Elements of Industrial Hazards

T0264855

Elements of Industrial Hazards

Health, Safety, Environment and Loss Prevention

Ratan Raj Tatiya

CRC Press
Taylor & Francis Group
Boca Raton London New York Leiden

CRC Press is an imprint of the
Taylor & Francis Group, an **informa** business

A BALKEMA BOOK

First issued in paperback 2017

CRC Press/Balkema is an imprint of the Taylor & Francis Group, an informa business

© 2011 Taylor & Francis Group, London, UK

Typeset by Vikatan Publishing Solutions (P) Ltd, Chennai, India

Published by: CRC Press/Balkema
P.O. Box 447, 2300 AK Leiden, The Netherlands
e-mail: Pub.NL@taylorandfrancis.com
www.crcpress.com – www.taylorandfrancis.co.uk – www.balkema.nl

Library of Congress Cataloging-in-Publication Data

Tatiya, Ratan.
Elements of industrial hazards : health, safety, environment, and loss prevention / Ratan Raj Tatiya.
  p. cm.
Includes bibliographical references.
ISBN 978-0-415-88645-1 (hardback) – ISBN 978-0-203-83612-5 (ebook)
1. Industrial safety. 2. Industrial accidents. I. Title.

T55.T296 2011
363.11– dc22

                                                                2010037681

ISBN 13: 978-1-138-11526-2 (pbk)
ISBN 13: 978-0-415-88645-1 (hbk)

To my Grandchildren Chavvi, Divya, Sambhav, Madhur, Aadesh, Parag & Parshav.

Contents

Preface   xix
Acknowledgements   xxv
Conversion tables   xxvii

I   Introduction   I

1.1   Introduction – industries & related issues   1
1.2   Industrialization – brief history   1
1.3   Current scenario   2
    1.3.1   Population growth   2
    1.3.2   Lifestyle   4
    1.3.3   Globalization   5
    1.3.4   Buyer's market   6
    1.3.5   Technological developments and renovations   6
    1.3.6   Mineral production and consumption trends, and rapid
        resources depletion   7
    1.3.7   Information Technology (IT) and its impacts   7
1.4   Industrial (Occupational) Health, Safety & Environment (HSE)   8
    1.4.1   Industry & environment   8
    1.4.2   Industry & safety   8
1.5   Impact of industrialization on society   11
    1.5.1   Mass balance system/equation   11
    1.5.2   Damage due to industrialization   12
    1.5.3   Birth of global issues   12
1.6   HSE – a critical business activity   12
1.7   Environmental policy   13
    1.7.1   Sustainable development   13
    1.7.2   Development of industrial technology   14
    1.7.3   Education   14
    1.7.4   The ultimate goal   15
Questions   15
References   16

2   Ecology, environment, mineral resources & energy   19

  2.1   Environment related issues   19
      2.1.1   Ecology   19
            2.1.1.1   Ecosystem   19
            2.1.1.2   Classification – ecology   20
  2.2   Earth's great spheres   21
      2.2.1   Biotic component of planet earth – biosphere   22
      2.2.2   Natural cycles   22
            2.2.2.1   Sulfur cycle   22
            2.2.2.2   Nitrogen cycle   22
            2.2.2.3   Carbon cycle   23
      2.2.3   Solar energy's contribution   25
  2.3   Food – food chains – food webs   26
  2.4   Abiotic   29
  2.5   Population   31
      2.5.1   Impacts of population growth   31
      2.5.2   Human population – important aspects   31
  2.6   Chemicals in motion: Cycles in the ecosphere   33
  2.7   Minerals – the non-renewable resources and their use in energy,
      goods and services production   34
      2.7.1   As we know without power & energy, world is dark,
            factories would come to halt, and services would be
            jeopardized; but is it electricity or minerals?   34
  2.8   Energy sources   36
      2.8.1   Classification of energy sources   36
      2.8.2   Green power and its purchasing options   37
      2.8.3   Energy sources and their merits and limitations   38
      2.8.4   Top 7 promising alternative energies   40
      2.8.5   GDP, energy consumption patterns and CO_2 emissions   40
      2.8.6   Risk of CO_2 emissions   44
      2.8.7   Coal for energy security   45
      2.8.8   Clean coal technology (CCT)   45
      2.8.9   Carbon capture & storage (CCS)   46
      2.8.10   Converting coal-to-oil   47
  2.9   Energy crisis   47
      2.9.1   Wayout/solution to the energy crisis   48
      2.9.2   Energy efficient lighting tips   53
      2.9.3   Energy conservation tips   54
      2.9.4   Things to remember/way forward   55
  Questions   55
  References   56

3   Air pollution   59

3.1   Introduction   59
3.2   Air pollution   59
        3.2.1   Clean and dry air composition   60
        3.2.2   Air pollutants   61
3.3   Air toxics   69
        3.3.1   Units of measuring concentration   69
3.4   Air quality standards   69
        3.4.1   The Air Quality Index (AQI) or Pollution Standard
                    Index (PSI)   70
        3.4.2   Determination of Pollution Standard Index (PSI) value   73
        3.4.3   Emission inventory estimation   74
3.5   Performance monitoring   74
        3.5.1   Air pollutant receptors and adverse impacts   75
3.6   Global warming – the greenhouse effect   76
        3.6.1   Greenhouse impacts   77
        3.6.2   Changing climate   79
        3.6.3   Calculation of CO_2 emission from hydrocarbons   80
3.7   Acid rain   83
        3.7.1   How acid rain is formed?   83
        3.7.2   pH   85
        3.7.3   Calculation related to acid rain   85
        3.7.4   Adverse impacts of acid rain   86
3.8   Ozone gas & photochemical smog (PCS)   88
        3.8.1   Ozone depletion process   90
3.9   Noise pollution   92
        3.9.1   Noise sources   92
        3.9.2   Industrial noise   92
        3.9.3   Important relations for sound/noise measurement   93
        3.9.4   Noise control techniques   94
        3.9.5   Noise related calculations   94
        3.9.6   Noise threshold limits   97
3.10   Vibrations   97
3.11   Dust   98
        3.11.1   Conditions for dust to become nuisance   98
        3.11.2   Factors affecting the degree of health risk   98
        3.11.3   Physiological effects of dusts   99
        3.11.4   Sources of dust   100
        3.11.5   Control of dust   100
3.12   Particulate Matter (PM)   101
        3.12.1   Grouping particulate matter   101

3.13	Air samples	103
	3.13.1 Gas detection techniques	103
3.14	Remedial measures	105
3.15	Concluding remarks	105
Questions		105
References		107

4   Water pollution — 109

4.1	Introduction	109
	4.1.1 Water cycle	110
4.2	Worldwide water resources – some facts	111
4.3	Water quality standards (WQS)	113
	4.3.1 Water quality standards based on receiving environment	114
4.4	Groundwater	117
	4.4.1 Sources and routes for groundwater contamination	119
4.5	Water use	119
	4.5.1 Industry	119
	4.5.2 Mining	121
	4.5.3 Agriculture	122
4.6	Water pollution (information taken from various sources)	123
	4.6.1 Point sources of pollution	123
	4.6.2 Non-point sources of pollution	123
	4.6.3 Types of water pollutants	125
	4.6.3.1 Biological agents	125
	4.6.3.2 Toxic substances	126
	4.6.3.3 Organic substances	126
	4.6.3.4 Thermal pollution (increase in temperature)	126
	4.6.4 Natural pollution	126
	4.6.4.1 Dissolved Oxygen (DO)	127
	4.6.4.2 Biological Oxygen Demand (BOD)	127
	4.6.4.3 Hardness	127
	4.6.4.4 Acidity	127
	4.6.4.5 Alkalinity	128
	4.6.4.6 Colors	128
	4.6.4.7 Radioactive	128
	4.6.4.8 Oils & petrochemicals	128
	4.6.4.9 Red tide	128
4.7	Sewage	128
	4.7.1 Suspended or sedimentary solids	129
	4.7.2 Polluted municipality water	130
4.8	Marine pollution	130
4.9	Water in subsurface (underground) areas	131

4.9.1	The main sources of water		132
4.9.2	Effects of subsurface water		132
4.10	Acid mine drainage		133
4.10.1	Chemistry		134
4.10.2	Yellow boy		134
4.11	Case study: Water pollution due to mining, petroleum products handling and industrial activities		135
4.11.1	Study areas		135
	4.11.1.1	Copper mining	135
	4.11.1.2	Areas adjacent to the petroleum refinery, PDO area and shell marketing area	135
	4.11.1.3	Industrial area	137
4.11.2	Sample collection and preparation		137
	4.11.2.1	Analysis	137
4.11.3	Observations		137
4.11.4	Inference drawn – physical and chemical properties		137
4.11.5	Heavy metal concentration		142
4.11.6	Conclusion		143
4.12	Bottled water		145
4.12.1	Bottled water – do we need it? Some facts		145
4.13	Concluding remarks		146
Questions			146
References			147
5	**Solid industrial waste & land degradation**		**149**
5.1	Introduction		149
5.2	Classification		150
5.2.1	Non-hazardous wastes		150
5.2.2	Hazardous wastes		150
5.3	The growth of waste		152
5.3.1	The waste problem		152
5.3.2	Lifestyle		153
5.4	Methods of waste disposal (information gathered from various sources)		154
5.4.1	Source reduction		155
5.4.2	Reuse		156
5.4.3	Recycling of waste		156
	5.4.3.1	Merits	156
	5.4.3.2	Limitations	157
5.4.4	Treatment		158
	5.4.4.1	Incineration	158
5.4.5	Responsible disposal		159
	5.4.5.1	Landfill	159

5.5	Land degradation due to industrial or domestic waste disposal		160
	5.5.1	Land degradation	160
	5.5.2	Soil degradation/pollution	161
5.6	Waste generation and its management in mining and excavation (civil) industries – some basics		161
	5.6.1	Surface excavations/mining	161
		5.6.1.1 Open pit elements	162
		5.6.1.2 Stripping ratio	162
	5.6.2	Dumping site	166
5.7	Waste management in petroleum industry – a case study		168
	5.7.1	Waste management	168
	5.7.2	Waste in the petroleum industry	168
		5.7.2.1 Audit	169
		5.7.2.2 Waste management plan	169
		5.7.2.3 Waste consignment note	170
	5.7.3	Waste minimization	170
	5.7.4	Inventory management	172
	5.7.5	Improved operation	172
	5.7.6	Material substitution	172
	5.7.7	Equipment modifications	173
	5.7.8	Waste reuse	173
	5.7.9	Waste recycling	174
	5.7.10	Waste treatment	174
	5.7.11	Incineration	174
	5.7.12	Waste disposal	175
5.8	Tips for reducing solid waste (as advised by EPA)		176
5.9	A classic example from Lord Buddha's disciple as how to reuse!!!!!		176
5.10	Concluding remarks		177
	Questions		177
	References		178
6	**Industrial hazards**		**179**
6.1	Industrial hazards		179
	6.1.1	List of hazards	180
	6.1.2	Disaster	181
	6.1.3	Health risk	181
6.2	Fires		181
	6.2.1	The fire triangle concept	181
	6.2.2	Concepts – mechanism of fire	182
	6.2.3	Ignition sources of major fires	183
	6.2.4	Classification of fires	184
	6.2.5	Fire protection	184
	6.2.6	Fire and emergency	186
	6.2.7	Fixed fire fighting equipment	187

	6.2.8	Fire fighting department	187
		6.2.8.1 Introduction	187
		6.2.8.2 Functions	187
6.3	Explosions (information from various sources)		190
	6.3.1 Classification		190
		6.3.1.1 Mechanical	190
		6.3.1.2 Detonation, deflagration and shockwaves	191
		6.3.1.3 Confined & unconfined explosions	192
		6.3.1.4 Air blast	192
		6.3.1.5 Pressure Vessel Ruptures (over-pressure)	193
		6.3.1.6 Rock burst and bumps	193
		6.3.1.7 Vapor Cloud Explosions (VCE)	193
		6.3.1.8 Physical explosion or eruption	194
		6.3.1.9 Methane and coal dust explosions	194
		6.3.1.10 Sulfide dust explosions	195
		6.3.1.11 Explodable dusts	195
6.4	Dow index to access degree of hazards		195
6.5	Incidents responsible for onset of hazards and accidents		197
	6.5.1	Spillage	197
	6.5.2	Leakage (also refer sec. 9.6)	197
	6.5.3	Unintended venting (also refer sec. 9.6)	197
	6.5.4	Failures at normal working pressure (also refer sec. 9.6)	198
	6.5.6	Equipment failure due to excessive pressure – (also refer sec. 9.6)	198
6.6	Losses in the chemical industry due to fires and explosions		198
6.7	Hazards with flammable liquids, and precautions		198
6.8	Static hazards associated with Ammonium Nitrate Fuel Oil mixture (ANFO) loading		199
	6.8.1	Blasting agent ANFO	199
	6.8.2	Case history: On static electricity hazards	201
6.9	Toxic gases		201
	6.9.1	Asphyxiate gases	201
	6.9.2	Irritant gases	201
	6.9.3	Poisonous gases	202
	6.9.4	Portal of entry	202
	6.9.5	Remedial measures	202
	6.9.6	Toxicology	202
	6.9.7	Summary: Classification – toxicity-related hazards	204
6.10	Hazards while using machinery		205
6.11	Surface or subsurface (underground) mine hazards		206
6.12	Classification of hazardous materials		207
	6.12.1	Explosive materials	208
	6.12.2	Compressed gases	208
	6.12.3	Flammable liquids and solids	209

6.12.4	Chemically reactive materials	210
6.12.5	Corrosive material	210
6.12.6	Flammable solids	210
6.12.7	Controlled materials	210
6.12.8	Workplace Hazardous Materials Information System (WHMIS)	210
6.13	Hazards analysis methods	211
6.14	Inherent safer design strategies	213
6.14.1	Minimize	213
6.14.2	Substitute/eliminate	215
6.14.3	Moderate	215
6.14.4	Simplify	216
6.14.5	Location/sitting/transportation	216
6.14.6	Change to inherent safety strategy	217
6.15	Breathing apparatus	217
6.16	The way forward	218
6.17	Vocabulary	219
Questions		220
References		221

7 Occupational Health & Safety (OHS)   **223**

7.1	Occupational Health and Safety (OHS)	223
7.2	Elements: Occupational Health (OH)	223
7.3	Industrial hygiene	224
7.3.1	Steps for managing industrial hygiene	224
7.4	Fundamental principles of industrial hygiene	224
7.4.1	Anticipation	224
7.4.2	Identified/recognized	226
7.4.2.1	Dust generation	226
7.4.2.2	Asbestos fibers	226
7.4.2.3	Noise generation	227
7.4.2.4	Vibrations	227
7.4.2.5	Welding	227
7.4.2.6	Hazardous salts	231
7.4.2.7	Diesel emissions	231
7.4.2.8	Foul gases	231
7.4.2.9	Metals	233
7.4.2.10	Extreme temperatures – heat & humidity	233
7.4.2.11	Radiation hazards	238
7.4.2.12	Vapors	245
7.4.2.13	liquids	245
7.5	Aqueous effluents – permissible quality & efficient discharge	245
7.5.1	Parameters concerning effluent discharge	245

	7.5.2	Performance standards	246
	7.5.3	Effluent discharges receiving environment	246
	7.5.4	Effluent discharge/disposal – surface water-bodies	246
		7.5.4.1 Water Quality Standards (WQS)	246
	7.5.5	Effluent discharges/disposal – marine	248
	7.5.6	Effluent discharges – sewage treatment systems	248
7.6	House keeping		249
	7.6.1	Aspects to be adhered to	250
	7.6.2	Dealing with spillage	251
	7.6.3	Administrative controls	251
	7.6.4	The 5S concept	251
	7.6.5	Sanitation	253
7.7	Working conditions		253
7.8	Ergonomics		254
	7.8.1	Introduction	254
	7.8.2	Making things user-friendly	255
	7.8.3	Impacts of poor ergonomics	256
	7.8.4	Impacts of good ergonomics	256
		7.8.4.1 Improved labor relations	256
		7.8.4.2 Safeguarding skilled and experienced human resources	256
		7.8.4.3 Offsetting limitations on age of employees	257
		7.8.4.4 Reduced maintenance downtime	257
	7.8.5	Work in neutral postures	257
	7.8.6	Identifying wasted activities	260
	7.8.7	Fresh insights on your operations	260
7.9	Occupational health surveillance		261
	7.9.1	Organizational culture and workplace stresses	261
		7.9.1.1 Organizational culture and commitment	261
		7.9.1.2 Workplace stress, its adverse impacts and ways to avoid	263
	7.9.2	Lost performance at work (presenteeism)	265
		7.9.2.1 Presenteeism	265
		7.9.2.2 Health Promotion Management (HPM) – what it is?	266
		7.9.2.3 Health risks and behavior	267
		7.9.2.4 Developing a health profile for businesses – a case study	267
	7.9.3	Occupational hygiene risk – exposure assessment and control measures	270
		7.9.3.1 Health-related variables influencing the working life of an industrial worker	270

		7.9.3.2	Periodic health surveillance based on exposure risk	271
7.10	Notified diseases and preventive measures			274
Questions				275
References				276

8   Industrial safety   279

8.1	Introduction		279
8.2	Safety elements and strategies		280
8.3	Safety elements		280
	8.3.1	People/industrial workers	280
	8.3.2	Systems developed to run the show	286
	8.3.3	The working environment	287
8.4	Strategies		287
8.5	Lifecycle approach		288
8.6	Layers of protection		290
8.7	Accidents		290
	8.7.1	Accident – a three-step process	290
	8.7.2	Accidents/incident analysis	291
	8.7.3	Accident-related calculations	293
	8.7.4	Degree (type) of injuries	294
	8.7.5	Causes of accidents	294
	8.7.6	Accident costs	295
	8.7.7	Remedial measures	295
8.8	Conceptual planning, detailed design and evaluation		296
8.9	Training and education		296
8.10	Personal Protective Equipment (PPE)		300
8.11	Risk analysis		300
8.12	Case study: Without a 'Sugar' Coat!: British Sugar		304
8.13	Substandard behaviour and workplace accidents, and ways to avoid		305
Questions			305
References			307

9   Loss prevention   309

9.1	Introduction		309
	9.1.1	Aims and objectives of an industrial set-up	309
	9.1.2	Input resources	309
9.2	Loss prevention		310
9.3	Loss prevention strategy		310
	9.3.1	Content employees	310
	9.3.2	Efficient systems	311

	9.3.3	Legal compliance	311
9.4	Human Resources (HR) – manpower – HR management		311
	9.4.1	Some basics of man-management	312
	9.4.2	Some basics of leadership	312
9.5	Managing plant, equipment, machines, tools and appliances		314
	9.5.1	Proper equipment selection	314
	9.5.2	Efficient utilization	314
	9.5.3	Effective maintenance	316
	9.5.4	Preventive maintenance	320
9.6	Abnormalities		321
9.7	Classification – losses		325
	9.7.1	Direct losses in various forms or types	325
	9.7.2	Indirect losses	325
	9.7.3	Losses in a manufacturing plant – reasons and suggested measures to minimize them	325
9.8	Wastage		325
9.9	Case-study illustrating computation of financial losses		332
9.10	Effective systems – best practices		333
	9.10.1	Quality Management System (QMS)	333
		9.10.1.1 Quality is a function of these factors	333
		9.10.1.2 Reasons for quality defects	333
	9.10.2	Six sigma	334
	9.10.3	Quality Control Tools (QC Tools)	334
	9.10.4	Benchmarking & standardization	334
	9.10.5	ISO 9000	335
	9.10.6	Other models of standards	336
9.11	Legal compliance including Environment Management Systems (EMS)		337
9.12	ISO 14000 and ISO 14001		338
	9.12.1	Procedure to develop these standards	338
	9.12.2	ISO 14001standard	338
	9.12.3	Potential benefits of an EMS based on ISO 14001	339
9.13	Effective training, competency and awareness		339
9.14	Effective communication		341
9.15	World Class Management (WCM)		342
9.16	Precision in operations		342
9.17	Emergency preparedness and response		343
9.18	Way forward		343
9.19	Health, Safety and Loss Prevention (HSLP) management system and its effectiveness		345
9.20	Case study – three pillars of equal strength for loss prevention		346
Questions			349
References			350

10 HSE management system   353

  10.1   Introduction   353
         10.1.1   HSE – a critical business activity   354
         10.1.2   Vision   355
  10.2   HSE leadership and commitment   355
         10.2.1   Visibility   355
         10.2.2   Target setting   356
         10.2.3   Culture   356
         10.2.4   Informed involvement   356
         10.2.5   Accountabilities   356
         10.2.6   Checklist   359
  10.3   HSE policy   360
  10.4   Organization, responsibilities, resources,
         standards & documents   360
         10.4.1   Training needs   361
         10.4.2   Resources required   361
         10.4.3   Roles and responsibilities   361
  10.5   Hazards and effects management   362
         10.5.1   Steps in hazards and effects management process   362
         10.5.2   Control of hazards and effects   364
         10.5.3   HEMP tools – risk analysis   365
         10.5.4   Recovery measures   365
  10.6   Planning and procedures   371
         10.6.1   Emergency measures/preparedness   371
  10.7   Implementation and monitoring   371
  10.8   HSE audit   372
  10.9   Review   373
  10.10 Management commitment   373
  10.11 Management: Occupational hazards (health & physique)   374
  10.12 Environment management   375
         10.12.1 Why pollution?   375
         10.12.2 Mass balance system/equation   376
         10.12.3 Environmental degradation in an industrial setup   376
         10.12.4 Main sources of pollution in air, water and land
                  environments   377
  10.13 Environmental management   377
  10.14 Sustainable development   380
  10.15 Concluding remarks   382
  Questions   384
  References   385

  Subject index   387

Preface

Industries are the backbone of the economy of any country. Natural resources such as minerals, fossil fuels, air, water and flora are the basic raw materials to run them. Since last century industrialization has grown very rapidly due to growing population and changes in our life style and living standards. This growth should continue.

The output from industries is not only goods and services but also the generation of gaseous emissions, liquid effluents, solid wastes, radiations, particulate-matters, heat and noise. Equally associated with them are hazards such as fires, explosions, inundation, accidents, disasters and few others. All these are detrimental to health not only to direct and indirect stockholders particularly when they exceed the allowable limits but also to the biotic and abiotic components of nature. Growing health-problems of the world's citizens and global issues (problems) such as acid rain, ozone depletion, photochemical smog, acid drainage and global warming are witnessing this fact.

Thus, the foregoing discussion reveals that we run Industries to produce goods and services of many kinds. Production is the bread and butter of those concerned (right from shop floor worker to the topmost executive) in any industrial set up. Productivity brings excellence to production; but both cannot be achieved if safety is jeopardized, pollution is at its paramount; workers' health is not looked after and welfare of the Society is neglected. Thus, HSE must be considered a 'Critical Business Activity' at par with 'Production' and 'Productivity'. And a thorough balance is required amongst these three critical business activities together with proper care for the society to achieve sustainable development, which is beneficial Socially, Economically and Ecologically to the present as well as future generations. There is no simple and straightforward solution for it; but this is for sure that *minimizing losses of various kinds could achieve this*. As such 'Loss Prevention strategy' should be an integral part of running any industry effectively. And to accommodate this feature, a chapter on 'Loss Prevention' has been added to this book. Thus, these are some of the themes that have been put together to make this book meaningful. For each theme (Industrial Hazards, Occupational Health, Safety, Environment, HSE-combined and Loss prevention) a rational approach has been tried to cover them reasonably.

The foregoing discussion would reveal that components for the subject areas chosen for this book are many, and they are very vital. And to write on these subject areas; one's Industrial background ensures that the material is industrially relevant and his academic background ensures that the fundamental and basics required to help readers are included.

The author had opportunity to have 40 years of experience in Mineral Industry and related disciplines including as a university professor to teach students of Chemical, Petroleum and Mining industries which are considered to be the most Risky and Polluting Industries. As a 'Consultant' he spent good amount of time in the Preparation of Environment Impact Assessment Statement (EIAS) and Management Plans for industrial establishments. He worked with multinationals from more than 40 countries; and multi-cultured environment; initially for a decade in the industry and then as senior university professor and industrial consultant for more than 25 years, and presently at the senior position with the industry that inspired him to write this book.

The book encompasses material that is useful for Engineering Disciplines such as: Chemical, Petroleum, Environment, Mining, Mechanical, Civil, Construction, Occupational Health and Safety (OHS) and Geology. As such it is intended to serve as a textbook for students of undergraduate level and beginning year of graduate level at Schools or Institutes having any of the above-mentioned disciplines. More material presented that can be accommodated in a 2-credit course, and it should be sufficient for a 3 credit-course. In addition, the officials, supervisor, engineers and professionals of these disciplines are intended to use this book.

Some will remark that contents are not complete or some details are missing. The purpose of this book, however, is not to be a complete but to provide starting point for those who wish to learn about this important area, hence it is an *elementary book on the subject areas* but one would find that it is a unique combination of Occupational Health, Safety, Environment and Loss Prevention. And it is not a distant future that these will be required subjects for all the college and university students. Knowing about these subjects is as important to college graduates as knowing about history, science, mathematics, language, or the arts.

In summary, this book contains a comprehensive text on its principal elements, which have been divided into 10 chapters.

'Introductory' chapter 1 introduces **Industries and their contribution** and impacts. It lists prominent air disasters, classifies hazards and introduces industrial evils, which have been dealt with in various chapters. While briefing prevalent global scenario it conveys a clear message that goods and services except that of strategic importance, are available at competitive price. It provides a way forward to address HSE related issues and illustrates consequences in the 'Working Life' of an Industrial Worker.

Chapter 2 includes basics of 'ecology', contribution of **minerals** in the production of goods, services and energy. It also includes **energy** sources and their impacts to the environment. It highlights 'Energy Crisis', which has already begun and suggests few way-outs including tips to conserve energy. It briefly covers technologies such as: Clean Coal Technology (CCT) and Carbon Capture & Storage and lists Promising Alternative Energies and suggest some 'Best Practices' to reduce CO_2 emissions.

Chapter 3 on 'Air Pollution' details principal air pollutants their sources and natural removals. It explains Air Quality Index (AQI) that is used to determine air pollution level. The culprits for air pollution result in: *Acid Rain, Global Warming, Ozone Depletion and Photochemical Smog*. Each one of them has been dealt with in detail. And ultimately how all these are causing damage to human beings, flora and fauna have been dealt with in this chapter. It also includes Noise pollution, airborne dusts and particulate matters. Numerical problems have been added to bring clear understanding of the concepts described in this chapter.

Chapter 4 on 'Water Pollution' describes Global Water Resources, Water Cycle, and Water Quality Standards. It gives fair treatment to the Water Pollutants. Problems of mine-water included. It covers 'Acid Mine Drainage' and its impacts. Case study on: Water pollution due to Mining, Petroleum Products' handling and Industrial activities provides a true picture as how water pollution can be harmful to the direct and indirect stockholders. It discloses the hidden facts of 'Bottled Water'.

Chapter 5 on: 'Solid Industrial Waste & Land Degradation' begins with classification of non-hazardous and hazardous wastes, their growth and resultant problems. Disposal techniques based on its management hierarchy then follow. Waste Management in Petroleum Industry – A Case Study gives insight as how effectively this task could be managed in a Petroleum Field, or alike. Lastly tips of EPA for reducing waste have been listed.

Chapter 6 on 'Hazards' describes fires, explosions, toxic and machinery hazards; as applicable to various industries. It outlines hazardous materials. It also highlights hazards of surface and underground mining. Important features of this chapter include: Application of 'Safer Design Strategies', overview of 'Hazards Analysis Methods' and description on 'Incidents' responsible for onset of hazards and accidents. The chapter ends with suggestions for accepting challenges of working in hazardous conditions and proposes way outs.

Chapter 7 has been devoted to 'Occupational Health (OH)'. It pertains to health of industrial workers as a result of exposure during their working life. It covers 4 elements of OH: Industrial Hygiene including efficient effluent discharge; Housekeeping & Working Conditions; Ergonomics and Occupational Health Surveillance. Its special features include: '5S Concept'; Organizational Culture and workplace stresses; Lost Performance at Work (Presenteeism); Risk – Exposure Assessment & Control Measures; and Making things user-friendly through good ergonomics. These features have been incorporated with case studies and input from professional practitioners.

Chapter 8 on 'Industrial Safety' describes its 3 elements: Workers, Systems and Working Environment. It advocates: *man is the most valuable resource to run industry* and he/she *must be protected*. And proposes use of '4Es Concept': *Education, Engineering, Enforcement and Engagement*. Its special features include: Strategies, Lifecycle approach, Layers of Protection; Training and Education, HSE Risk Analysis (based on concepts used at a Petroleum Field), substandard behaviour and workplace accidents.

It gives fair coverage to Accidents/Incidents On: Process, Analysis, Causes, Consequences, Calculations, Controls, Costs and Remedial Measures. Through a case study it explains how change in behavior could allow any individual to leave his family (home) happy and return happier. This could be achieved (as explained) by effective training and education using concept of: *Why, Who, When, Where, What and How*. Outline for a Three-Day Supervisory Program -Total Accident Control Training (TACT) has been included.

Chapter 9 describes 'Loss Prevention Strategies' and suggests: Three Pillars of equal strength: Content Employees, Efficient Systems and Legal Compliances. It reasonably covers ways to effectively manage: Plant, Equipment, Machines, Tools and Appliances.

- *Its important features* include: Diagnosis of Abnormalities and Remedial Measures to overcome them. It lists direct and indirect losses and suggests measures to minimize them. With the help of a Case-Study Computation of Financial Losses has been illustrated.

- *Its special features* include: Effective Systems & Best Practices, which have been explained through: Quality Management System (QMS); Six Sigma; Quality Control Tools (QC Tools); Benchmarking & Standardization; ISO 9000 & 14000; Environment Management Systems (EMS); World Class Management (WCM), Autonomous Maintenance System (AMS) and Precision in Operations.

- *It emphasizes need* for Effective Training, Competency, Awareness and Communication to run any industry smoothly. Ways to effectively manage OHS and Loss Prevention has been explained by a Case Study – Three Pillars of Equal Strength for Loss Prevention.

- *It proposes* an effective model to accomplish continual improvements in *not only HSE but equally to Production and Productivity.*

- *This chapter emphasizes need of using our 5 SENSES: EAR, NOSE, EYES, FEEL AND TOUCH, to remove abnormalities, which is an effective dose (tool) to minimize losses of all kinds.*

The concluding chapter 10 **'HSE Management System'** attempts to cover Leadership, Commitment and Policies. It also details steps for the preparation of Environment Impact Assessment Statements (EIAS) and Management Plans. It includes emergency preparedness, risk analysis and guidelines for *'Sustainable Development'*. As this chapter ends so does this book with the concluding remark: 'It is the teamwork that could minimize losses, which could approach zero-illness, zero-injury, zero-incident, zero-waste, zero-emissions and zero-rework' – A goody-goody situation for everybody, particularly those concerned. Should we not strive for it?

Thus, the 'Book', features Inclusion of 'Best Practices'; 'Case Studies', 'Latest Trends and Practices' wherever feasible. And top of all it advocates for **'Prevention of Losses of all kinds'**, which is a noble way of working as it results in *producing goods and services with maximum productivity and least costs safely*. It provides solution as how an industry can survive in the scenario of *'Global Competition and Recession'*. Each chapter ends with either 'Way-Forward' or Concluding Remarks' to conclude it. With each chapter 'Questions' have been listed at the end to help readers for better understanding of the subject matter. However, Instructors/Professors could design 'Quizzes and Questions', using this material and by providing/suggesting supplementary material on the subject matter.

In the end this book is a result of appreciation from students and colleagues, support from my family members and cooperation from professional societies and government agencies (SME, ICMM, Wikipedia, AIMEX, AIME, SPE, EPA) and academic institutes including Sultan Qaboos University, Oman.

Companies such as: Centennial-coal, CSIRO, Rio Tinto, Newmont Mining Corporation, BHP Billiton, Lonmin, Center for Applied Macroeconomics, Pacific Strategy Partners, AngloGold Ashanti, CaterPillar, PDO, and a few others who encourage me by providing valuable information, on my request and also granted permission to use published material. I wish to express my sincere gratitude to them, and all those who helped me directly or indirectly in this endeavor. I wish to extend special thanks to Aditya Birla Group; as while working with this group I learnt a lot pertaining to the subject area.

My sincere thanks to Jane Olivier who helped me with great enthusiasm and professionalism, and advised me on the book-proposal submitted to SME, that the material is worth publication through a publisher with wide audiences.

I sincerely acknowledge my family members: my wife Shashi and sons Anand and Gaurav and their wives Smirita and Pallalvi; and Daughter Sapna and her husband Roopam who provided patience, understanding and encouragement throughout preparation of this book.

Ratan Raj Tatiya
June, 2010

Acknowledgements

The author wishes to express his sincere gratitude to the educational institutes, professional societies, manufacturing companies, publishers, business houses, government bodies (agencies), Non-Government Organizations (NGOs) and other institutes, as listed below, for their permission, facilities and release of technical data, literature, information and material.

AIMEX Australia
American Institute of Chemical Engineers U.S.A.
AngloGold Ashanti Southafrica
BHP Billiton & BHP Billiton-Mitsubishi Alliance Australia
Canadian Standards Association, Canada
CaterPillar, USA
Centennial Coal Australia
Center for Chemical Process Safety of the American Institute of Chemical Engineers,
    New York
Climatic Research Unit, University of East Anglia
CRC Mining Australia
CSIRO Australia
Environment Health Center, National Safety Council. USA
Environmental Protection Agency (EPA) U.S.A.
Essel Mining and Industries – Aditya Birla Group, India
Hella Australia
International Council on Mining and Metals (ICMM) UK
International Organization for Standardization
International Social Security Association
Lonmin, Southafrica
Mineral Information Institute (MII), Denver, Colorado
Mining, Metallurgical, and Petroleum Engg., USA (AIME)
Ministry of Environment and Forests; Govt. of India
Modular Mining System, USA
NASA Earth Observatory USA
Newmont Mining Corporation Australia
Pacific stetegy Partners, Australia
Petroleum Development Oman (PDO), Sultanate of Oman

Rio Tinto Australia/Southafrica
Royal Dutch/Shell Group International.
SME/SMME – Society for Mining, Metallurgy, and Exploration, Inc., USA
Society of Petroleum Engineers (SPE) USA
Sultan Qaboos University, Oman
Wikipedia, the free encyclopedia

Conversion tables

Multiply Metric unit	By	To obtain English unit
kilometer (km)	0.6214	mile (mi)
meter (m)	1.0936	yard (yd)
meter (m)	3.28	foot (ft)
centimeter (cm)	0.0328	foot (ft)
millimeter (mm)	0.03937	inch (in)
sq kilometer (km^2)	0.3861	sq mile ($mile^2$)
hectare (ha)	2.471	acre
sq meter (m^2)	10.764	sq. foot (ft^2)
sq meter (m^2)	1550	sq inch (in^2)
sq centimeter (cm^2)	0.1550	sq inch (in^2)
cu centimeter(cm^3)	0.061	cubic inch (in^3)
cubic meter (m^3)	1.308	cubic yard (yd^3)
liter (l)	61.02	cubic inch (in^3)
liter (l)	0.001308	cubic yard (yd^3)
km/h	0.621	mph
liter (l)	0.2642	US gallon
liter (l)	0.22	imperial gallon
metric ton (t)	0.984	long ton (lg ton)
metric ton (t)	1.102	short ton (sh ton)
kilogram (kg)	2.205	pound advp. (lb)
gram (gm)	0.0353	ounce advp. (oz)
kilonewton (kn)	225	pound (force)
newton (n)	0.225	pound (force)
cu centimeter(cm^3)	0.0338	fluid ounce
kg/m^3	1.686	pounds/yd^3
kg/m^3	0.062	pounds/ft^3
kg/cm^2	14.225	pounds/in^2
kilocalorie (kcal)	3.968	Btu
kilogram-meter (kg.m)	7.233	foot-pound
meter- kilogram- (m.kg)	7.233	pound-foot
metric horsepower (cv)	0.9863	Hp
kilowatt (kw)	1.341	Hp
kilopascal (kpa)	0.145	psi
bar	14.5	psi
tons/m^3	1692	pounds/yd^3
decaliter	0.283	bushel

Multiply English unit	By	To obtain Metric unit
mile (mi)	1.609	kilometer (km)
yard (yd)	0.9144	meter (m)
foot (ft)	0.3048	meter (m)
inch (in)	25.4	millimeter (mm)
sq mile (mile^2)	2.590	sq kilometer (km^2)
acre	0.4047	hectare (ha)
sq. foot (ft^2)	0.0929	sq meter (m^2)
sq inch (in^2)	0.000645	sq meter (m^2)
cubic yard (yd^3)	0.7645	cubic meter (m^3)
cubic inch (in^3)	16.387	cu centimeter(cm^3)
cubic foot (ft^3)	0.0823	cubic meter (m^3)
cubic inch (in^3)	0.0164	liter (l)
cubic yard (yd^3)	764.55	liter (l)
mph	1.61	km/h
Ton-mph	1.459	tkm/h
U.S. gallon	3.785	liter (l)
U.S. gallon	0.833	Imperial gallon
long ton (lg ton)	1.016	metric ton (t)
short ton (sh ton)	0.907	metric ton (t)
pound advp. (lb)	0.4536	kilogram (kg)
ounce advp. (oz)	28.35	gram (gm)
pound (force)	0.0045	kilonewton (kn)
pound (force)	4.45	newton (n)
fluid ounce (fl oz)	29.57	cu centimeter(cm^3)
pounds/yd^3	0.5933	kg/m^3
pounds/ft^3	16.018	kg/m^3
pounds/in^2	0.0703	kg/cm^2
Btu	0.2520	kilogram calorie
foot-pound	0.1383	kilogram-meter (kg.m)
horsepower (hp)	1.014	metric horsepower
horsepower (hp)	0.7457	kilowatt (kw)
psi	6.89	kilopascal (kpa)
psi	0.0689	bar
pounds/yd^3	0.0005928	tons/m^3
pounds (no. 2 diesel fuel)	0.1413	U.S. gallon
bushel	3.524	decaliter

Note: Some of the above factors have been rounded for convenience. For exact conversion factors please refer to the International System of Units (SI) table. (Courtesy: Caterpillar)

Metric unit equivalents

1 km	=	1000 m
1 m	=	100 cm
1 cm	=	10 mm
1 km^2	=	100 ha
1 ha	=	10,000 m^2
1 m^2	=	10,000 cm^2
1 cm^2	=	100 mm^2
1 m^3	=	1000 liters
1 liter	=	100 cm
1 metric ton	=	1000 kg
1 quintal	=	100 kg
1 N	=	0.10197 kg.m/s^2
1 kg	=	1000 g
1 g	=	100 mg
1 bar	=	14.504 psi
1 bar	=	427 kg.m
1 cal	=	427 kg.m
		0.0016 cv.h
		0.00116 kw.h

Torque unit		
1 CV	=	75 kg.m/s
1 kg/cm^2	=	0.97 atmosph.

English unit equivalents

1 mile	=	1760 yd
1 yd	=	3 ft
1 ft	=	12 in
1 sq. mile	=	640 acres
1 acre	=	43,560 ft^2
1 ft^2	=	144 in^2
1 ft^3	=	7.48 gal liq.
1 quart	=	32 fl oz
1 fl oz	=	1.8 in^3
1 sh ton	=	2000 lb
1 lg ton	=	2240 lb
1 lb	=	16 oz, avdp
1 Btu	=	778 ft lb
		0.000393 hph
		0.000293 kw.h

Torque unit		
1 mechnical hp	=	550 ft-lb/s
1 atmosph.	=	14.7 lb/in^2

Power unit equivalents

kW	=	kilowatt
hp	=	Mechanical horse power
CV	=	Cheval Vapeur (steam horsepower)
		French designation for horsepower
PS	=	German designation for horsepower
1 hp	=	1.014 CV = 1.014 PS = 0.7457 kW
1 PS	=	1 CV = 0.986 hp
		0.7355 kW
1 kW	=	1.341 hp
		1.36 CV
		1.36 PS

Note: Some of the above factors have been rounded for convenience. For exact conversion factors please refer to the International System of Units (SI) table. (Courtesy: Caterpillar)

Chapter 1

Introduction

'Industries generate wealth. Industries also damage environment, and hurt men and properties by virtue of accidents. These adverse impacts cannot be eliminated but certainly efforts could be made to minimise them. Zero accident rate and clean environment should be our ultimate goal'.

Keywords: Industrialization, Air Polluting Industries, World's Population, Life Style, Globalization, Buyer's Market, Mineral Production and Consumption Trends, Occupational Health, Safety & Environment, Industry & Environment, Industry & Safety, Hazard Classification, Mass Balance System, Environmental Policy, Sustainable Development, Industrial Evils.

1.1   INTRODUCTION – INDUSTRIES & RELATED ISSUES[9, 11]

An industry adds value to the input material. It makes it usable. It produces commodities of our basic needs and provides services of various kinds and earns profits. It provides employment. We cannot do away with it and it is thus, indispensable. It has its limitations too. Every industry has hazards and so has every profession. Only their magnitude (quality and quantity) and direction (as to whom they influence) differs. Figure 1.1 briefly outlines all these aspects. Figure 1.6 depicts general classification of hazards, and chapter 6 deals with them in detail. These hazards endanger safety of men, machines, sets of equipment and property. They degrade surroundings (the environment). All these impacts are detrimental to our health.

1.2   INDUSTRIALIZATION – BRIEF HISTORY[1, 2]

Industrialization that began more than a century ago, particularly in developed countries, is at present at its peak in many developing countries. The undeveloped countries are also anxious to have it. Thus, most of the countries in Europe and North America have already experienced the fruits and damages (as briefly described in Table 1.1, sec. 3.2 and Figure 3.18) of this culture.

The reason behind growing industrialization is the phenomenal growth in world's population during the past years (twentieth century) as shown in Figure 5.2. To understand this growth; let us analyze the current/prevalent scenario, which is outlined below:

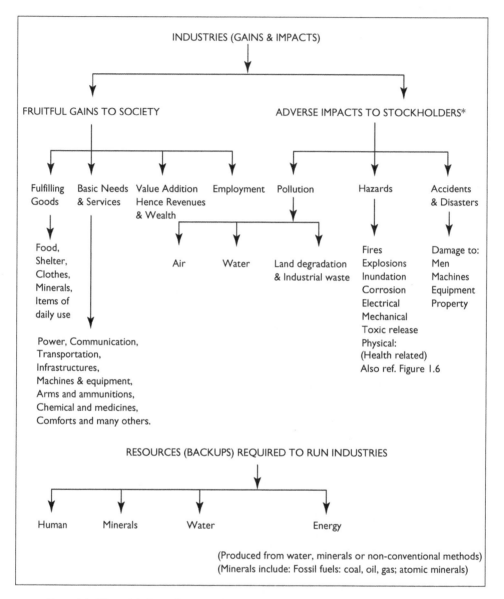

Figure 1.1 (Upper) Industrial gains and adverse impacts.

Direct stockholders – Industrial workers; and indirect stockholders – those affected by the industrial activities. (Lower) Resources required to run industries.

1.3 CURRENT SCENARIO

1.3.1 Population growth[7, 15]

With the present growth rate of about 1.8%, the world's population is going to be doubled in the next 39 years, while using the formula [equations (2.1) and (2.2)].

Table 1.1 Brief history of impacts of air polluting industries.[1,2]

Period	Impact of air polluting industries	Remarks
Centuries Preceding Industrial Revolution	Dust, fumes, foul gases and smell from many sources.	Metallurgy, ceramic and preservation of animal products.
Industrial revolution	As detailed below:	Producing steam for generating power and running machinery.
Nineteenth century	Smoke and ash from burning coal or oil in the boilers.	
Twentieth century 1900–25	Cities and factories grew in size, the severity of problem.	National Air Pollution control legislation enacted.
1925–50	Present day air pollution problems and solutions emerged.	
1950–90	Prominent disasters:	In Europe research activities expanded.
Dec. 1952	London; Low temperature inversion, thick fog, stagnant air, smoke and SO_2 accumulation. 4000 deaths and thousands hospitalized.	
Jan. 1956	London, similar to 1952; exceeding 1000 deaths.	These are some air pollution related disasters.
Dec. 1962	London; almost similar to 1952 and 1956; causing more than 700 deaths.	
Nov. 1966	New York; due to excessive SO_2 concentration; exceeding 176 deaths.	
Jun. 1974	Flixborough, England; failure of temporary pipe section replacing a reactor; 28 deaths and 36 injured.	These are Chemical Industry related disasters; and there are many more in other industries particularly in Mining and Petroleum Industries.
10-7-1976	Seveso, Italy; unintended formation and release of few kg of super-toxin causing wide spread illness and lasting environment damage.	
3-12-1984	Bhopal India, Release of 30–40 tonne. of toxic isocynate vapor in a pesticide factory; 2000 deaths and 50000 people affected.	
23-10-1989	Pasadena, Texas; a massive explosion causing 23 fatalities, 314 injuries and capital loss of over $715 million.	
1950–2008	Serious air pollution problems at large cities the world over.	
Future	Much would depend upon the use of fossil fuels and nuclear fuels; and the success of using solar, photovoltaic, geo-thermal, wind, non-fossil fuel (biomass) and oceanic (thermal gradient, tidal, and wave) sources of energy. Refer sections 2.8.4 and 2.8.5 also.	

The population problem is not just numbers but it has various environmental impacts. Providing ever-increasing numbers with clean air, water, nourishing food and shelter is becoming more and more difficult.

It is also an established fact that population control is not simple; use of birth-control technologies and measures can help this aspect to some extent. It has its religious, political, social and economical impacts. Some of the Asian countries are thickly populated whereas at some western countries, the growth rate is below the world's average. As detailed in section 2.5, the current level of population is already having problems on the following accounts:

- Food
- Shelter
- Water (Refer Figures 2.11 and 2.12)
- Education
- Employment
- Environment & health.

1.3.2   Lifestyle[7, 8, 13]

World's culture, customs and lifestyles must be considered in dealing with environment-related issues. People's lifestyle also plays an important role in degrading the environment. In some of the developed countries people's habits and lifestyle require large consumption of goods, services and energy. Wastage of these commodities is very high, thereby resulting more environment- related problems. In some under-developed or developing countries, even though their population is considerable, their simple living and moderate lifestyle gives very little environment-related problems.

Large urban areas consume vast quantities of natural resources, which may be transported from hundreds or even thousands of kilometers away. They also produce vast quantities of waste (refer sec. 5.3.2). People living in these urban areas often have little contact with the natural environment and may have little knowledge of, or concern for, their impact on it. Table 1.2 illustrates as how consumption patterns of minerals have drastically changed within 200 years in the USA. This is mainly due to

Table 1.2 A change within two centuries in the pattern of mineral consumption in U.S.A[8]—A developed country.

Few minerals and metals	Per capita consumption in pounds by American citizen		In 200 years how many times increase
	1776	1996	
Al	0	93	93
Cement	12	742	62
Coal	40	7581	190
Clay	100	326	3.26
Cu	1	23	23
Glass	1	150	150
Iron	20	603	30
Lead	2	13	6.5
Phosphate	0	340	340
Potash	1	44	44
Salt	4	404	101
Sand, Gravel, Stone	1000	19061	19
Sulfur	1	111	111
Zinc	0.5	12	24

changes in the lifestyle and living standard of people. Table 1.3 illustrates the lifestyle of an average middle-class American citizen. Is it not detrimental to the environment, as comparing the same with citizens from Asian and African continents whose lifestyle is very simple (mild) but their populations are higher?

1.3.3   Globalization[15]

Business has crossed the barriers/boundaries. Most of the countries are welcoming foreign investment which was closed by many countries in the past. Multinational Companies (MNCs) are looking for acquisition of natural resources such as oil, gas, minerals, etc., and establishing industries of various kinds world over. They are also quitting their rights and possessions where conditions cease to remain favorable. Uneconomical industrial units are no more continuing. Foreign investors bring different systems, culture and managerial skills. They look for political and fiscal stability, prevalent regulations and industrial culture at the country of operation. Globalizations' major driving force is the concern to create a single global market place. Its major characteristics include:

- A single market place where free trading would be possible.
- It encourages use of more and more computerization and automation in industries and trade.
- It establishes new information and communication channels.
- It results in massive urbanization and population migration between countries.

MNCs invest huge sum of money and expect its recovery at the earliest, and hence, the higher rate of returns. Could it be possible to operate industries in an

Table 1.3 (a – upper) & (b – lower): An illustration of the lifestyle of an average middle-class American citizen. Is it not detrimental to the environment?

Consumption pattern during lifetime	Imagine impact to the environment!!!
Drive 700,000 miles	in a dozen cars, using more than 28,000 gallons of gasoline.
Read and throw away 27,500 newspapers	a rate of seven trees a year.
Add 110,250 pounds of trash	to the nation's garbage heap.
Wear 115 pairs of shoes.	and throw away to the nation's garbage heap.
Automobile 63% Light bulb 54% Telephone 42% Television 22% Aspirin 19% Microwave oven 13% Hair-dryer 8% Personal computer 8%	In polling 1,000 Americans, an MIT study found these *essential inventions* that people said they could not do without
There are more than 134,000,000 cars in the United States	More than 8 million new cars are made every year for use in the U.S.A**

Source: Life's Big Instruction Book.

**The consumption pattern varies from one country to another.

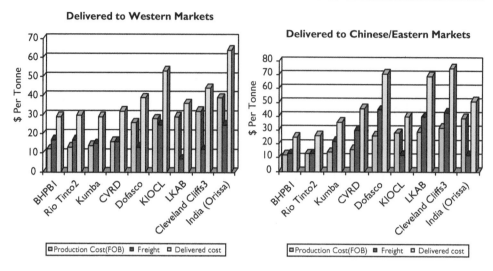

Figure 1.2 A prevalent competition in the delivery price of iron ore to eastern as well as western markets by the leading MNCs.

eco-friendly manner and achieve *a thorough balance amongst production, productivity and safety?* In the prevalent scenario globalization is considered as a system that is having both positive and negative implications for sustainable development.

1.3.4   Buyer's market

Except for minerals, goods and services of strategic importance, the rest of the minerals, goods and services are available at a competitive price. No monopoly in any sector exists any more and thus a quality product at a cheaper rate (competitive price) is readily available; but what about the damage to the environment? Are they producing these goods and services with least pollution so that not only the present generation but also future generations would be benefited?

To understand the impact of competition, Figure 1.2 (left) illustrates the delivery price of iron ore to eastern as well as western markets by the leading MNCs. Companies such as BHP Billiton, Rio-Tinto, Kumba and CVRD are having delivery (production + freight costs) of iron ore just half of some of the Companies in developing countries or in Europe; who are trying hard to overcome shortcomings in the areas such as basic infrastructures, systems, technical know-how etc. and also for bringing further technological excellence to perform better even in adverse natural conditions (such as depth and type of deposits etc.). But to compete in the International market one will have to get rid of all these hurdles.

1.3.5   Technological developments and renovations[13, 15]

Looking at the global scenario, the progress that has been made in manufacturing and production of goods of mass consumption in the last five decades is far ahead of what could have been achieved during the last five centuries. Since the 19th century there have been many important events, inventions and developments that have resulted in new techniques, methods and equipment. This includes use of giant-sized equipment, application of automation, modular systems, and remote control techniques. In tomorrow's industries the use of robotics to carry out repetitive tasks in hazardous and risky locales would be part of the process. The application of lasers for precise survey, measurements and monitoring would play an important role. Sensors installed at strategic locations would help to monitor factories' atmospheres. There will be thorough technology transfer from one field of engineering to another.

1.3.6 Mineral production and consumption trends, and rapid resources depletion[8, 13]

Minerals are the basic raw materials, which are used to produce value added products, and energy (coal, oil, atomic), arms and ammunitions. Their use has been established both during war and peace times. After sun, air and water they are our basic need. Even food cannot be produced without their aid/contribution. Development of any country is measured as per-capita mineral consumption and production. Figure 1.3, illustrates this aspect. Thus, mining is a basic industry that supplies raw materials to numerous industries. It is indispensable but it has adverse impacts on air, water and land environments. It damages flora and fauna.

1.3.7 Information Technology (IT) and its impacts[13, 15]

Use of IT and computers in all the industrial sectors is growing and it has compelled producers to compete in international markets. Many manufacturing companies/

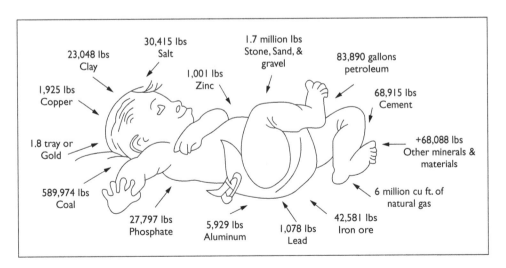

Figure 1.3 A newly born baby would require 3.75 million pounds of minerals, metal and fuels throughout his/her life (or 60 kg/day with life span of 80 years) in a developed country, such as U.S.A What about this requirement in your own country?

units are merging. MNCs are looking for out-sourcing and wish to take advantage of cheaper energy/power, human resources and know-how.

Based on the above description it could be said that consumerism and materialism drive industrialization. Global competition, computerization, automation and mature markets also exercise an influence. These factors all contribute to higher levels of natural resource exploitation and increased levels of pollution and waste with negative impacts on the environment and human health, as described in chapters 2 to 10.

1.4   INDUSTRIAL (OCCUPATIONAL) HEALTH, SAFETY & ENVIRONMENT (HSE)[9, 10, 11, 13]

HSE stands for Health, Safety and Environment. (Here health addresses the health of industrial workers (the direct stockholders), which could be affected by the occupational diseases of many types that are associated with industries (refer chapter 8). It also addresses the health of those affected by the industrial operations – the indirect stockholders. Is there any interrelationship between these three elements?

1.4.1   Industry & environment[13, 14]

Every one of us likes to have precious metals such as gold, platinum, or precious stones such as diamond. Have you ever thought as how we could obtain them? And is there any relation with the environment in getting them. To understand this aspect, let us take example of gold, how do we get it? Gold is mined if its average grade (concentration) is 5 gms/tonne. Or more. It means while mining 1 tonne (1000 kg) of gold-ore (gold + associated rocks), only 5 gms (on an average) would be of useful and the remaining 999.995 kg of rocks that would be generated are waste. In recovering this 5 gms gold; it has gone through Mining (breaking rock into small fragments from in-situ); concentration (crushing, grinding into powder and separation from the rest of the rock-mass using chemicals), smelting and lastly refining and casting into bars or any other shape (Figure 1.4). One can imagine how much energy it required, materials of different kinds it consumed, foul gases it produced and land it required to dispose of the wastes generated. This is the reason mining is blamed for pollution. Figure 1.5 outlines impacts of gold production to the environment. It is not only the gold but minerals are our basics needs and we cannot do away with them. One should not forget that more than 75% power is generated using fossil fuels (coal, oil and gases). Automobiles are run by fossil fuels. Fossil fuels have greenhouse effect (sec. 2.8.5 and 3.6). Exhaust from automobiles is responsible for bad ozone (sec. 3.8). Industrial pollutants are causing acid rain (sec. 3.7), for example in a recent survey it was found that in China it is affecting about one-third of its land.

1.4.2   Industry & safety[13, 14]

Not only the environment, in the process of producing gold, imagine! As at how many stages the direct industries (mining through to – mineral beneficiations, smelting, refining, casting and production of finished goods) and indirect industries (those providing services such as water, power, communication, transport etc.) men have saved

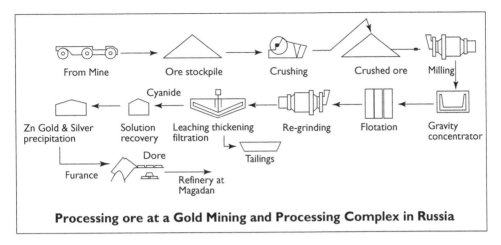

Processing ore at a Gold Mining and Processing Complex in Russia

Figure 1.4 Stages mining through to finished product (metal) involving air and water pollution and land degradation but who should be blamed? Consumers (society as a whole), or the miners!!!

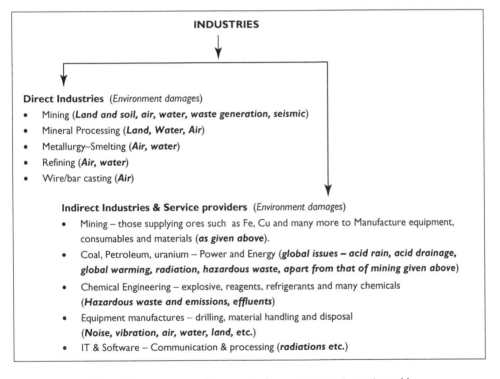

Figure 1.5 An account of damage to the environment in getting gold.

themselves from accidents. The accidents damage the men, machines and equipment, structures, and surroundings. They incur huge financial losses.

The preceding sections and illustrations (Figures 1.6 and 1.7) could explain how industries are responsible for creating pollutions of various kinds, and are also liable

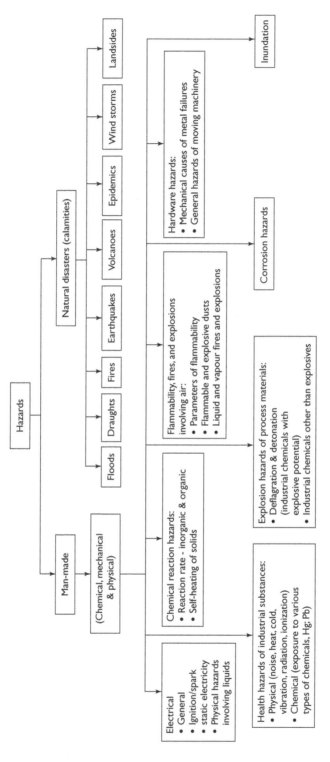

Figure 1.6 Classification of hazards.

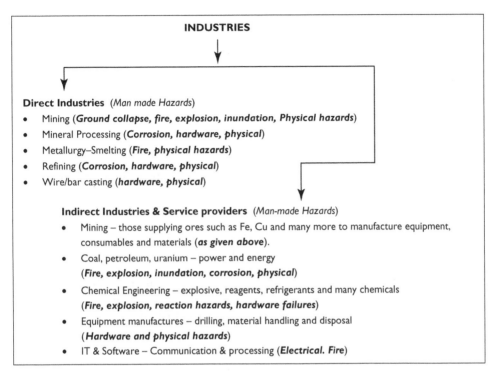

Figure 1.7 An account of hazards and risks involved in getting gold. (Physical hazards relate to damage to health due to occupational diseases of various types, as described in Chap. 6).

for hazards of many kinds that could damage the health of those who are involved directly or indirectly.

1.5 IMPACT OF INDUSTRIALIZATION ON SOCIETY[1, 2, 4, 9, 14]

1.5.1 Mass balance system/equation

To understand the impact of industrialization and pre-industrialization on society let us understand a basic equation, which is known as: **Mass Balance System/Equation.**

$$\text{ACCUMULATION} = \text{INPUT} - \text{OUTPUT}$$
$$\text{Or}$$
$$\text{ACCUMULATION RATE} = \text{INPUT RATE} - \text{OUTPUT RATE} +/- \text{TRANSFORMATION RATE}$$

'Imagine if a river becomes stagnant (doesn't flow), sky smoke shrouded (covered), dumping ground odoriferous and unsightly (ugly). Populations can ignore all these things till they have not ill effects on their health and well-being. It went on for many years but ultimately man started experiencing negative impacts on health, aesthetic and cultural pleasures and economic opportunities'.

ADD or NOT ADD Approach:

Basic Questions:
Can we Avoid POLLUTION?
Can we Minimize POLLUTION?
Can we Avoid ACCIDENTS?
Can we Minimize ACCIDENTS?

The answer lies in the fact that pollution and accidents cannot be avoided but they can certainly be minimized.

1.5.2   Damage due to industrialization[1, 2, 4, 9, 10]

The damages that have been caused by industrialization are cropping up. Let us understand what has happened in the past.

It is very difficult to calculate damages caused by air pollution and the remedy is even more expensive. But the following estimates are instructive. They relate to sales in air pollution combat equipment (1980–2000): US$ 4 billion in North America, US$ 4.2 billion in Europe and US$ 4.5 billion in the rest of the world. Damages to agricultural products, forests and human health exceeded US$ 10 billion. Annual expenditure to decrease 55%-65% of the remaining sulfur emissions from states of the European Community (EC) during the 1980–2000 ranges between US$ 4.6 billion to 6.7 billion, and the cost of monitoring nitrogen emissions by 10% annually till the year 2000 ranges between US$ 100,000 to 400,000. Cost for Health and safety for British Industry is in the range of 4 to 9 billion pounds sterling a year.[10]

1.5.3   Birth of global issues[7]

Damages that have been caused by industrialization include many disasters, as described in Table 1.1 and chapter 3. Besides these damages to health and environment it has given birth to some of the following issues:

• Global warming and green house effects
• Ozone depletion
• Bad ozone (photo-chemical smog)
• Acid rain and acid drainage.

1.6   HSE – A CRITICAL BUSINESS ACTIVITY[9, 12, 14, 15]

An accident that could result due to unsafe acts and unsafe conditions adversely affects the production, costs and productivity of any industrial setup (ref. Chap. 8). It spoils reputation and demoralizes workers. Many a times it results int court inquiries and disputes. It could damage the environment. The foregoing discussions also reveal that any occupational disease that could be due to degradation in the environment

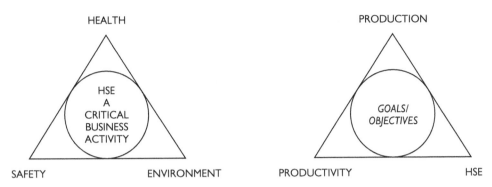

Figure 1.8 (a) (left) Interrelationship of occupational health, safety and environment. (b) (right) Interrelationship of critical business activities.

(pollution) could reduce working efficiency of an industrial worker, his morale and ultimately reputation of the company. All these issues are of public concern. Ultimately they result in a deviation from the laid out objectives and goals of a company. Thus, the logical approach is a thorough balance amongst the production, productivity and safety, giving equal weight to each of these components, like three sides of an equilateral triangle (Figure 1.8(b)).

Occupational Health, Safety and Environment (HSE) are amongst the critical business activities in achieving the set goals and objectives of a company. Any imbalance amongst them can jeopardize business (Figure 1.8(a)). It could be achieved by considering HSE as one of the critical business activity, at par with production and productivity as shown in Figure 1.8(b); and managing it by following the basic principles of business management (chapter 10) that include:

•   Organization
•   Policy
•   Procedures
•   Supervision
•   Management review and appraisal.

1.7   ENVIRONMENTAL POLICY[12, 15]

In addition to above concepts and criterion, the important aspect that need to be addressed is 'Sustainable Development' as briefly described in the following paragraphs, and in chapters 9 and 10, in detail. Together with HSE the welfare of society is equally important for achieving sustainable development as shown in Figure 10.1.

1.7.1   Sustainable development[4, 14, 15]

The eaning of the word 'sustain' is to keep in existence. Sustainable development is the one which keeps in existence the ecological cycles that sustain renewable natural resources.

Past development has emphasized exploitation and optimization of mineral and natural resources, manufacturing of goods and commodities of mass consumption and developing utilities and services with little concern for long-term environmental impacts. It has largely ignored constraints arising from the finite character of non-renewable natural resources and the ecological cycles that sustain renewable natural resources.

It is the industrialization in the immediate past, say within the last two centuries that has given rise to global issues such as: acid rain, ozone depletion, bad ozone, global warming, air and water pollutants that are causing huge expenditure on health care and remedial measures. It is this negligence and carelessness of the past that is compelling every one towards 'Sustainable Development'. Sustainable development is beneficial *socially, economically and ecologically* to the present as well as future generations.

Let us put it in a simple way; sustainable development is what takes care of environmental concerns. If environmental concerns are ignored, growth and development may lead to short-term improvements in overall living standards. However, they will lower the quality of life for many people, particularly poorer people who already face degraded living environments. Failure to address the sustainable use of natural resources will degrade the resource base on which we depend. To avoid this, environmental policy must set us on a course that will achieve the goal of sustainable use, where the environmental impacts of society are in harmony with natural ecological cycles of renewal. To achieve this, sustainable development must ensure that the direction of investments, the orientation of technological developments, and institutional mechanisms work together towards the goal of sustainable use that will meet present and future needs. For detail also refer sec. 10.14.

1.7.2   Development of industrial technology[15]

In order to achieve sustainable development, there must be compatibility between economic development and the maintenance of the environment. It also warrants development of industrial technology that can contribute to improved management of environmental impacts.

Industry has a key role to play in reconstruction and development. It has become more sophisticated, with advanced technological capabilities in some areas and has increased its contribution to GDP and employment creation but it should become eco-friendly; that would allow sustainable development.

1.7.3   Education[15]

In the present scenario, the role of Occupational Health, Safety and Environment-related education is of the utmost importance due to the fact that it improves our everyday lives by:

- Protecting Human Health
- Advancing Quality Education
- Creating Jobs in the Environmental Field

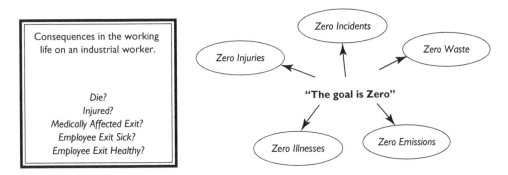

Figure 1.9 (Left) Consequences in the working life on an industrial worker. (Right) Aiming at "The goal is Zero" of industrial evils could result in a scenario of healthy and content employee when she or he retires.

- Promoting Environmental Protection along with Economic Development
- Encouraging Stewardship of Natural Resources.

1.7.4 The ultimate goal[5, 6, 14]

Industries require human resources, which are ample (Figure 5.2) but if the current trend of consumption pattern of other resources such as minerals, water and energy continues they would soon be depleted and their scarcity would be felt. The present boom in the mineral market and oil prices are already influencing world's economy. Water scarcity has begun, and it is being felt world over.

All these factors are appealing for producing goods and services with the optimum utilization of available resources: men, machines and equipment and energy. *This scenario is also asking for better recoveries, minimum wastage and cost reduction in every sector. HSE will have to be considered as a 'Critical Business Activity' to achieve Sustainable Development*, described in the above sections/paragraphs, and rest of the chapters.

Working in industry could cause, as shown in Figure 1.9 (left); deaths, loss of man-shifts due to injuries, illness when medically affected and loss of productivity due to inefficiency on health grounds, resulting in a sick employee when he or she retires from industry. But following the strategy (as shown in Figure 1.9 (right)) of zero incidents, zero injuries, zero fatalities, zero illness, zero waste and zero emissions could result in a scenario of a healthy and content employee when she or he retires. Should we not strive for this strategy? Please also refer to sections 10.14 and 10.15.

QUESTIONS

1 Is there any relation between industry and safety; also industry and environment? Describe them.
2 Can we do away with minerals? What is their contribution to our society? How do they influence environment?

3 How can we produce goods and services of mass scale consumption/utilization with least pollution?
4 In the prevalent worldwide recession and competitive market situation what strategy would you propose for achieving the best results?
5 List resources required to run industries. List gains and adverse impacts from industries.
6 Name some air pollution disasters of the past and what had been the main cause for most of them? How industries are responsible for pollution.
7 What are the impacts of globalization? List global issues that have adverse impacts on the world's citizens? Are they a result of industrialization during the recent past?
8 What are industrial evils? Why should we aim at 'Zero-Goal of industrial evils'? What should be the ultimate goal to achieve the best results from industries?
9 What are the impacts of population growth? How does the lifestyle of people influence the problem of growing population? How can population be controlled?
10 Why is Occupational Health Safety and Environment (OHSE)-related education essential? What should be our environment policy?
11 Write the mass balance equation i.e., Accumulation rate and its significance.

REFERENCES

1 Boubel, R.W., Turner, D.B. & Fox, D.L. (1984) *Fundamentals of air pollution*. Academic press Inc. pp. 1–15.
2 Crowl, D.A. & Louvar, J.F. (2002) *Chemical process safety*. New Jersey, Prentice Hall PTR. pp. 23–29.
3 Environment Health Center, National Safety Council. Reporting on climate change: Understanding the Science, 2nd. Edt. June 2000.
4 EPA (Environmental Protection Agency, USA) URL:http://www.epa.gov/ebtpages/air.html data and information for the period 1980–2000.
5 Jager, K.D. (2006) Wellness in the Workplace. *International conference focusing on safety and health* 14–16 November 2006, Johannesburg, South Africa. London ICMM.
6 Jansen, J. (2006) First Steps on the Journey to Zero Harm. *International conference focusing on safety and health* 14–16 November 2006, Johannesburg, South Africa. London ICMM. (Permission: Lonmin).
7 Kupchella, C.E. & Hyland, M.C. (1993) *Environmental Science*. Prentice-Hall International Inc. pp. 61–68, 112–14, 191–93.
8 Mineral Information Institute (2001) *Web site*, Denver, Colorado.
9 Petroleum Development of Oman – HSE management, course material, reports and documents and interaction with those concerned during period: 1995–2004. Also incorporating: Shell International Exploration and Production Company – HSE Management system.
10 Ridley, J. & Channing, J. (1999) *Workplace safety*. Butterworth Heinemann. pp. ix–x.
11 Tatiya, R.R. (2005) *Civil Excavations and Tunneling – A Practical Guide*. London, Institution of Civil Engineers/Thomas Telford Ltd. pp. 273–305.

12 Tatiya, R.R.: *Course material for Petroleum and Natural Resources Engg; Chemical Engg. And Petroleum and Mineral Resources Engg.* Sultan Qaboos University, Oman, 2000–2004.

13 Tatiya, R.R. (2006) Health, Safety and Environment (HSE) Management – Where do you stand? *Souvenir: 17th Environment Week Celebration, Indian Bureau of Mines.*

14 Tatiya, R.R. (2005) *Surface and underground excavations – methods, techniques and equipment.* Netherlands A.A. Balkema/Taylor and Francis. pp. 1–17.

15 White Paper on: Environmental Management Policy, Department of Environment Affairs and Tourism, South Africa, July 1997.

Chapter 2

Ecology, environment, mineral resources & energy

"The people have the right to clean air, pure water, and to the preservation of natural, scenic, historic and aesthetic values of the environment. Pennsylvania's public natural resources are the common property of all the people, including generations yet to come. As a trustee of these resources, the Commonwealth shall conserve and maintain them for the benefits of all the people. (Article 1, Section 27)[16]"

Keywords: Ecology, Great Spheres, Ecosphere, Biosphere, Biotic & Abiotic Components, Natural Cycles, Food Chain, Ecological Pyramid, Population & its Growth, Mineral Resources, Nonrenewable and renewable Resources, Energy Sources, Green Power, Clean Coal Technology, Energy Crisis, CO_2 emissions.

2.1 ENVIRONMENT RELATED ISSUES

2.1.1 Ecology

Meaning of Greek word *'Oikos'* is *'house* or *place of residence' and 'logy' means study.*[24] Thus, Ecology is a study of components that exists or resides on the 'Planet Earth'. There are two components that reside on earth:

1 Biotic (those having lives i.e. organisms) and
2 Abiotic (those not having lives), also known as physical components.

2.1.1.1 Ecosystem

A self-regulating natural unit consisting of all plants, animals and micro-organisms (biotic component) in an area functioning together with all of the non-living physical (abiotic component) of the environment.[19] For example: An ant farm inside a glass aquarium is an ecosystem. Maggots and microbes causing decay of garbage in a collection container constitutes an ecosystem, a pond, or valley, or the planet itself can be considered an ecosystem.[18]

Biotic components of an ecosystem includes all living organisms and their products including the waste products and decaying matters such as urine, feces, decaying leaves and twigs, bones and flesh. Whereas abiotic or physical components of ecosystem includes water, air, sunlight, minerals and their interaction to produce climate, salinity, turbulence and other conditions that influence the physical conditions under which the organism survives.[18]

2.1.1.2   Classification – Ecology

Ecology is a broad discipline comprising many sub-disciplines. A common, broad classification, moving from lowest to highest *complexity*, where complexity is defined as the number of entities and processes in the system under study, as shown in Figure 2.1.[28]

Ecology can also be sub-divided according to the species of interest into fields such as animal ecology, *plant ecology, insect ecology, marine ecology and so on*[9, 19, 20]

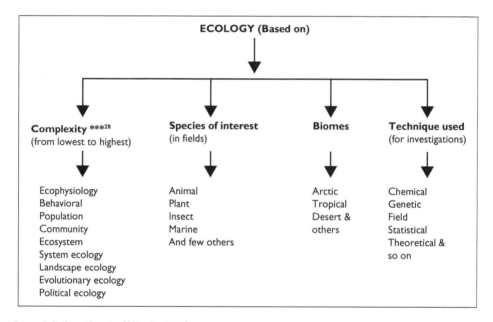

Figure 2.1 Classification*** – Ecology[9]

- **Ecophysiology** examines how the physiological functions of organisms influence the way they interact with the environment, both biotic and abiotic.
- **Behavioral** examines the roles of behavior in enabling an animal to adapt to its environment.
- **Population** studies the dynamics of populations of a single species.
- **Community** focuses on the interactions between species within an ecological community.
- **Ecosystem** studies the flows of energy and matter through the biotic and abiotic components of ecosystems.
- **System ecology** is an interdisciplinary field focusing on the study, development, and organization of ecological systems from a holistic perspective.
- **Landscape ecology** examines processes and relationship in a spatially explicit manner, often across multiple ecosystems or very large geographic areas.
- **Evolutionary ecology** studies ecology in a way that explicitly considers the evolutionary histories of species and their interactions.
- **Political ecology** connects politics and economy to problems of environmental control and ecological change.

Another frequent method of subdivision is by biomes. Biomes are the major community types. Biomes are characterized by specific climatic conditions (such as temperature, rainfall and the evaporation rate), topography and lithology of the region that results varying plants types. Plant types in turn govern the animals that can exist in a particular biome. Major biomes include: Tundra, Taiga, Temperate Forest, Temperate Grasslands, Tropical Savannas, Tropical Forest and Deserts. Ecologically they are known as: Arctic or polar ecology, tropical ecology, desert ecology etc. Thus, a biome is a community of interacting vegetation, soil-types, and animal populations, which are adapted to the physical environment of a region.[18] Land biomes include rainforests, grassland, desert, tundra, coniferous forest and deciduous forest whereas water biomes include marine and freshwater.

Biomass production is production of grams of organic matter per square meter in any region and biomass-productivity is defined as the production of grams of organic matter (carbon) per square meter per day in that region.

An ecosystem such as that found in taigas may be high in biomass, but slow growing and thus low in productivity. Ecosystems are often compared on the basis of their turnover (production ratio) or turnover time, which is the reciprocal of turnover.

Total productivity of ecosystems is sometimes estimated by comparing three types of land-based ecosystems and the total of aquatic ecosystems. Biomes differ in production and productivity while compared amongst themselves (Earth's land area) and also when compared with the rest in the ocean, as under:[9]

- One third of earth's land area is occupied by dense biomass; one third by Savannas, meadows and marshes which contain less biomass but they are productive and represent the regions which meet human needs for food. Deserts, semi-deserts, tundra, alpine meadows and steppes occupies 1/3 of earth's land area but they are less productive and have sparse vegetation.
- Lastly marine and fresh water ecosystems (3/4 of earth's surface) are characterized by very sparse biomass except at the coastal zones/areas.

Techniques such as chemical ecology, genetic ecology, field ecology, statistical ecology, theoretical ecology and so on and so forth, (Figure 2.1), which are used for investigation also subdivide this discipline into groups.

2.2 EARTH'S GREAT SPHERES

The outer layer of the Planet Earth is accommodating the following envelopes, known as Earth's great spheres:[9, 16]

- Atmosphere (House of Earth's gases)
- Lithosphere (House of Earth's soils, rocks and minerals)
- Hydrosphere (House of Earth's water)

Ecosphere is the grand system that includes 'all life forms' on earth together with the parts of the earth (non-living components) on which, and in which, living things exist. The Ecosphere comprises all living organisms and some parts of all the three spheres up to which living organisms could reach. But other parts of all these spheres are inaccessible to living things.

2.2.1 Biotic component of planet earth – Biosphere

Biosphere is the sphere within ecosphere where life can survive without the aid of man-made protective devices. Relative to the volume of the Earth, the biosphere is only the very thin surface layer that extends from 11,000 meters below sea level to 15,000 meters above it.[18] It encompasses deserts, lakes, forests etc. within it. Within Biosphere, as known to us there are about[23]:

- 330 thousand species of green plants
- 930 thousand species of Animals
- 80 thousand species of bacteria and fungi
- It is believed that total number these species is about 3 millions.

Thus, Biosphere is very large and complex. It encompasses air, light, heat, food and habitats (shelter). All the ecosystems make 'Biosphere'.

The biosphere contains great quantities of elements such as C, N, H_2 and O_2. Other elements, such as P, Ca, and K, are also essential to life, but they are present in smaller amounts. At the ecosystem and biosphere levels, there is a continual recycling of all these elements, which alternate between the mineral and organic states. In the following sections, cycles of S, N and C have been briefly described to illustrate how natural cycles function and adverse impacts due to imbalance, if any.

2.2.2 Natural cycles

2.2.2.1 Sulfur cycle[16]

- It is an important element in protein and carbohydrate complexes.
- Organic sulfur is converted into inorganic sulfate by common decomposers (bacteria and fungi). Plants take it as sulfates.
- The sulfur cycle requires numerous species of bacteria and fungi that inter-convert sulfates, sulfides and elemental sulfur.
- Sulfur may enter the atmosphere as SO_2 or H_2S. SO_2 is toxic to plants and animals. And it reacts in the atmosphere to form sulfuric acid. This falls to the ground with various forms of precipitation (Acid Rain Figure 3.8) (ref. sec. 3.7)
- SO_2 is formed by the combustion of organic material – leaves, coal etc.

2.2.2.2 Nitrogen cycle

Nitrogen constitutes 78.08% of the gaseous envelope (Table 3.1) that surrounds the planet earth. Nitrogen is inaccessible to most of the living organisms. Nitrogen must

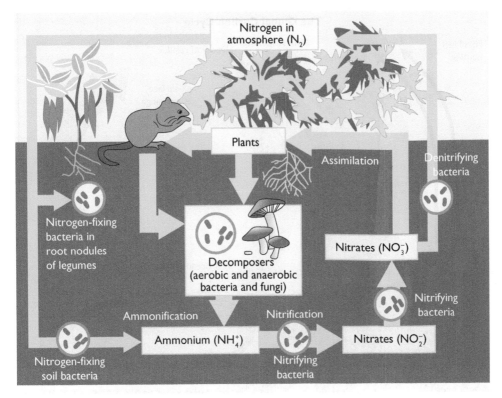

Figure 2.2 Nitrogen cycle.[11] Courtesy: EPA, USA.

be fixed by soil bacteria living in association with the roots of a particular plant such as legumes, clover, alfalfa, soybeans, peas, peanuts and beans. Living on nodules around the roots of legumes, the bacteria chemically combine nitrogen in the air to form nitrates (NO_3) and ammonia (NH_3) making it available to plants.[11] Organisms whose feed are plants ingest nitrogen and release it in their organic wastes. Denitrifying bacteria frees the nitrogen from the wastes returning it to atmosphere;[11] this is how the cycle is completed, as shown in Figure 2.2.

2.2.2.3   Carbon cycle[6, 8, 16]

Carbon constitutes 0.03 to 0.04% of the gaseous envelope (Table 3.1) that surrounds the earth. It is the fourth most abundant element on the earth and the key element of the living organisms including human beings. Conversion of carbon dioxide into living matter and then back is main pathway of carbon cycle (Figure 2.3).

Most of carbon is available from within inorganic mineral deposits such as limestone and organic fossil fuel deposits such as coal and petroleum. Due to combustion of fossil fuels, weathering and dissolution of carbonate rocks, and volcanic activities, some of the bound carbon (locked in the geologic reservoir as shown in Figure 2.3) returns to the atmosphere as carbon dioxide or carbonic acid.[23]

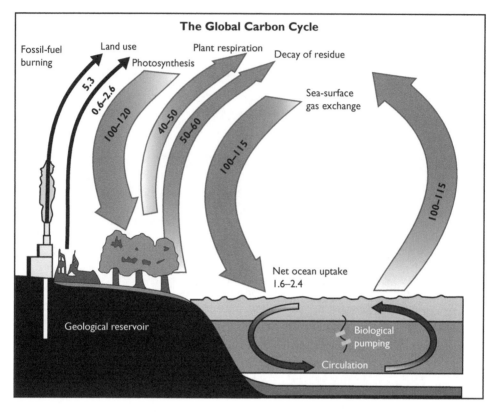

Figure 2.3 Global carbon cycle.[6]

Source: Oak Ridge National Laboratory, http://www.esd.ornl.gov/iab/iab2-2.htm

Plants draw about one quarter of the carbon dioxide out of the atmosphere and photosynthesize it into carbohydrates. Some of the carbohydrates are consumed by plants' respiration and the rest are used to build plants' tissues and growth. Animals consume the carbohydrates and return carbon dioxide to the atmosphere during respiration. Carbohydrates are oxidized and returned to the atmosphere by soil microorganisms decomposing dead animals and plant remains (soil respiration).

The world's oceans through direct air-water exchange absorb another quarter of atmospheric carbon dioxide. Interchange of CO_2 between atmosphere and the oceans though diffusion and the oceanic reservoir tends to regulate the atmospheric CO_2 concentration.[23]

It is worth noting that the percentage of CO_2 in the atmosphere has been increasing since the last century. The main source is anthropogenic sources like burning fossil fuels etc. It is ultimately resulting into global warming. The solution is not simple and straightforward. There are number of parameters that are contributing the heavy emissions, as shown in Figures 2.20 and 2.31. Companies having a proactive attitude started taking positive steps as illustrated in Figure 2.30 to reduce emissions and also to conserve energy.

2.2.3   Solar energy's contribution[11, 19, 22]

There is a slight input of geothermal energy; the bulk of the functioning of the eco-system is based on the input of solar energy. As shown in Figure 2.4; 99.9% of Sun's energy reaching earth is reflected to space, or absorbed, or evaporates water and the rest 0.1% is used by plants. About 20% of this is used in plants' respiration; and 50–80% used in plants' growth.[18]

Plants and photosynthetic microorganisms convert light into chemical energy by the process of *photosynthesis*, which creates glucose (a simple sugar) and releases free oxygen. Glucose thus becomes the secondary energy source that drives the eco-system. Some of this glucose is used directly by other organisms for energy. The rest of the sugar molecules can be converted to molecules such as amino acids. Plants

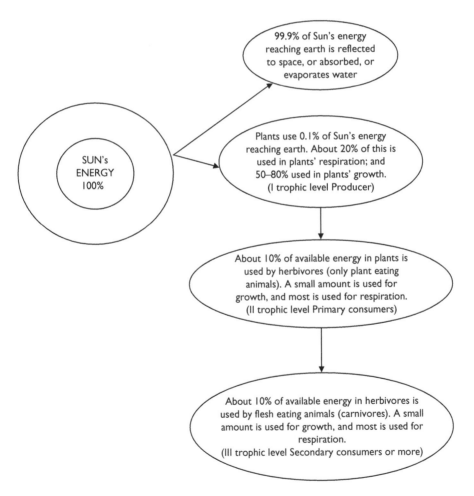

Figure 2.4 Solar energy and its contribution.

use some of this sugar, concentrated in nectar, to entice pollinators to aid them in reproduction.

$$6CO_2 + 12\ H_2O \rightarrow ^{\text{LIGHT (solar) ENERGY}} \rightarrow C_6H_{12}O_6 \text{ (Glucose)} + 6O_2 \text{ (Oxygen)} + 6H_2O \text{ (Water)}$$

Simplified presentation of the process 'Photosynthesis'[16]

Cellular precipitation is the process by which organisms (like mammals) break the glucose back into its constituents: water and CO_2, thus regaining the stored energy the sun originally gave to the plants.

$$6O_2 + C_6H_{12}O_6 \text{ (Glucose)} \rightarrow ^{\text{ENERGY}} \rightarrow 6CO_2 + 6HO_2$$
Simplified presentation of the process 'Cellular Respiration'[16]

The proportion of photosynthetic activity of plants and other photosynthesizers to the respiration of other organisms determines the specific composition of the Earth's atmosphere, particularly its oxygen level. That's why ecologists have used two terms: *Gross production* and *net production* (Figure 2.6). Global air currents mix the atmosphere and maintain nearly the same balance of elements in areas of intense biological activity and areas of slight biological activity.[9]

Water is also exchanged between the hydrosphere, lithosphere, atmosphere, and biosphere in regular cycles. Solar energy causes water to *evaporate*. When water evaporates from plants it is called *Transpiration*. All the water vapor ultimately undergoes cooling; this process is called *Condensation*. This converts water into droplets or ice crystals, and return to the earth surface; this process is known as *Precipitation*. (Refers Sec. 4.1.1 also)

An ocean is a large tank that stores water, ensures thermal and climatic stability, and facilitates the transport of chemical elements. This is how there is a large contribution of the oceans and their oceanic currents.

As described in the preceding sections living things need shelter, food, energy, water, and the ability to maintain constant internal body temperature.

The living space (shelter) where an organism lives is called '*habitat*'. A group of similar organisms is called '*Population*'. A group of different populations is called '*Community*'. Thus, an ecosystem has community and abiotic features. As stated above in an ecosystem living things interact with themselves as well as with the non-living things in a particular region. The ecosystem cannot satisfy all the needs of the living things in a habitat, particularly food and water. As such only a limited number of the same types of living populations can survive there and that is why many animals defend their areas which they consider as a suitable space for their living. This creates '*Competition*' which is a struggle amongst the living things to get proper and adequate space, food, water and energy. This competition persists within species and also between species.

2.3   FOOD – FOOD CHAINS – FOOD WEBS[9, 11, 13, 16, 18, 24]

Organisms that make their own food are known as autotrophs (*Producers*). Plants get energy from sun to make food (sugar) from CO_2 and water and that is why they are known as: '*Producers*'. All green plants are producers. They take in chemicals from air, water and soil and heat energy from the sun. *Plants do not produce from nothing.*

Animals cannot produce their food and they have to depend on grass or other animals, and they are known as '*Consumers*'. There are three types of consumers.

1   Animals that eat only plants are known as '*Herbivores or Primary consumers*'; for example: deer, elk, mouse and beaver, grasshopper etc. who eats grass.
2   Animals that eat other animals are known as '*Carnivores or secondary consumers*'. And those carnivores that eat other carnivores are known as '*tertiary consumers*' *(lions, leopards, jackals, hyenas, timber wolves etc.).* In a marine-food web: Phytoplankton is eaten by the small fishes that are eaten by seals and seals are the food of killer whales.
3   Animals and human beings who eat both animals and plants are known as '*Omnivores*' (coyotes, badgers etc.).
4   The organisms that get their food from dead organisms are called '*decomposers*'. Bacteria, fungi and insects are the decomposers who degrade organic matter of all types and restore the nutrients to the environment; and producers then consume these nutrients, and this '*completes the cycle*'.

In this manner energy from the sun is passed from one organism to another. This is called a '*Food chain*'. Any organism rarely eats one type of food having the same energy level; rather they have a few choices, and similarly they too serve as food for a variety of other creatures that prey on them. These interconnections create a '*food web*'.

Based on the above it could be inferred that energy transfer is about 10% from one trophic level to the one above it. Net production is only a fraction of the gross production as the organisms have to consume it for their own survival and the rest is transferred.

Water: Living things are made of water for example human beings (70–80%); Chickens (75%); Pineapple (80%) and tomato (95%); Illustrating the fact that how important is water for living organisms.[10]

Homeostasis is the ability of an organism to keep conditions inside the body the same even though conditions in its external environment change. Such animals are called warm-blooded animals.[13]

The mass of all organisms in an area or ecosystem is known as '*Biomass*'. While assessing mass at each trophic level and then plotting it, it forms a '*Pyramid*' in which the mass of producers is far greater than each of the consumers (primary, secondary and tertiary).

Whenever an organism eats another, not all the energy is transferred but only about 10% of the energy of a producer is transferred to the consumer that eats it. This leads to progressive loss of energy at each level of a food chain. While assessing energy level at each level of an ecosystem or an area and plotting it, the resultant shape of the plot is also a pyramid, as shown in Figures 2.5 and 2.6. This is known as '*Pyramid of energy*'.

Similarly, by counting the number of producers and consumers at each trophic level in an area or ecosystem we can draw the '*Pyramid of numbers*'. We would notice that there is a progressive drop in numbers while moving from producers towards consumers. This is because of the drop in energy that could be transferred from one level to another. H. T. Odum analyzed the flow of energy through a river ecosystem in Silver Springs, Florida. His findings are shown in Figure 2.6. The units are given in kilocalories per square meter per year (kcal/m²/yr).[12]

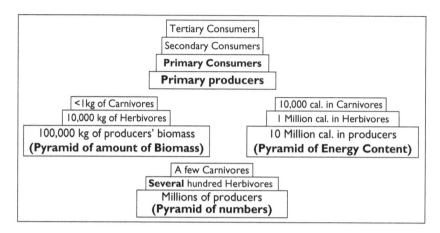

Figure 2.5 An ecological pyramid – Numbers, amount, and energy levels (Not to scale but only illustrative).[11, 19, 22, 28]

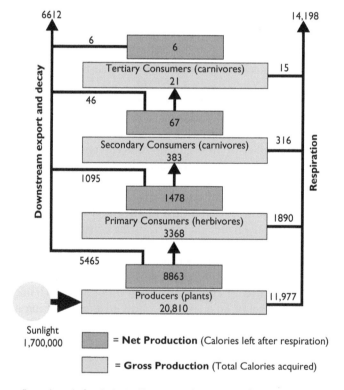

Figure 2.6 Energy flow though food chain illustrating formation of pyramid. After H.T. Odum who analyzed energy flow through a river ecosystem in Silver Springs, Florida. Numbers are shown in kilocalories per square meter per year. Adapted from (http://users.rcn.com/jkimball.ma.ultranet/BiologyPages/F/FoodChains.html).[12]

| Corn & Grains to Feed 1 Cow to produce beef for 1 man for period X | = | Corn & Grains to Feed 20 men for period X |

Figure 2.7 Impact of eating-habits on conservation of energy and resources.[18] (After Moore, G.S. 2007).[18]

These concepts lead to the idea of biomass (the total living matter in an ecosystem), primary productivity (the increase in organic compounds), and secondary productivity (the living matter produced by consumers and decomposers in a given time).

These last two ideas are keys; since they make it possible to evaluate the *carrying capacity*–the number of organisms that can be supported by a given ecosystem. Since in any food network, the energy contained in the level of the producers is not completely transferred to the consumers; the higher up the chain, the more energy and resources are lost.[9]

Thus, from a purely energy and nutrient point of view, as illustrated in Figure 2.7, it is more efficient for humans to be primary consumers (to subsist from vegetables, grains, legumes, fruits, etc.) than to be secondary consumers (consuming herbivores, omnivores, or their products) and still more so than as a tertiary consumer (consuming carnivores, omnivores, or their products). An ecosystem is unstable when the carrying capacity is overrun.

2.4 ABIOTIC[19, 22, 28]

Abiotic components of planet earth are geological, geographical, hydrological, and climatological parameters. These nonliving parts of an ecosystem have physical and chemical features. Physical features include wind, terrain, soil moisture, water current, temperature, soil porosity and light intensity. Chemical components are all the materials, minerals, gases, water and a wide variety of chemicals. All chemicals could be part of a living thing and also that of a nonliving component. Figure 2.8 illustrates an ecosystem's various combinations.

POINTS TO NOTE[2, 16, 19, 21, 24, 28]

- Living components affect the physical and chemical features. Chemicals can be part of living components at one moment and part of nonliving components a later moment, for example the carbon molecule is a part of a biotic system when consumed by an organism but it becomes abiotic when exhaled as a molecule of CO_2 and becomes part of the abiotic environment.
- Living organisms manipulate (handle in an unfair way) the constituents of earth: air, water and soil.
- All human environmental problems have their roots in fundamental principles of ecology. An ecological crisis can cause extinction and well-being of the remaining individuals.
- Technology and population growth are worsening the situation day by day.
- Chemical and physical features of an environment: heat (temperature), moisture (water), light (which provides energy to the ecosystem through photosynthesis),

Figure 2.8 Ecosystem – various combinations. Courtesy: Wikpedia. Sources for some of the photographs are references.:[9, 11, 16, 18, 21, 24]

* For atmospheric gases (air's composition) also refer sec. 3.2.1.
** Mountain Ecosystems: Mountain ecosystems play a key role in providing forest cover, feeding perennial river systems, conserving genetic diversity, and providing an immense resource base for livelihoods through sustainable tourism. At the same time, they are among the most fragile of ecosystems in terms of susceptibility to anthropogenic shocks. There has been significant adverse impact on mountain ecosystems by way of deforestation, submergence of river valleys, pollution of freshwater sources, despoliation of landscapes, degradation of human habitat, loss of genetic diversity, retreat of glaciers, and pollution.

soil composition (pH, salinity, nitrogen and phosphorus content, ability to retain water, density, etc.) influence the organisms that live there. But the reverse is also true that living organisms also influence the physical and chemical features of the prevalent environment. As it can be seen by the impact of global warming on Swiss Alps in the following photographs (Figure 2.9).[9]

Figure 2.9 The retreat of Aletsch Glacier in the Swiss Alps (situation in 1979, 1991 and 2002), due to global warming.[9] Courtesy: Wikpedia.

- Generally, an *ecological crisis* occurs due to imbalance between the biotic and abiotic components of an ecosystem. The impact could be local or global. Abnormal emission of CO_2 is a global phenomenon which has started showing its impacts, as evident by these photographs (and as shown in Figures 3.4 to 3.6, 2.18 to 2.22 and sections 2.8.5, 3.6; Table 3.11).

2.5   POPULATION [16, 24, 25]

It is a group of interbreeding members of the same species.

2.5.1   Impacts of population growth

In the calculation of the doubling period, the following mathematical relation could be applied.[16]

Doubling period $T_d = 70/r$ (2.1)
r = Percentage growth rate = b – d + m   (2.2)
b = Percentage birth rate
d = Percentage death rate
m = Percentage emigration rate

Exponential (Geometric) Growth Rate:
Exponentially means some constant raised to power. This means as population gets larger and larger, it grows faster and faster in succeeding intervals of time. Doubling time can vary from minutes in case of bacteria to many years in case of elephants (750 years).
Zero Population Growth:
Crude birth rate = crude death rate; the result is zero population growth.

2.5.2   Human population – important aspects[9, 16, 18, 24, 25]

Environmental problems (adverse impacts) in any region are influenced not only to numbers of people but also on impact of their lifestyle. A few people with a highly consuming lifestyle can have a much greater impact on the ecosphere than a large number of people with a simpler lifestyle. For example USA having 5% of the world's population consumes about 1/3 of mineral resources. Asian and African communities though they are large in numbers, have moderate lifestyle. Also refer sec. 1.3.1.

1   Population control is much more than birth control technology. Economical, social, religious and political factors also come into play.
2   In the past, war famine and disease have controlled population. Today the logic that regulates population growth in humans is economic, social and instrumental.
3   When the fertility rate fall below the replacement rate (as in USA), it will take a number of years to reach zero population growth.
4   Present annual growth rate of the human population is about 1.8% and it will take about 39 years *to double it*. With the rise in population, lifestyle has also changed as shown in Figure 2.10.
5   Mechanisms to bring more food to the world market are limited and it will require extensive energy inputs. Developing and underdeveloped countries cannot afford meat as one of the main foods consumed and that is why Asian and African Countries, having huge populations, the vegetarians are in majority.
6   The hungriest people are usually the poorest.
7   Traditional farming is based on ecological principles and cycles. It is more sustainable than chemical farming.
8   It has yet to be determined whether the human population growth curve is *J-shaped* or *S shaped.*

The population problem is just not numbers, but has various environmental impacts. Providing ever-increasing numbers with clean air, water (sec. 4.2), nourishing food and shelter is becoming more and more difficult. The world's culture, customs and lifestyles must be considered in dealing with it. Figures 2.11 and 2.12 depict the influence of growth of population and increasingly use of water due to industrialization and life style that could lead to a water crisis particularly in Africa and central Asia, where together with rising population the rivers are continuously getting dried. This is based on the study undertaken by UNEP.

Figure 2.10 With the change in population – lifestyle is also changing.[9] (Courtesy: http://en.wikipedia.org).[9]

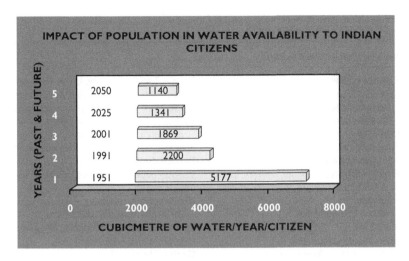

Figure 2.11 Impact of population on water resources.[25]

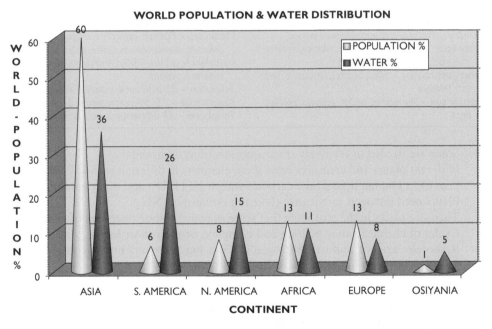

Figure 2.12 Water resources and world population – an imbalance between the two, and the Impacts that would be significant in the days to come.[25]

2.6 CHEMICALS IN MOTION: CYCLES IN THE ECOSPHERE[9, 16, 20]

- Living things are made of about 20–30 elements (out of 100) and out of which some are needed in relatively large amounts–*Macronutrients (greater than 1% dry organic weight)*. Refers Table 2.1 also.

Table 2.1 Minerals' contribution in making goods for our daily use. Minerals once mined are depleted and they cannot be replaced.[17, 26]

Industrial minerals used around the house

Carpet – *Calcium carbonate, limestone*	Paint – *Titanium dioxide, kaolin clays, calcium carbonate, mica, talc, silica, wollastonite*
Glass/Ceramics – *Silica sand, limestone, talc, lithium, borates, soda ash, feldspar*	Concrete – *Limestone, gypsum, iron oxide, clay*
Linoleum – *Calcium carbonate, clay, wollastonite*	Wallboard – *Gypsum, clay, perlite, vermiculite, aluminum hydrate, borates*
Glossy paper – *Kaolin clay, limestone, sodium sulfate, lime, soda ash, titanium dioxide*	Spackling – *Gypsum, mica, clay, calcium carbonate*
Cake/Bread – *Gypsum, phosphates*	Pencil – *Graphite, clay*
Plant fertilizers – *Potash, phosphate, nitrogen, sulfur*	Carbon paper – *Bentonite, zeolite*
Toothpaste – *Calcium carbonate, limestone, sodium carbonate, fluorite*	Ink – *Calcium carbonate*
Lipstick – *Calcium carbonate, talc*	Microwavable container – *Talc, calcium carbonate, titanium dioxide, clay*
Baby powder – *Talc*	Sports equipment – *Graphite, fiberglass*
Hair cream – *Calcium carbonate*	Pots and pans – *Aluminum, iron*
Counter tops – *Titanium dioxide, calcium carbonate, aluminum hydrate*	Optical fibers – *Glass*
Household cleaners – *Silica, pumice, diatomite, feldspar, limestone*	Fruit juice – *Perlite, diatomite*
	Sugar – *Limestone, lime*
Caulking – *Limestone, gypsum*	Drinking water – *Limestone, lime, salt, fluorite*
Jewelry – *Precious and semi-precious stones*	Vegetable oil – *Clay, perlite, diatomite*
Kitty litter – *Attapulgite, montmorillonite, zeolites, diatomite, pumice, volcanic ash*	Medicines – *Calcium carbonate, magnesium, dolomite, kaolin, barium, iodine, sulfur, lithium*
Fiberglass roofing – *Silica, borates, limestone, soda ash, feldspar*	Porcelain figurines – *Silica, limestone, borates, soda ash, gypsum*
	Television – *35 different minerals & metals*
Potting soil – *Vermiculite, perlite, gypsum, zeolites, peat*	Automobile – *15 different minerals & metals*
	Telephone – *42 different minerals & metals*

- Some are needed in relatively trace amounts (micronutrients).
- Different plants and creatures need these elements in different forms. Plants need C as CO_2. Human beings take carbon as sugar ($C_6H_{12}O_6$) and many eatable items. Plants need nitrogen as Nitrates (NO_3) or Ammonia (NH_3).
- There are places in the ecosphere that serve as major abiotic reservoirs of chemicals.
- Cycles of chemicals may be long and complex, or even short loops.
- Reservoirs are not uniform (nitrogen gas is a big reservoir) but as compound in soil its presence is limited.

2.7   MINERALS – THE NON-RENEWABLE RESOURCES AND THEIR USE IN ENERGY, GOODS AND SERVICES PRODUCTION[17, 26]

2.7.1   As we know without power & energy, world is dark, factories would come to halt, and services would be jeopardized; but is it electricity or minerals?

The bulb shown in the illustration below (Figure 2.13) is glowing by electricity. Now, try to analyze its various components. It includes filament, lead wires, fuse,

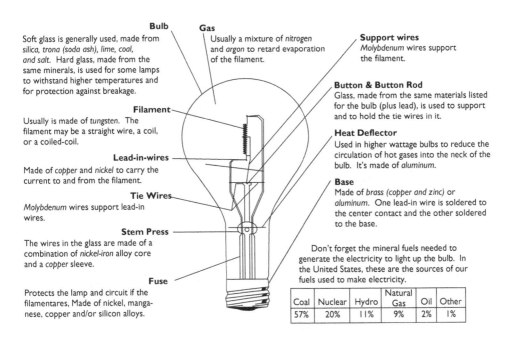

Bulb
Soft glass is generally used, made from *silica, trona (soda ash), lime, coal, and salt.* Hard glass, made from the same minerals, is used for some lamps to withstand higher temperatures and for protection against breakage.

Filament
Usually is made of *tungsten.* The filament may be a straight wire, a coil, or a coiled-coil.

Lead-in-wires
Made of *copper* and *nickel* to carry the current to and from the filament.

Tie Wires
Molybdenum wires support lead-in wires.

Stem Press
The wires in the glass are made of a combination of *nickel-iron* alloy core and a *copper* sleeve.

Fuse
Protects the lamp and circuit if the filamentares, Made of nickel, manganese, copper and/or silicon alloys.

Gas
Usually a mixture of *nitrogen* and *argon* to retard evaporation of the filament.

Support wires
Molybdenum wires support the filament.

Button & Button Rod
Glass, made from the same materials listed for the bulb (plus lead), is used to support and to hold the tie wires in it.

Heat Deflector
Used in higher wattage bulbs to reduce the circulation of hot gases into the neck of the bulb. It's made of *aluminum.*

Base
Made of *brass (copper and zinc)* or *aluminum.* One lead-in wire is soldered to the center contact and the other soldered to the base.

Don't forget the mineral fuels needed to generate the electricity to light up the bulb. In the United States, these are the sources of our fuels used to make electricity.

Coal	Nuclear	Hydro	Natural Gas	Oil	Other
57%	20%	11%	9%	2%	1%

Figure 2.13 Is it electricity/power that is providing energy to us, or the coal and other minerals?[17]

stem press, tie wires, base, heat deflector and support wires. You would find that many minerals have been used to make them. A mixture of argon and nitrogen gas has been filled in this bulb and the current that lights it is generated from minerals, which are known as fossil fuels–coal, oil, or natural gas (as described in the following sections/paragraphs). It can also be produced from atomic minerals. In USA 57% (Figure 2.9) and India 68% of electricity is *generated by coal.* More than 90% of electricity all over the world is generated using these minerals. So what do you say?

- The world is dark without electricity/power or, is it without coal and minerals?
- Factories would come to a halt without electricity or, is it without coal and minerals?
- Services such as Transportation, Communication, Information-Technology and others would be jeopardized without Electricity or, is it without coal and other minerals?

The foregoing discussion reveals that '**Minerals**' are the basic raw materials, which are used to produce not only energy (coal, oil, atomic) but value added products, arms and ammunitions. Their use has been established both during war and peace times. *After sun, air and water they become our basic need.* Even food cannot be produced without their aid/contribution. The development of any country

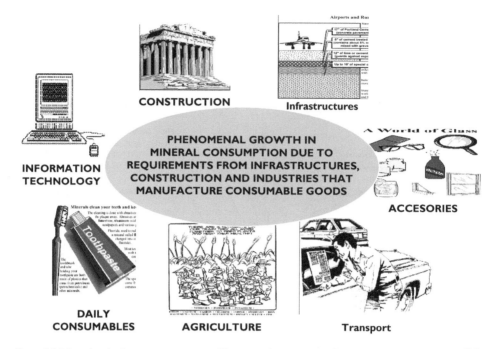

Figure 2.14 In a developing economy we would require them more and more in the years to come.[21, 30]

is measured as per capita mineral consumption and production, as illustrated in Figure 2.14.

Table 7.6 lists elements/minerals in human body; illustrating the fact that each element fulfills a critical purpose to keeps us fit. Thus, as described in sec. 1.3.6 and Figures 1.3, 2.14, Tables 2.1 and 7.5 minerals are our basic need and we cannot do away without them. Equally important is their role in providing us the energy in any of the forms they are available: Solid, Liquid or Gaseous.

2.8   ENERGY SOURCES[3, 7, 22, 27]

Energy is valuable resource on our planet; save it before it is too late. Saving energy helps to reduce pollution, thereby making a better tomorrow–a sustainable mechanism.

2.8.1   Classification of energy sources

Energy (Figure 2.15) can be considered in two categories–primary and secondary. Primary energy is the energy in the form of natural resources, such as wood, coal, oil, natural gas, natural uranium, wind, hydropower, and sunlight. Secondary energy is the more useable form to which primary energy may be converted, such as electricity and petrol. Primary energy can be renewable or non-renewable.

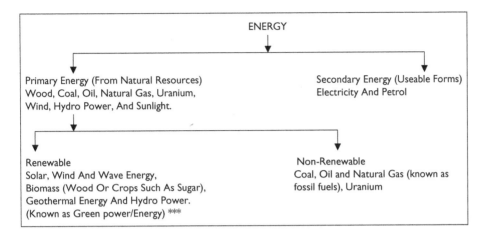

Figure 2.15 Classification of energy sources.

*** Refer Sec. 2.8.2.

Renewable energy sources include solar, wind and wave energy, biomass (wood or crops such as sugar), geothermal energy and hydropower. Non-renewable energy sources include the fossil fuels–coal, oil and natural gas, which together with uranium provide over 80% of our energy at present. Table 2.2, briefly compares various features of each energy source.

2.8.2   Green power and its purchasing options

Green Power: Green power can be defined as energy from indefinitely available resources and whose generation has zero/negligible environmental impacts, whether through reduced emissions or minimal environmental disruption. Such sources of energy include: Wind, Wave, Tidal, and Small Scale Hydropower, Biomass, Landfill gas, Geothermal and Solar energy.

Purchasing Options: Green power can either be purchased from off-site providers or produced via onsite generation. Purchasing delivered green power/RECs from an off-site provider currently costs slightly more than purchasing conventional power. This cost can be minimized, however, by entering into long-term contracts with the provider, which allows the purchaser to lock in a fixed-rate contract and increase opportunities for reduced pricing. Alternatively, green power generators can be installed on the site of the consuming building. This has a high initial cost, but produces "free" energy and, consequently, lower energy bills after the payback period. Additionally, federal and state governments offer incentives to reduce the initial cost of a green power installation. Moreover, as with any new technology, the price for green power generators has fallen and will continue to fall. As the market for green power grows, equipment will be manufactured in larger volumes, thus meeting economies of scale.[7]

2.8.3   Energy sources and their merits and limitations[4, 7, 22, 27]

Table 2.2 Energy sources and their merits and limitations.

Primary Energy Source	Merits	Limitations
Hydro-electric They harness the energy of falling water, which can occur naturally, but more often has to be engineered by the construction of large dams with lakes behind them.	• Generating facilities have the attraction of providing electricity without polluting the atmosphere. • The advantages of hydro-electricity have long been appreciated and today it provides 18% of the world's power.	• In many countries most of the suitable dam sites have already been used, thus limiting further major development of this source.
Solar The sun pours abundance on to our planet each day. We see this energy in a variety of forms, ranging from solar radiation, through wind and waves, to trees and vegetation, which convert the sun's rays into plant biomass. In addition, there is an enormous amount of energy in the materials of the earth's crust, the fossil fuels also storing energy from the sun.	• On a small scale (and at relatively high cost) it is possible to store electricity. • The main role of solar energy in the future, as considered by the scientist will be that of direct heating. • Solar energy has considerable logical and popular appeal.	• Solar input is interrupted by night and by cloud cover, which means that solar electric generation plant can typically only be used to a small proportion of its capacity. • Also, there is a low intensity of incoming radiation and converting this to high-grade electricity is still relatively inefficient • The capital, energy and materials costs of conversion, maintenance and storage are extremely high.
Wind, like the sun, is 'free' and is increasingly harnessed for electricity.	• About 20,000 megawatts capacity is now installed around the world. • Nevertheless, costs have come down, and contracts for electricity from some new US plants are as low as for conventional sources.	• However, in meeting most electricity demand, similar backup issues arise as for solar. It is not always available when needed, and some means is required to provide substitute capacity for windless periods.
Geothermal This energy derived from natural heat below the earth's surface. Where hot underground steam can be tapped and brought to the surface it may be used to generate electricity.	• There are also prospects in other areas for pumping water underground to very hot regions of the earth's crust and using the steam thus produced for electricity generation.	• Such geothermal sources have potential in certain parts of the world, and some 8000 MW capacity plants are operating.
Biomass Most forests and agricultural crops are technically capable of being converted into some form of energy, even if the primary purpose of the crop is to provide food.	• Biomass does provide a useful and growing source of energy, especially for rural communities in third world countries. Organic waste and water plants can be used to produce methane or 'biogas'.	• There are also some 'energy farms', where crops are produced solely for energy production. Such farms however compete with other crops for water, fertilizer and land use, thus requiring some choice between fuel and food.

(Continued)

Table 2.2 (Continued)

Primary Energy Source	Merits	Limitations
Gas Natural gas is widely used alongside coal and oil, as a very versatile fuel **Landfill Gas** It is extracted from landfills using a series of wells and a blower/flare (or vacuum) system. This system directs the collected gas to a central point where it can be processed and treated, depending upon the ultimate use for the gas. From this point, the gas can be flared or used to generate electricity.	• After the oil price shocks of the 1970s, increased exploration efforts revealed huge deposits of natural gas in many parts of the world and today these are extensively used for power stations. • Heat values of natural Gas = 39 MJ/m^3	• Gas can be seen in the same way as oil, as being too valuable to squander for uses such as large-scale electricity generation.
Oil Oil is a convenient source of energy. Heat value of Crude Oil = 45–46 MJ/kg	• It has the great advantage of being a portable fuel and remains vital for transportation. • It played an important role in the economic development of many countries during the past century.	• Oil has generally become too expensive to use for electricity.
Coal It is abundant and world production is about 3.5 billion tonnes per year, most of this being used for generation of electricity. Its heat values are as under: Brown coal = 9 MJ/kg Black coal (low quality) = 13–20 MJ/kg Black coal = 24–30 MJ/kg	• **Fossil fuels** have served us well. • It is the most abundant of fossil fuels—and is a reliable, secure and affordable fuel for both power generation and industrial applications (as described in the following section).	• The future of coal-using industry is largely pinning its hopes on the Clean Coal Technologies and on policy of security of energy supply, as describe in following sections.
Uranium Uranium is an energy source, which has been locked into the earth since before the solar system was formed, billions of years ago. It is also abundant, and technologies exist which can extend its use 60-fold if demand requires it. Uranium* – in light water reactor; the heat values = 500,000 MJ/kg	• World mine production is about 35,000 tonnes per year, but a lot of the market is being supplied from secondary sources such as stockpiles, including material from dismantled nuclear weapons.	• One of the most dangerous consequences of running a nuclear reactor is that it can release radioactive material into the environment. A lethal radioactive byproduct of a nuclear reactor is plutonium, produced in large quantities. • Less than one-millionth of a gram of plutonium is carcinogenic. If one pound of plutonium is uniformly distributed through the earth, it will cause lung cancer to every inhabitant of this planet.

2.8.4 Top 7 promising alternative energies[13]

Table 2.2 gives an account of various sources and options that are available for energy. As per the study by *Mark Jacobson (2009 through* a report in New Scientist; under atmosphere and energy program at Stanford University, US), the *most promising types of alternative energies*, based on their total ecological footprint and their benefit to human health are listed in descending order:

1 *Wind,*
2 *Concentrated solar power (mirrors heating a tower of water),*
3 *Geothermal energy,*
4 *Tidal energy,*
5 *Solar panels,*
6 *Wave energy, and*
7 *Hydroelectric dams.*
8 Bio-fuels from corn and plant waste*
9 Nuclear power** and
10 "Clean" coal. ***
11 Oil and gas***

  *Biomass would have problems in common with agriculture. Burning bio-mass would generate hydrocarbons and has the potential to generate carcinogenic compounds.
  **Nuclear fission–its environmental impacts are high; its impacts are primarily on land but air and water are also impacted. The problem is the disposal of radioactive waste for which a viable solution has not been found so far.
  ***Environmental impacts are high, as described in the following sections

Thus, this study found wind power to be by far the most desirable source of energy. Bio-fuels from corn and plant waste came right at the bottom of the list, along with nuclear power and "clean" coal, oil and gas. This study inferred that not just the quantities of greenhouse gases that would be emitted from the alternative sources of energy but also the impact the fuels would have on the ecosystem – taking up land and polluting water, for instance.

2.8.5 GDP, Energy consumption patterns and CO_2 emissions[29]

Figures 2.16 and 2.17 give an account of energy consumption country-wise, and the trend, which is on the rising side. There is a relationship amongst Gross Domestic Products (GDP), energy consumption and CO_2 emissions, as shown in Figures 2.18 and 2.19; which illustrate these trends in the USA and Japan. It depicts the fact that in Japan energy consumption was more than 2.5 times which resulted rise in CO_2 emissions 3 times and multiplied its GDP about 4 times comparing to 1965 levels. In USA these Figures are 1.4, 1.6 and 2 times respectively. Thus, energy use plays a dominating role in bringing prosperity to any country but at the cost of phenomenal growth in CO_2 emissions.

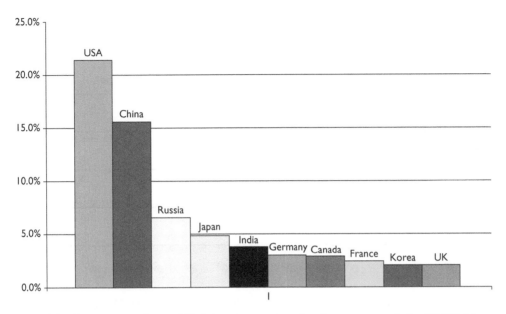

Figure 2.16 Country-wise share of Global energy consumption (in percentage during 2006)[29]. After Warwick J. McKibbin (2007).

Source: BP Statistical Review of World Energy.

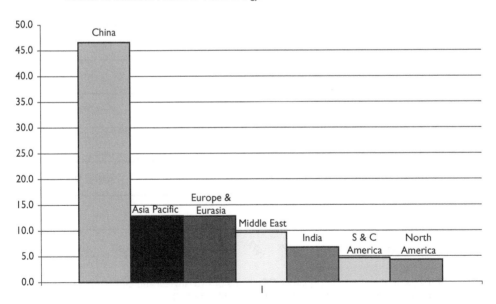

Figure 2.17 Contribution to growth in global energy consumption from 2000 to 2006; illustrating countries responsible for adding (percentage share) the CO_2 emissions.[29] After Warwick J. McKibbin (2007).

Source: BP Statistical Review of World Energy 2007 and author's calculation.

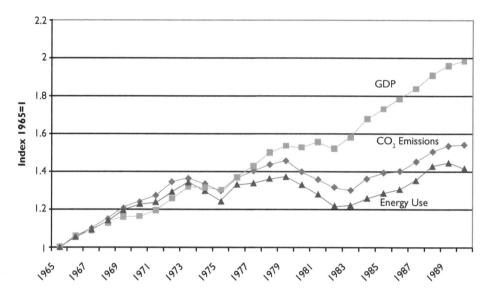

Figure 2.18 Growth of energy use (squares), CO_2 emissions (triangles) and GDP (rectangles) in USA from 1965 to 1990.[29] After Warwick J. McKibbin (2007).

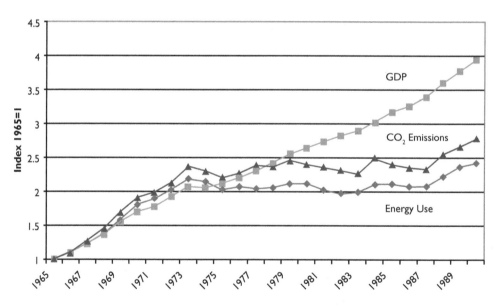

Figure 2.19 Growth of energy use (squares), CO_2 emissions (triangles) and GDP (rectangles) in Japan from 1965 to 1990.[33] After Warwick J. McKibbin (2007).

The sectors responsible for CO_2 emission (Figure 2.20 and Figure 2.31) are transportation (range: 13 to 35%), Industry and Construction (range: 12 to 30%) and generation power/electricity (range: 13 to 40%) of which the later dominates. Considering world's average it accounts for 38.3%; next is transportation (24.1%), and Industry and construction accounts for 18.5%.

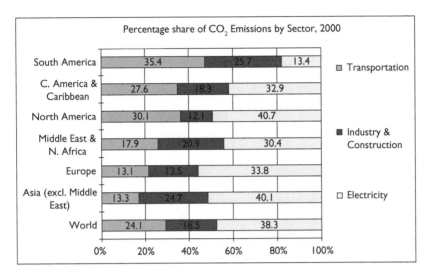

Figure 2.20 Percentage share of CO_2 Emissions by sector (as illustrated in the bar-chart: Transportation (left); Industry and Construction (middle); Electricity (right); in the year 2000.[29] After Warwick J. McKibbin (2007).

Source: World Resource 2005, World Resources Institute, International Energy Agency, UNFCCC.

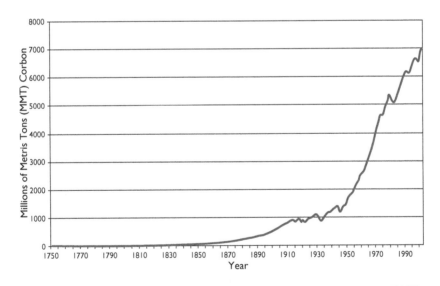

Figure 2.21 Global CO_2 emissions from fossil fules.[29] After Warwick J. McKibbin (2007).

Figure 2.21 shows the growth of CO_2 emissions which has been very steep particularly after 1950. The projections country-wise have been shown in Figure 2.22; describing the fact that the fossil fuels use would be rising and so is the emission of CO_2.

The impact of temperature change in Australia–an example illustrating it is not just a climate risk but could influences other aspects also[29]; as illustrated in Figure 3.6.

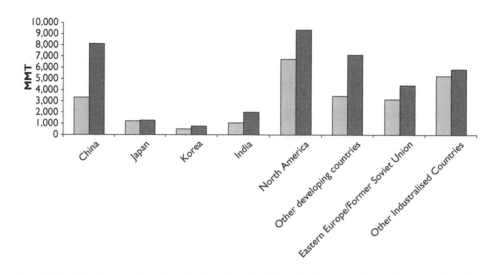

Figure 2.22 Global carbon dioxide emissions from fossil fuels. (as illustrated in the bar-chart:Year 1990 (left);Year 1925 (right).[29] After Warwick J. McKibbin (2007).

Source: Energy Information Administration/International Energy Outlook.

2.8.6   Risk of CO_2 emissions[29]

Prof. Warwick J. McKibbin (2007)[3] lists following risks due to CO^2 emissions:

- Risk of environmental damage
  - from climate variability
  - from bad policy design
- Risk of severe economic damage
  - from not taking action
  - from taking the wrong action
- Risk of collapse of a policy regime that needs to last many decades

Thus, it is not just climate risk that matters but other aspects should also be looked into to arrive at a viable solution to this problem. In addition to the above, the following facts need consideration:

- As we enter into the new century, the energy resources available for economic large-scale electricity generation are likely to be fossil fuels and nuclear.
- Today in the industrial countries of the world, we use between 150 and 350 gigajoules* per person each year, an increasing proportion of it in the form of electricity.

*Joule (J) = A unit of energy
Mega joule (MJ) = 10^6 Joules
Gigajoule (GJ) = 10^9 Joules.

- The challenge today is to move away from our heavy dependence on fossil fuels and utilize non-carbon energy resources more fully.
- Concerns about global warming are a major reason for this. (Refer sec. 3.6 also)
- Sec. 2.8.1 describes top 7 promising alternative energies, based on their total ecological footprint and benefits to human health. It clearly advocates 'clean coal' enabling us to have the secure supply of energy, as described in sections 2.8.7 and 2.8.8.

2.8.7 Coal for energy security[4, 7, 22, 30]

The world recoverable coal reserve base is estimated at 1.001×10^{12} tonnes with a life of about 180 years at current consumption level. Four countries account for 67% of this reserve base: USA (27%), Russia (17%), China (13%), and India (10%) and they also accounted for 63% of the world's production in 2003. Globally, coal accounted for 24% of the total energy consumption in 2003. About 67% of the coal was used for power generation, 30% for the industrial sectors (steel, cement etc.) and 3% for the residential and commercial sectors.[4]

Coal may have a unique role in providing the demand for a secure supply of energy. It is the most abundant of fossil fuels–and is a reliable, secure and affordable fuel for both power generation and industrial applications. Thus, the world needs coal for energy security. The *International Energy Agency* (IEA) has stated: World reserves of coal are enormous and, compared with oil and gas, widely dispersed".

Coal can have significant environmental impacts at every stage of its production and utilization. This, however, can be mitigated. The coal industry is continuing to improve its environmental performance by working to ensure that coal is produced and used efficiently and that the opportunities for technological advancement, called *Clean Coal Technology*, are fully and vigorously pursued.

2.8.8 Clean coal technology (CCT)[4, 7, 26, 34]

Clean coal technology (CCT) is the technology employed and being developed to meet coal's environmental challenges. It represents a continuously developing range of options to suit different coal types, different environmental problems, and different levels of economic development. CCT reflects the coal industry's concern over the issues facing coal, but also its optimism about the future, given progress already made and the technological options available for successfully addressing those issues. Coal's technical response to its environmental challenges is to have three core elements:

1 Reducing emissions of pollutants such as particulate matter and oxides of sulphur and nitrogen
2 Energy efficiency is necessary for at least three major reasons: Growing concern for climate change implies a reduction of greenhouse gas emissions; the expectation of persisting high prices for oil and gas increases the economic value of efficiency improvements which could become a serious competitive advantage; reduction in energy consumption should lessen market tightness and therefore improve the volume and price dimensions of security of energy supply.
In 2005, the EU published a report 'Green Paper on Energy Efficiency" that outlines European initiatives in the field of energy efficiency, including R&D in increasing the efficiency of fossil fuel-based power production (European

Commission, (EC), 2005). Concerning energy efficiency in coal utilization, a new coal-fired power plant facility in Germany–called BoA technology–has a thermal efficiency up to 43%. This is higher than the current average efficiency in Europe (38%) and in the world (30%). High thermal efficiency will reduce coal consumption per energy unit generated.

3   Reducing CO_2 emissions to near zero levels through carbon capture and storage (CCS) Source: *International Energy Agency (IEA) (2002)*

With substantial quantities of CO_2 being emitted into the atmosphere by the coal-based power generating stations across the globe and the consequent international effort to mitigate its impact, there is a world-wide stress on making use of coal cleaner by focusing attention on the role of carbon capture and storage (CCS) during the combustion process in the sustainable use of coal. It will not be out of place to declare that use of coal, as a fuel in 21st century will largely depend upon how best CCS can be made more effective.[34]

2.8.9   Carbon capture & storage (CCS)[34]

This system involves capturing the CO_2 being emitted by a coal-based thermal power plant, transporting the same either directly to a user industry or indirectly from a storage point in the form of large below the earth underground reservoirs.

Thermal power industry globally is pitched to grow current level of 3500 GW to 7000 GW(Giga-watt) by 2030 with resultant CO2 emission increasing from 9000 Mtpa to 1700 Mtpa (Million tonnes per annum) indicating the quantum of the problem. Of this generating capacity coal is contributing 71%; as such stress on coal has to be maximum. (See Figures 2.23 left and right).[30]

In addition to the above, CCT also aims to achieve the following:

•   Create a green mining technology system of coal resources to turn harm into treasure. This concept aims at:
    •   Establishing suitable green mining technology system of coal resources taking into consideration the prevalent condition at the mine site.

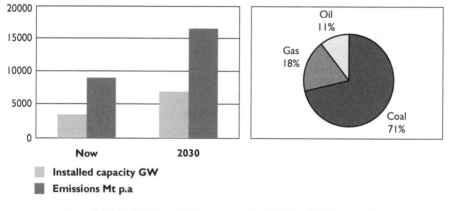

Figure 2.23 (Left) Why CCS for power. (Right) Why CCS for coal.[34]

- Carrying out the comprehensive development and utilization of coal waste, slime, Coal-bed-methane (CBM), mine water and accompanying resources with coal
- Strengthening the protection of ecological environment and water resources in mining areas
- Harnessing the abandoned material and subsidence area and
- Researching the compensation mechanism for recovery of ecological environment in mining areas.

Using this concept when applied, for example in China, by 2010 the gas quantity drainaged on surface and from underground will reach 10.0 billion m^3, the comprehensive utilization rate of coal waste and slime will reach 75%, the discharge rate of qualified mine water will be 100%, main pollutants from large and medium coal enterprises will be totally qualified and discharged; and pollutant discharge from small mines will be controlled.

2.8.10   Converting coal-to-oil[22]

Efforts to carry out the comprehensive utilization of coal resources and ecological improvement also include efforts to convert coal-to-oil. Work on such projects across the globe is going on. For example in China: The Shenhua project is aiming at producing the annual production of about 20.0 million tonnes and 2.0 million liters of synthesis fuel and chemicals per day by 2020. In addition, the production facility of coal-based synthetic oil is also aimed at the Lu'an Mine Group.

Coal gasification[14] (Figure 2.28), economic approaches to carbon capture and storage (as described in the preceding sections), coal to liquids transformation and production of hydrogen from coal are the major areas of emphasis currently to continue to develop coal into a viable energy source across the globe.

2.9   ENERGY CRISIS[4, 5, 7, 16, 22, 30]

The energy crisis has already begun in many countries. Frequent power cuts and inadequate supply of electricity are routine features. Some parts of world are dark even today. With the growth in population, this trend would continue further. The salient points that emerge in the prevalent scenario are summarized here below:

- Within a brief period of 5 centuries we will have consumed all the coal, gas and oil formed over a period of 500 million years.[16]
- We are running out of some of the energy sources we depend upon the most. Oil and gas are at their peak production rates and this trend would continue in the future also, as shown in Figure 2.24.
- There is uncertainty due to economic, political, social and other reasons on the energy front. There is no global policy on this issue.
- Energy is wasted worldwide.

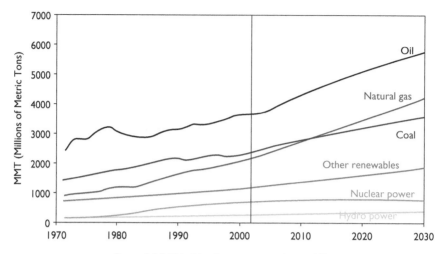

Figure 2.24 World primary energy demand.[30]

Source: IEA WEO 2005.

- Almost every environmental problem is directly related to the production and consumption of energy (refer chapter 3). Combustion of fuels account for air pollution. The extraction, processing and use–water pollution (chapter 4). Mining–land degradation. Energy is also responsible for noise and hazardous waste (sec. 5.2). The adverse impacts due to excessive use of energy to the ecosystem could be as illustrated in the Figures 2.25 and 2.26.

2.9.1   Wayout/solution to the energy crisis

'Past industrial experience taught us the lesson of sustainable development–a development that is not harmful to present as well future generations–should we not strive for it?' Goals commonly expressed by experts in the subject area include addressing the following issues:[1]

- The establishment of nature and biosphere reserves under various types of protection; and, most generally, the protection of biodiversity and ecosystems upon which all human and other life on earth depends.
- Very large development projects – megaprojects – pose special challenges and risks to the natural environment. Major dams and power plants are cases in point. The challenge to the environment from such projects is growing because more and bigger megaprojects are being built, in developed and developing nations alike.
- Reduction and clean up of pollution: Pollution cannot be eliminated fully but efforts must be made to minimize it. For example taking measures as shown in figure 2.27 that illustrates 'Clean Industrial Operations' by making use of enclosures to protect against noise & dusts.

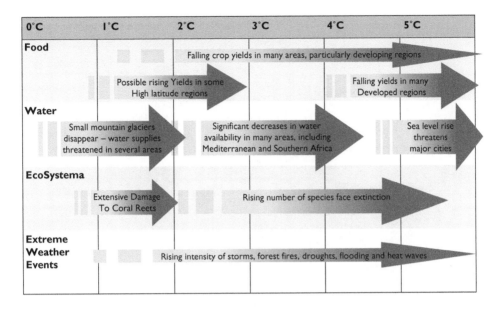

PACIFIC STRATEGY PARTNERS

Figure 2.25 Impact of rise in temperature in Global prospective.[1] After Bart de Hann, (2007).[1]

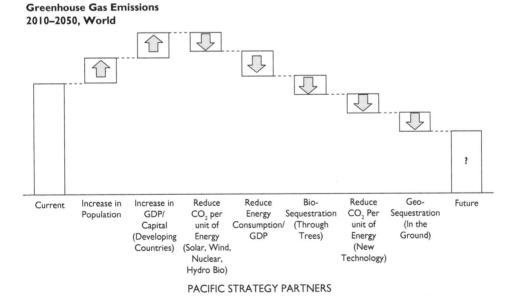

PACIFIC STRATEGY PARTNERS

Figure 2.26 An account of Green house gases during period 2010–2050 and projected strategy to reduce/overcome it.[1] After Bart de Hann, (2007).

Source: IPCC, IEA.

Figure 2.27 Clean industrial operations: Enclosures to protect against Noise & Dust.[30]

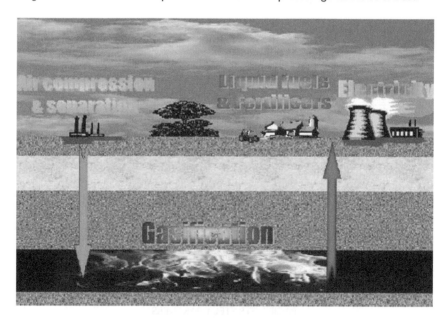

Figure 2.28 Gasification of poor quality, uneconomical to extract by conventional mining methods – An example of cleanly converting non-cyclic materials into energy through direct combustion. (After Jock Cunningham, 2007).[14]

Table 2.3 Best Practices through which conservation and sustainable use of scarce resources such as water, land, and air; could be achieved.[1] After Bart de Hann, (2007).

Best practices	Details
1 Expand Product Driven Benefits	• Extend product life – improve quality • Reduce consumption – thinner coatings, new alloys. Figure 2.29 illustrates this concept.
2 Reduce Operating Footprint	• Lower power use – better processes and new technology • Capture emissions – new technology

3 Recycle More	• Lower transport emissions
	    – More secondary feed/closer to plants
	    – More small scale production/closure to feed source
	• More secondaries – new technology to treat old residues and more complex secondaries

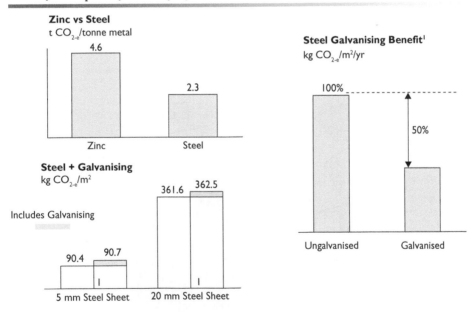

Life Cycle CO$_2$ Example – Company X

PACIFIC STRATEGY PARTNERS

Figure 2.29 An example illustrating reduction in consumption of large CO$_2$ emiters – thinner coatings, new alloys (as descibed in table 2.3).[1] After Bart de Hann, (2007).[1]

    [1]   15 mm Steel Sheet, Zinc layer 10 μm, Corrosion Condition C3.
      Source: CSIRO, Pacific Strategy Partners.

BHP Billiton Climate Policy (June 2007)

- **Goal of Zero Harm**
  - Understand life cycle emissions
  - Improve energy and emissions per unit produced
  - AUD$300 M over 5 years to R&D and awareness
  - Emissions trading
- **Reduction Targets (2007–2012)**
  - Energy: 14%
  - Greenhouse Gas: 14%
- **Site-based targets** and reporting
- **Carbon pricing sensitivities** to large capital projects

PACIFIC STRATEGY PARTNERS

Figure 2.30 An example illustrating 'Reduce Operating Footprint'. World-class companies has already initiated (for example BHP Billiton in this illustration), as described in table 2.3.[1] After Bart de Hann, (2007).[1]

Source: BHP Billiton.

Transforming heavy emitters is difficult

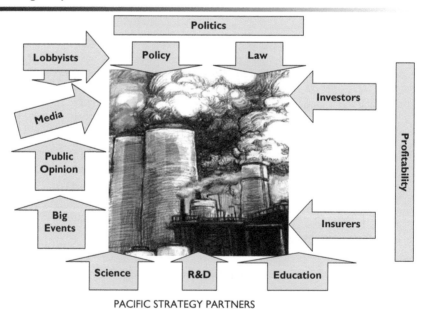

PACIFIC STRATEGY PARTNERS

Figure 2.31 An illustration showing reduction in CO_2 emissions is a difficult task.[1] After Bart de Hann, (2007).[1]

- Cleanly converting non-cyclic materials into energy through direct combustion, for example there are coal seams that are practically un-mineable and lying at great depths or their mining is uneconomical by the conventional methods. Applying non-conventional methods such as Coal-gasification (figure 2.28), coal-bed-methane (CBD) drainage; the useful components could be converted into useful energy.[2, 14]
- Reducing societal consumption of non-renewable fuels.
- Development of alternative, green, low-carbon or renewable energy sources (Ref. Sec. 2.8.4)
- Conservation and sustainable use of scarce resources such as water, land, and air; as described in the preceding sections, also by adapting best practices as shown in Figure 2.29/Table 2.3.
- Protection of representative or unique or pristine ecosystems.
- Preservation of threatened and endangered species extinction.
- Conservation of non-renewable resources such as minerals could minimize pollution. Mining and processing require huge amounts of energy. As such it becomes all the more important to consume them as minimally as possible. Table 2.1 lists goods of our daily use that are manufactured using number of minerals. Should we not try to consume them judiciously?

2.9.2   Energy efficient lighting tips[1, 5, 11]

Followings are some of the considerations that could bring improvement while using electricity for lighting purposes.

- One kWh is enough energy to run one 100-watt bulb for 10 hours. Consider that the average home has approximately 34 sockets (which may vary from one country to another). If each socket is filled with a 100-watt bulb and is powered for 5 hours each day, 133 kg of Carbon Dioxide are emitted into the atmosphere in one week!
- Compact fluorescent bulbs are about 3 to 4 times more efficient than incandescent bulbs. Therefore if all 34 bulbs in our example were replaced with 30-watt compact fluorescent bulbs, only 43 kg Carbon Dioxide (which is one-third) would be emitted into the atmosphere.
- By switching from incandescent lighting to compact fluorescent lighting the average consumer can save 50% to 80% in electricity costs without any loss in lighting quality. The average compact fluorescent bulb lasts 8 to 10 times longer than any incandescent bulb.
- Wasting water wastes electricity. Why? Because the biggest use of electricity in most cities is for pumping and supplying water. About 75 percent of the water we use in our homes is used in the bathroom. If we have a high flush toilet, we use about 15 liters to 25 liters of water with every flush! A leaky toilet can waste more than 40000 liters of water a year. Another simple way to save water and energy is to take shorter showers. We will use less hot water; as water heaters account for nearly 1/4 of our home's energy use and increases electricity bills.

2.9.3 Energy conservation tips

- Turning lights off when we leave the room, or installing occupancy sensors.
- Installing energy-saving floodlights outdoors.
- Using solar-powered accent lights outdoors.
- Buying energy-efficient lighting equipment.
- Making use of natural daylight
- Reducing background light levels and rely more on task lighting
- Switching to compact fluorescent lamps. Using tube fluorescent lighting. Also using incandescent lights wisely.
- Cooking Tips (efficient cooking can save considerable energy)

    - Covering pans while cooking to prevent heat loss.
    - Making sure that pan covers the flame/coil of the stove/ cooking range. If flame/coil peeping out from the sides of pan, energy is lost.
    - Using lesser liquid and fat results in quicker cooking. Use of Pressure cooker is always advantageous. It really saves on energy. Try using a solar box cooker also.

Fuel saving tips (while using automobiles)[7]

- Switching to radial-ply tires because they offer less rolling resistance and have a longer life than the cross-ply variety.
- Avoiding driving with under-inflated tires because a tire pressure that is too low not only increases consumption, it also markedly reduces a tire's life.
- Avoiding stop-start driving
- Accelerating slowly
- Not speeding too much, as faster driving consumes more petrol.

Recycling: for example consider; recycling aluminum cans saves how much energy, compared to making new aluminum? Recycling aluminum cans saves 95 percent of the energy required to make the same amount of aluminum from raw aluminum ore, bauxite. For other recycled products, the amount of energy saved differs by material, but almost all recycling processes achieve significant energy savings compared to production using virgin materials. Harvesting, extracting, and processing the raw materials used to manufacture new products is an energy-intensive activity. Reducing or nearly eliminating the need for these processes, therefore, achieves huge savings in energy.[7] EPA Q & A. [Accessed Aug. 2009]

Vegetation and energy saving–Shade trees around our house can reduce our air conditioner bill by up to how much? Answer is 75%. Plants can significantly reduce a building's energy needs since it's cooler in the shade of trees and warmer behind plants that block the winter winds. The Department of Energy (in USA) predicts that proper placement of as few as three shade trees will save an average household $100-$250 in energy costs each year. And a Pennsylvania study found that air conditioning needs could be reduced by up to 75 percent by shading a house with trees![7] EPA Q & A. [Accessed Aug. 2009]

Tire-derived fuel: How can scrap tires help save the environment? Answer is burn them for fuel. In 2003, 130 million scrap tires were used as fuel, up from 25.9 million in 1991. Tires can be used as fuel either in shredded form–known as tire derived fuel (TDF)–or whole, depending on the type of combustion device. Scrap tires are typically used as a supplement to traditional fuels such as coal or wood. There are several advantages to using tires as fuel. For example, tires produce the same amount of energy as oil and 25 percent more energy than coal.[7] EPA Q & A. [Accessed Aug. 2009]

2.9.4 Things to remember/way forward

Efforts should be made to create and sustain a culture in which everybody shares his/her commitment to save energy. Such a culture often results into making us aware of the facts that:

- Inefficient use of energy is like burning money
- Leaks make our future bleak
- Energy serves us the way we deserve
- Saving one unit or more a day, enable us to keep power-cut (problem) away
- The less we burn, the more we earn
- Where conservation fails, pollution starts
- We should conserve oil for a clean environment
- We should practice to manage energy well to avoid damage & hell.
- Ecology and Energy are twin concerns of development
- Use a portfolio of solutions to reduce emissions[1] (After Bart de Hann, 2007)

  - Reduce energy use: now
  - More clean energy: now
  - Bio-sequestration: now
  - Carbon offsets: now
  - Geo-sequestration: when affordable
  - Clean coal technology: when affordable
  - Let's start conserving energy instead of only talking about conserving energy.

  Concluding remarks: In addition to above

- *Nature is beautiful; we should Preserve it and Respect it.*

QUESTIONS

1 Briefly describe 'Green Power' and its Purchasing Options.
2 Classify Ecology.
3 Comment on: An account of Greenhouse gases during period 2010–2050 and projected strategy to reduce /overcome it as projected by Bart de Hann in this chapter.
4 Define terms: Ecology; Environment; Ecosphere; Organisms
5 Draw simplified cycles of: Carbon, Sulfur and Nitrogen and write briefly important features of each of them.
6 How useful is Carbon Capture and Storage technology?

7 How helpful is Clean coal technology (CCT) in solving problems of fossil-fuels' adverse impacts to the environment?

8 How would you advise to conserve energy and what impacts it can have?

9 Illustrate relation amongst GDP, Energy Consumption Pattern and CO_2 Emissions.

10 Is the bulk of the functioning of the ecosystem based on the input of solar energy? Explain it.

11 List 'Best Practices' through which conservation and sustainable use of scarce resources such as water, land and air could be achieved?

12 List different sources of energy. What are energy sources in your own country? Do you think consumption pattern of energy varies from country to country, if so, who are major consumers? Compare potential environment impacts of different energy resources including the future ones.

13 Name Earth's great spheres. Illustrate them showing position of ecosphere and biosphere within them. Where do you find most of the life on the earth? What is the structure of biosphere?

14 Name some producers, consumers and decomposers and also write their function. How many species of plants, animals and bacteria are known to be within the biosphere? Draw ecological pyramids for: Numbers, amount, and energy levels.

15 Name types of energy that are involved in ecosystems. How are nutrients cycled in nature, give a simple diagram. What is photosynthesis? Illustrate the use of this system in the production of fruits.

16 Of how many elements are living things made of? What is nutrient? Name major and minor macronutrients.

17 What is the impact of a rise in temperature in Global perspective?

18 What strategy should be followed to achieve reduction in consumption of large CO_2 emitters?

19 World is dark without Electricity/Power or, is it without coal and minerals? Comment.

20 World's birth rate of human population is 38 per thousand and death rate is 12 per thousand. Calculate doubling period. How do you determine the growth rate of population? A population control is much more than birth control technology; explain it.

REFERENCES

1 Bart de Haan. (2007) Corporate Climate Change Policies. Presentation in: *Mining 2020, The AIMEX International Mining Conference, Sydney, Australia,* Sept. 5–6.

2 Carter, P. (2007) New mine information technologies. Presentation in: *Mining 2020, The AIMEX International Mining Conference, Sydney, Australia,* Sept. 5–6.

3 Christian, J.J. & Davis, D.E. (1964) Endocrines, Behavior and Populations. *Science* 146:1550–1560.

4 Chugh, Y; Patwardhan, A. & Moharana, A. (2006) US Coal Industry Trends and Challenges: An Overview. *International Coal Congress & Expo 2006 on Sustainable Development of Coal for Energy Security, 11–13 December, New Delhi, India.*

5 *Energy Saving Tips (2007)* Safety Week Celebration, Publicity and Propaganda, Directorate General of Mines' Safety, Chaibasa Region, India.

6 Environment Health Center, National Safety Council. Reporting on climate change: Understanding the Science, 2nd. Edt. June 2000. (Ref. Oak Ridge National Laboratory. http://www.esd.ornl.gov/iab2–2.htm)

7 EPA Q & A. [Accessed Aug. 2009]

8 http://earthobservatory.nasa.gov/Library/CarbonCycle/carbon_cycle4.html/ [Accessed Sept. 2009]

9 http://en.wikipedia.org/wiki/Ecology; http://en.wikipedia.org/[Accessed Aug. 2009]

10 http://johnson.emcs.net/life/food_web] [Accessed in April 2010]

11 http://www.epa.gov/maia/html/nitrogen.html/ [Accessed Aug. 2009]

12 (http://users.rcn.com/jkimball.ma.ultranet/BiologyPages/F/FoodChains.html) [Accessed in April 2010]

13 J, Mark. (2009) Atmosphere and Energy Programme at Stanford University, US–ZEE TV News, India–15th Jan.

14 Jock, Cunningham. (2007) **Transforming the future mine.** Presentation in: *Mining 2020, AIMEX Conference; 5 September, Sydney, Australia.*

15 Kormondy, E.J. (1984) *Concepts of Ecology*, 3rd ed. Englewood Cliffs, N.J.: Prentice Hall.

16 Kupchella, C.E. & Hyland, M.C. (1993) *Environmental Science*. Prentice-Hall International Inc. pp. 3–85.

17 Mineral Information Institute (2001), Web site, Denver, Colorado. [Accessed Aug. 2001]

18 Moore, G.S. (2007) *Living with the earth*. 3rd edition. Boca Raton, FL, CRC Press. pp. 10–28.

19 Odum, E.P. (1971) *General Principles of Ecology*, Third Edition W. B. Suanders Company. pp. 17–20

20 Odum, E.P. (1983) *Basic Ecology*. Philadelphia: Saunders.

21 Pitocchelli, J.: http://www.anselm.edu/homepage/

22 Qiuliang, D; Hongzhu, Z. & Jianhua, D. (2006) Status and Development of Coal Industry in China. *International Coal Congress & Expo 2006 on Sustainable Development of Coal for Energy Security, 11–13 December, New Delhi, India.*

23 Rao, C.S. (1991) *Environment Pollution Control Engineering*. New Delhi, Wiley Eastern Ltd.

24 Saxena, N.C; Singh, G. & Ghosh, R. (2002) *Environment management in mining areas*. Scientific Pub. India. pp. 110–120.

25 Singh, R. (2009) *Rajasthan Patrika* (Daily Newspaper) – Water and its global scenario based on UNEP Report; March.

26 Tatiya, R.R. (2009): Exploration to Exploitation in Indian Mining Sector – Proposed strategy. *International Conference on Advances in Exploration to Exploitation of Minerals, Association of Mining Engineers of India, Jodhpur, India, 14–16, Feb.*

27   Uranium Information Center. www.uci.com [Accessed in 2003]
28   Warming, E. (1990) *Oenology of Plants – an introduction to the study of plant-communities*. Clarendon Press, Oxford.
29   Warwick J. McKibbin. (2007) The Future of Mining in a carbon-constrained world. Presentation in: *Mining 2020, AIMEX Conference; 5 September, Sydney, Australia*. (Sponsored by Hella).
30   Webb, M. (2006) International Policy Perspective on Clean Coal. *International Coal Congress & Expo 2006 on Sustainable Development of Coal for Energy Security, 11–13 December, New Delhi, India*.

Chapter 3

Air pollution

'Fresh air, safe drinking water, unadulterated food and clean environment are our basic needs.' When air is clean, calm, clear and cool; it is fresh and unpolluted.

Keywords: Pollution, Air Pollution, Air Pollutants, Air Composition, Air Toxics, Air Quality Standards, Air Quality Index, Global Warming, Greenhouse Effect, Changing Climate, Acid Rain, Ozone Depletion, Noise Pollution, Industrial Noise, Gas Detection Techniques, Particulate Matter, Vibrations, Dust.

3.1   INTRODUCTION[29, 31]

When air is *Clean, Calm, Clear and Cool*; it is fresh and unpolluted air. A clean and dry air has composition as shown in Table 3.1. **Pollution** has its origins in the Latin word *polluere*,[31] means 'contamination of any feature of the environment'. Pollution has the following characteristics:

1   It is the addition of substance at a faster rate than the environment can accommodate. Certain pollutants like Pb, Hg and few others as shown in Table 3.2 have natural levels in nature, but when their levels exceed allowable limits, they become pollutants.
2   Pollutants are not only chemicals, but also forms of energy like heat, sound (so called noise pollution) and radioactive rays.
3   To be a pollutant, a material has to be potentially harmful to life.

3.2   AIR POLLUTION

The layer of air which supports life extends about 8 km above the Earth's surface, and is known as the troposphere (Figure 3.11). The composition of air remains remarkably constant except there may be small-localized variations. The onset of Air Pollution has been since humans began to use fire. The problem became significant since the industrial revolution in the 19th century, as shown in Table 1.1. Almost all air pollutants are the result of burning fossil fuels, either in the home, by industry or in internal combustion engines, as shown in Figures 3.1 and 3.3. Fossil fuels include: gasoline, natural gas, coal and petroleum products (oil).

Figure 3.1 classifies air pollutants, which includes organic and inorganic gases, aerosols and energy. Table 3.3 details salient features of principal air pollutants including five major air pollutants namely ground-level ozone, particulate-matter, carbon monoxide, sulfur dioxide and nitrogen dioxide. It also includes hydrocabons (excluding methane) and VOCs (Volatile Organic Compounds) pollutants.

3.2.1 Clean and dry air composition[4, 5, 26, 30]

Table 3.1 Clean and dry air average composition at the mean sea level.

Composition of air	Symbol	Content by volume	Density gms/Lit	Remarks
Air at STP			1.2928	(1 ppm = 0.0001%)
Air at 59°F			1.2256	
Nitrogen	N_2	78.084%	1.2506	These gases totals
Oxygen	O_2	20.947%	1.4290	99.998%
Argon	Ar	0.934%	1.7837	
Carbon dioxide	CO_2	0.033%	1.9770	
Neon	Ne	18.2 ppm	0.8999	This average is for the
Helium	He	5.2 ppm	0.1785	*clean and dry air* at
Krypton	Kr	1.1 ppm		the Mean Sea Level
Sulfur dioxide	SO_2	1.0 ppm		
Methane	CH_4	2 ppm		
Hydrogen	H_2	0.5 ppm	0.0899	
Nitrous oxide	N_2O	0.5 ppm		
Xenon	Xe	0.09 ppm		
Ozone	O_3	0.0 to 0.07 ppm		
Ozone winter	O_3	0.0 to 0.02 ppm		
Nitrogen dioxide	NO_2	0.02 ppm		
Iodine	I_2	0.01 ppm		
Carbon monoxide	CO	0.01 to trace	1.2500	
Ammonia	NH_3		0.0 to trace	

Table 3.2 Permissible exposure limit 8 hrs time weighted average as adopted by EPA USA.[4, 5]

Parameters/Variable	PEL[A]	Remarks
Asbestos[B]	0.1 fiber/cc	A = PEL = Permissible exposure
Benzene[B] (organic)	1 ppm	limit 8 hrs time weighted average
Bromine	0.1 ppm	B = Confirmed human carcinogen
Cadmium (all forms)[C]	5 µg/m³	C = suspected human carcinogen

(Continued)

Table 3.2 (Continued)

Parameters/Variable	PEL[A]	Remarks
Carbon dioxide	5000 ppm	D = acceptable ceiling
Carbon disulfide	20 ppm	concentration
Carbon monoxide	50 ppm	E = simple asphyxiant
Carbon tetrachloride[C]	10 ppm	Abbreviations
Chlorine	1 ppm[D]	cc – cubic centimeter
Chloroform	50 ppm[D]	$\mu g/m^3$ – microgram per cubic
Cresol	5 ppm	meter
Ethyl alcohol (ethanol)	1000 ppm	mg/m^3 – milligram per cubic meter
Fluorine	1 ppm	ppm – parts per million
Formaldehyde[C]	0.75 ppm	
Gasoline	300 ppm	
Hydrogen cyanide	10 ppm	
Iodine Magnesium	0.1 ppm[D]	
Iron oxide (Fume)	10 mg/m^3	
Isopropyl alcohol	400 ppm	
Lead (all forms)	50 $\mu g/m^3$	
Manganese compounds as Mn	5 mg/m^3	
Mercury	1 $mg/10\ m^3$ [D]	
Methyl alcohol (methanol)	200 ppm	
Nitric oxide	25 ppm[D]	
Propane[E]	1000 ppm	
Selenium compounds as Se	0.2 mg/m^3	
Sulfur dioxide	5 ppm	
Sulfuric acid	1 mg/m^3	
Tellurium compounds as Te	0.1 mg/m^3	
Tetraethyl lead as Pb	0.075 mg/m^3	
Toluene	200 ppm	
Turpentine	100 ppm	
Vinyl chloride[B]	1 ppm	
Zinc oxide fumes	5 mg/m^3	
Zinc oxide dust	15 mg/m^3	

Source: OSHA (Occupational Safety and Health Administration – USA).

3.2.2 Air pollutants

All VOCs contain carbon (C), the basic chemical element found in living beings. Carbon-containing chemicals are called organic. Volatile chemicals escape into the air easily. Many VOCs, such as the chemicals listed in the Table 3.4, are also hazardous air pollutants, which can cause very serious illnesses.

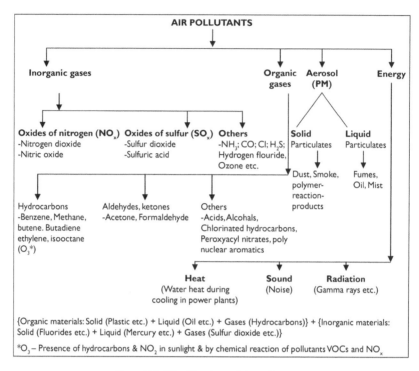

Figure 3.1 Classification of air pollutants.

Index Values	Levels of Health Concern	Cautionary Statements (Color code)
0–50	Good	None (Green)
51–100*	Moderate	Unusually sensitive people should consider reducing prolonged or heavy exertion outdoors. (Yellow)
101–150	Unhealthy for Sensitive Groups	Active children and adults, and people with lung disease, such as asthma, should reduce prolonged or heavy exertion outdoors. (Orange)
151–200	Unhealthy	Active children and adults, and people with lung disease, such as asthma, should avoid prolonged or heavy exertion outdoors. Everyone else, especially children, should reduce prolonged or heavy exertion outdoors. (Red)
201–300	Very Unhealthy	Active children and adults, and people with lung disease, such as asthma, should avoid all outdoor exertion. Everyone else, especially children, should avoid prolonged or heavy exertion outdoors. (Purple)
301–500	Hazardous	Everyone should avoid all physical activity outdoors. (Maroon)

Figure 3.2 AQI – Air Quality Index and its related aspects.

Source: On U.S. EPA.[6] http://cfpub.epa.gov/airnow/index.cfm?action=health_prof.main.[6]

Table 3.3 Details (taken from various sources) of principal air pollutants including five major air pollutants: ground-level ozone, particulate-matter, carbon monoxide, sulfur dioxide and nitrogen dioxide. [3,4,20,24,25,29,30,31,32]

Pollutant composition	Toxic* rank	Major source	Natural sources	Major sinks	Stay period	Major damages — Human health; plants and materials
• Carbon monoxide CO. • Colorless, odorless, slightly soluble in water.	9	• Fuels' incomplete combustion (in autos and other places). • Coal mines. • Industrial processes • Boilers • Incineration.	• Forest fires • Photochemical reactions	• Taken by soil where microbes convert into CO_2.	0.09–2.7 yrs.	Human health • Heart, Brain, Lungs • Reduces ability of blood to bring oxygen to body cells and tissues; cells and tissues need oxygen to work. • Particularly hazardous to people who have heart or circulatory (blood vessel) problems, or damaged lungs or breathing passages
• Hydrocarbons (excluding methane) & VOCs* (volatile organic compounds); smog-formers. VOCs include chemicals such as benzene, toluene, methylene chloride and methyl chloroform	8	• VOCs are released from fossil fuels; solvents, paints glues and other products used at work or at home. • Automobiles, • Evaporation from storage tanks of gasoline (petrol and diesel)	• Total non-CH_4 Hydrocarbons • Biogenic processes in soil and water	• Oxidized to CO_2. Absorption in soil then microbial degradation. • Photochemical degradation	1.5–2 yrs.	Human health • Fat below skins • Liver • In addition to ozone (smog) effects, many VOCs can cause serious health problems such as cancer and other effects Plants • For unsaturated hydrocarbons • Chlorosis (leaf tip and margin yellowing),

(Continued).

Table 3.3 (Continued)

Pollutant composition	Toxic* rank	Major source	Natural sources	Major sinks	Stay period	Major damages
						Human health; plants and materials
						• Leaf abscission, decreased yield. • In addition to ozone (smog) effects, some VOCs such as formaldehyde and ethylene may harm plants.
Hydrogen sulfide H_2S. • Colorless, rotten eggs odor, • Slightly soluble in water. It behaves like a weak acid.	7	• Chemical industries, • Mines, • Oil wells and • Refineries.	• Volcanoes, • Biogenic processes in soil and water	• Oxidation to SO_2.	0.08–2 Days.	Human health • Lungs Materials • Tarnish Cu, Ag. Leaded paints get blackened.
• Metals (heavy metals such as Cu, Zn, etc.) • Lead	6	• Mineral dressing, • Smelting and • Refining. • Leaded gasoline (being phased out), • Paint (houses, cars), • Manufacture of lead storage batteries				Human health • Bronchus, Bronchioles, Small intestine, Skin, Peripheral arteries, mid brain, Lungs, Kidney, Bones. • Brain and other nervous system damaged; Children are at special risk.

Pollutant	No.	Sources	Formation	Residence time	Effects
Nitrogen oxides [$NO_x(g)$, where x may be 1 or 2] • NO is radish brown. Somewhat soluble. • Nitrogen dioxide – an oxidizing agent when react with water forms nitric acid. • It also involves in the production of ozone in the atmosphere.	5	• Automobiles engines • Fertilizers • Blasting in mines • Burning of fossil fuels.	• The most natural way of forming NO_x is lightning. Atmospheric nitrogen reacts with nearby oxygen to form nitrogen monoxide. Nitrogen monoxide further reacts with oxygen to form **nitrogen dioxide**. Also • Biogenic processes in soil.	• NO – Atmospheric oxidation to NO_2; Precipitation scavenging then oxidation to nitrates. 3–5 days.	• Some lead-containing chemicals cause cancer in animals. • Lead causes digestive and other health problems. Lead can harm wildlife Human health • Lungs damage • Illnesses of breathing passages and lungs (respiratory system) Plants • Brown spots on leaf, suppression of growth. • Nitrogen dioxide is an ingredient of acid rain (acid aerosols), which can damage trees and lakes. Acid aerosols can reduce visibility Materials • Dyes fading. • Acid aerosols can eat away stones used on buildings, statues, monuments, etc.

(Continued)

Table 3.3 (Continued)

Pollutant composition	Toxic* rank	Major source	Natural sources	Major sinks	Stay period	Major damages
Particulate matter less than 10 μm dia. (PM) • Solid particles or liquid droplets including smoke, dust and aerosols.	4	• Process industries • Mines • Construction sites • Automobile traffic • Fires • Refuse • Incineration • Burning of wood, diesel and other fuels • Agriculture (plowing, burning off fields) • Unpaved roads	• Forest fires			Human health; plants and materials Human health • Nose and throat irritation, lung damage, bronchitis, early death. • Formation of scar in lung tissues by inorganic fibers and dusts such as: • Silica-silicosis • Asbestos – asbestosis • Talk – talcosis • Coal – black lung, • Organic fibers and dusts such as: Cotton, flax & hemp – brown lung. • Metallic fumes: • Tin oxide – stannosis; • Iron oxide – siderosis Plants • Haze that reduces visibility Materials • Soiling building materials, painted surface and clothes. • Discolor structures and other property, including clothes and furniture

Pollutant		Sources	Residence time	Effects
Sulfur dioxide SO_2: • Colorless, pungent smell, soluble in water. This acid forming oxidizing agent can form sulfurous and sulfuric acid. • SO_x	3	• Combustion of sulfur containing fuels (coal, oil), petroleum refining, metal smelting, paper industries. • Volcanoes • Reaction of biogenic S emissions	20 minutes to 7 days. • Precipitation scavenging then oxidation to sulfate particles.	Human health • 0.01% – uncomfortable; 1.0% bronchitis after exposure for half an hour. • Breathing problems, may cause permanent damage to lungs Plants • Bleached spots on leaf, Chlorosis, • Decreased yield and growth. • SO_2 is an ingredient in acid rain (acid aerosols), which can damage trees and lakes. Acid aerosols can also reduce visibility Materials • SO_x • Corrosion to Cu, Al; Leaching and weakening of building materials; embitterment of leather, paper; fabric's reduces strength.
Hydrogen fluoride HF. • Colorless, pungent, soluble in water. • These are highly reacting reducing agents.	2	• Manufacturing of fertilizers, • Ceramic • Al refining.		Human Health • Fluoride, I • Small intestine, Thyroid, Bones Plants • Chlorosis • Dwarfing, Leaf abscission, decreased yield Materials • Put etch marks on glass & impair opacity.

(Continued)

Table 3.3 (Continued)

Pollutant composition	Toxic* rank	Major source	Natural sources	Major sinks	Stay period	Major damages Human health; plants and materials
Ozone O_3: • Pale blue gas, sweetish, unstable, water soluble. • Ozone at ground level (principal component of smog)	I	• Presence of Hydrocarbons & NO_2 in sun light. • Chemical reaction of pollutants; vocs and nox		• Photochemical reaction (reversion to O_2) in atmosphere.	2 hrs–3 days.	Human health • Breathing problems, reduced lung function, asthma, irritates eyes, stuffy nose, reduced resistance to colds and other infections, may speed up aging of lung tissue Plants • Reddish brown flecks on upper surface of leaf • Bleaching, • Leaf abscission, • Decreased yield and premature aging. • Smog can cause reduced visibility Materials • Cracking and weakening of rubber and fabrics. • Dyes fading.
Carbon dioxide CO_2		• Burning of fossil fuels.	• The natural one is from the respiration of organisms.	• Dissolves in oceans. • Taken up by plants during photosynthesis ensuring it does not accumulate.		Human Health • Global warming impacts Materials • Deterioration of building stones particularly limestone

* Toxic rank as adapted by NAAQS of USA.

3.3   AIR TOXICS[3, 4, 14, 17, 25]

The emission of toxic substances into the air can be damaging to human health and to the environment. Human exposure to these pollutants at sufficient concentrations and durations can result in cancer, poisoning, and rapid onset of sickness, such as nausea or difficulty in breathing.[4] Other less measurable effects include immunological, neurological, reproductive, developmental, and respiratory problems. Pollutants deposited onto soil or into lakes and streams affect ecological systems and eventually human health through consumption of contaminated food.[4] The contributors and their impacts have been illustrated in Figure 3.18.

3.3.1   Units of measuring concentration

Table 3.5 details 'Units' used to measure concentration of the pollutants, and Table 3.2 lists Permissible Exposure Limit (PELs) 8 hrs time weighted average as adopted by EPA, USA.[4]

3.4   AIR QUALITY STANDARDS[4, 6]

Air Quality Standard is a measure of the concentration of contaminants in the atmosphere following an initial period of dispersion, and is primarily concerned with ambient concentrations of contaminants. Air Quality Standards are based on the premise that the environment can tolerate defined concentrations of contaminants without

Table 3.4 Toxic air pollutants as per EPA. Damage to human's body caused by major pollutants shown.[3, 4, 14, 17, 25]

Toxic air pollutants (As per EPA)	Damage to human body by pollutants[17, 25]
• Fuel combustion • Non-ferrous metals processing • Ferrous metals processing • Mineral products processing • Petroleum and natural gas production and refining • Organic liquids and gasoline distribution • Surface coating processes • Waste treatment and disposal • Agricultural chemicals production • Fibers production processes • Food and agriculture processes • Pharmaceutical production processes • Polymers and resins production • Production of organic and inorganic chemicals • Halogenated solvent and dry cleaning operations • Chromium electroplating • Chromic acid anodizing • Commercial sterilization facilities • Others	Hydrocarbons – Fat below skins, Liver SO_2 – Bronchioles CO – Heart, Brain, Lungs H_2S – Lungs NO – Lungs Ozone – Lungs Fluoride, I – Small intestine, Thyroid, Bones Metals – Bronchus, Bronchioles, Small intestine, Skin, Peripheral arteries, Mid brain, Lungs, Kidney, Bones Also refer Table 3.3.

Table 3.5 Units of measuring concentration.

Units of concentration

$mg/m^3 = ppm \times mol.wt/22.4$ (at STP 0°C & 1 atm. pressure.)
$mg/m^3 = (ppm \times mol.wt/22.4) \times (T0/T_1) \times (P_1/P_0)$
$(\mu g/m^3 = (mg/m^3)/1000$

Whereas:
ppm – parts per million i.e. volume.
mol.wt – Molecular weight of the gas under consideration.
T_0 – Standard Temperature 273, K
T_1 – Absolute temperature, in kelvins (K = °C + 273)
P_0 – Standard pressure which 101.325 kPa or 1 atmosphere
P_1 – Absolute pressure in kPa or atmospheres

- Ideal Gas Law:
$\rho = PM/RT$
ρ = Density of gas in kg/m^3
P = Absolute pressure in kPa
M = Molecular mass in gms./mole
R = Universal gas constant = 8.3143 J/K. mole
T = Absolute temperature, K

presenting a hazard. The potential effects of atmospheric emissions depend not only on the concentrations present at a particular point in time, but also on the duration of exposure. This is taken into account by specifying periods of time over which contaminants should be measured, referred to as the averaging period. In general, the Air Quality Standards specify short-term averaging periods for substances with high acute toxicity, while longer-term averaging periods are used to address chronic effects.

Air Quality Standards are listed in this Specification as limit values and guide values (Ref. Table 3.6). Limit values are intended to protect public health. Exceeding the limit values presents a risk to human health and should be considered as a non-compliance with this Specification. Action should be taken to reduce exposure concentrations in the event of a limit value being exceeded.

All Air Quality Standards in Table 3.6 have been expressed in terms of mass per standard unit volume, as specified by the World Health Organization (WHO) and European Union (EU).

3.4.1 The Air Quality Index (AQI) or Pollution Standard Index (PSI)[4, 6]

It is an easy-to-read report of daily air quality. It tells us how clean or polluted our air is, and which associated health effects might be a concern for us. The AQI focuses on health effects you may experience within a few hours or days after breathing polluted air. The AQI reports on five major air pollutants,[4, 6] including

Table 3.6 Foul gases and their permissible limits.[28]

Species	Averaging period	Limit value ($\mu g/m^3$)	Guide value ($\mu g/m^3$)	Source
Nitrogen dioxide (NO$_2$)	1 hour	400	–	WHO-EU
	4 hours	–	95	WHO-EU
	24 hours	150	–	WHO-EU
	1 year	–	30	WHO-EU
Sulphur dioxide (SO$_2$)	10 minutes	500	–	WHO-EU
	1 hour	350	–	WHO-EU
	24 hours	125	125	WHO-EU
	1 year	50	30	WHO-EU
Hydrogen sulphide (H$_2$S)	30 minutes[1]	–	7	WHO-EU
	24 hours	150[2]	–	WHO-EU
Carbon monoxide (CO)	1 hour[3]	40	–	Netherlands
	8 hours[4]	6	–	Netherlands
Benzene	1 hour	–	7.5	WHO-EU
	1 year	10	5	Netherlands
Total suspended particulate matter	1 year	120	–	WHO-EU
Particulates (products of incomplete combustion)	24 hours	125[5]	–	WHO-EU
	1 year	50[5]	–	WHO-EU

Notes:
1 odor prevention
2 not applicable to emergency situations
3 99.99 percentile
4 98 percentile
5 particulate matter diameter less than 10 ìm

ground-level ozone, particle pollution (extremely fine dust, also known as "particulate matter"), carbon monoxide, sulfur dioxide, and nitrogen dioxide. For each of these pollutants, EPA has established national air quality standards to protect public health.[4]

Each category corresponds to a different level of health concern. The six levels of health concern and what they mean are:

"Good" The AQI value between 0 and 50. Air quality is considered satisfactory, and air pollution poses little or no risk.

"Moderate" The AQI between 51 and 100. Air quality is acceptable; however, for some pollutants there may be a moderate health concern for a very small number of people. For example, people who are unusually sensitive to ozone may experience respiratory symptoms.

"Unhealthy for Sensitive Groups" When AQI values are between 101 and 150, members of sensitive groups may experience health effects. This means they are likely to be affected at lower levels than the general public. For example, people with lung disease are at greater risk from exposure to ozone, while people with either lung disease or heart disease are at greater risk from exposure to particle pollution. The general public is not likely to be affected when the AQI is in this range.

"Unhealthy" Everyone may begin to experience health effects when AQI values are between 151 and 200. Members of sensitive groups may experience more serious health effects.

"Very Unhealthy" AQI values between 201 and 300 trigger a health alert, meaning everyone may experience more serious health effects.

"Hazardous" AQI values over 300 trigger health warnings of emergency conditions. The entire population is more likely to be affected.

Example 3.1 illustrates how PSI value could be determined. Example 3.2 details steps to be followed to estimate 'Emission Inventory'. Table 3.7 depicts how monthly emission levels should be recorded. It also gives sources from where emissions are emitted.[28, 34]

Table 3.7 Illustration – how monthly emission level is recorded in a Petroleum Company in the Gulf region. The sources from where emission is emitted are also shown.[28, 34]

Emission type	Parameter	Unit
Flares (Includes AP, LP and HP Flares)	Flaring efficiency	%
Gas fired Combustion Systems	Monthly total emissions by component:	
	CO_2	Tonnes
	CO	Tonnes
	NO_x	Tonnes
	N_2O	Tonnes
	SO_x	Tonnes
	CH_4	Tonnes
	VOC	Tonnes
Diesel fired combustion systems	Monthly total emissions by component from:	
	Stationary	
	Mobile	
	CO_2	Tonnes
	CO	Tonnes
	NO_x	Tonnes
	N_2O	Tonnes
	SO_x	Tonnes
	CH_4	Tonnes
	VOC	Tonnes
Fugitive Emissions	CH_4	Tonnes
	VOC	Tonnes
CFCs	Inventory by end of Month	Kg
(R12, R502)	Total monthly releases	Kg
	Recovered for re-use	Kg

(Continued)

Table 3.7 (Continued)

Emission type	Parameter	Unit
HCFCs	Inventory by end of Month	Kg
(R22)	Total monthly releases	Kg
	Recovered for re-use	Kg
HFCs	Inventory by end of Month	Kg
(R134a, R404a, R407c)	Total monthly releases	Kg
	Recovered for re-use	Kg
Halons-1211	Inventory by end of Month	Kg
	Total monthly releases	Kg
Halons-1301	Inventory by end of Month	Kg
	Total monthly releases	Kg

3.4.2 Determination of Pollution Standard Index (PSI) value

Example 3.1: Determination of PSI Value

Determination of PSI Value

Suppose on a given day the following maximum concentration are measured:

1 hr. O_3 – 250 µg/m^3

8 hr. CO – 10 mg/m^3;

24 hr TSP – 50 µg/m^3

24 hr SO_2 – 100 µg/m^3

Find Pollution Standard Index (PSI) and indicate descriptive indicator that is to be used to characterize the day's air quality.

Pollutant Standard Index (PSI) breakpoints.

Index	1-Hr O_3; µg/m^3	8-Hr CO mg/m^3	24-Hr TSP µg/m^3	24-Hr SO_2 µg/m^3	TSP × SO_2 10^3 (µg/m^3)^2	1-Hr NO_2 µg/m^3
0	0	0	0	0	–	–
50	118	5	75	80	–	–
100	235	10	260	365	–	–
200	400	17	375	800	65	1130
300	800	34	625	1600	261	2260

PSI Value and Air Quality Descriptor.

1 hr. O_3 – 250 µg/m^3 – index +100

8 hr. CO – 10 mg/m^3; – index 100

24 hr TSP – 50 µg/m^3; – index less than 50

24 hr SO_2 – 100 µg/m^3; index less than 100

So for O_3 it is to be interpolated. The base is 100 and interpolation can be obtained by referring the values in the Table.

Referring the PSI Table
For a difference in concentration of $(400 - 235 = 165)$; the index difference = $(200 - 100 = 100)$
So for a concentration difference of $(250 - 235 = 15)$, the index value is?

PSI = $100 + [(250 - 235)/(400 - 235)] \times (200 - 100) = 109$ = Unhealthful.

PSI value	Descriptor
0–50	Good
50–100	Moderate
101–199	Unhealthful
200–299	Very unhealthful
Exceeding 299	Hazardous

3.4.3 Emission inventory estimation

Example 3.2: Steps to be followed to estimate Emission Inventory

Estimation of Emission Inventory

– List the source type – such as homes, automobiles, factories, etc.
– List type of air pollutant emission from the source – such as CO; SO_2 etc.
– Emission factor, listed in various ways i.e. 250 gms./1000 liter of coal consumption etc.
– Calculations then follow

For example given: Emission factor for CO from fuel oil = 50 g/1000 liters;
Sales of oil/day = 40,000 liters.
Hence, CO emission/day = $250 \times 40000/1000$ kg = 9.6 kg

– Likewise emission from every source can be calculated, and adding them gives total emissions/day.

3.5 PERFORMANCE MONITORING[3, 4, 6, 25]

An air emission-monitoring program should be developed, implemented and maintained to demonstrate compliance with the laid out norms, standards and regulations. In Table 3.8 air pollutant sources and their natural removals have been shown. Figure 3.3 illustrates sources of air pollution – Mobile as well as stationary. Table 3.9 gives an account of air pollutant receptors. It also mentions adverse impacts due to these pollutants.

Table 3.8 Air pollutant sources and their natural removals.[4]

Air pollutant sources		Air pollutant removals	
Man made	*Natural*	*Natural sinks*	*Mechanism*
Mobile sources: • Automobiles, trains, airplanes; any equipment, which is portable and run by fossil fuel.	• Plant and animal respiration and their decay after death. • Volcanoes • Forest fires	• Soil • Vegetation • Structures • Water-bodies such as oceans.	• Oxidation of NO, NO_2, H_2S, SO_2, SO_3

Stationary:
- Power plants.
- Industry – chemical plants, steel mills, oil refineries etc.
- Mines.
- Agriculture and Construction sites.
- Solid waste disposal and incineration.
- Accidental release including leakages and spills.

3.5.1 Air pollutant receptors and adverse impacts

| Mobile sources | Stationary | Forest fires | Accidental release including Leakage & spills |

Figure 3.3 Sources for air pollution – Mobile as well as stationary.

Table 3.9 Air pollutant receptors and adverse impacts.[17, 24, 25, 32]

Receptor – something which is adversely affected by polluted air	Adverse impacts
Person or animal	Health effects – internal and external organs, as shown in Table 3.4
Trees, plants, vegetation	Growth and appearance is affected to the extent they may die
Material – cloth, paper, leather, stone, metals, paints etc.	Appearance and strength
Aquatic life in lakes and soils	Affected by acidic deposition
Properties of atmosphere	Ability to transmit radiant energy

3.6   GLOBAL WARMING – THE GREENHOUSE EFFECT[2, 3, 6, 7, 11, 13]

In addition to the usual three ways of heat transfer in the troposphere: evaporation-condensation, conduction, and convection, the fourth way is the greenhouse effect.

Solar energy (light radiation) absorbed by the earth is converted into heat energy and emitted into space as long-wave (heat) radiation (infrared) as shown in Figure 3.4. Although water vapor and carbon dioxide are transparent to short-wave radiation, they are nearly opaque (obstacle) to long-wave radiation. Thus, much of the earth's radiation is retained, raising the temperature of the atmosphere. This phenomenon is known as the '**Greenhouse Effect**', taking its name from the principle of greenhouse construction, where the glass operates in a fashion similar to carbon dioxide and water vapor, allowing solar rays to pass unhindered into the greenhouse, but blocking reverse radiation (Figure 3.4).[13]

This can even be experienced in any automobile including a car, which is much warmer inside than the outside atmosphere. This allows transmission of short waves but not the long waves.

The combination of water vapor, ozone, carbon dioxide, and the trace gases (such as methane, ammonia, sulfur dioxide and certain chlorinated hydrocarbons) behaves like a glass in a greenhouse, and known as 'Greenhouse Gases'. The heat these materials absorb and hold accounts for the earth's surface temperature. Table 3.10 details of greenhouse gases and their impacts. Table 3.11 depicts impacts of greenhouse gases on the ecosystem. Also refer sec. 2.8.5.

In the last century, human activities have added more than 25%, about 360 billion tonnes of carbon dioxide in the atmosphere. This has caused an increase in earth's temperature by 1°F. The trend of venting carbon dioxide to the atmosphere is increasing.

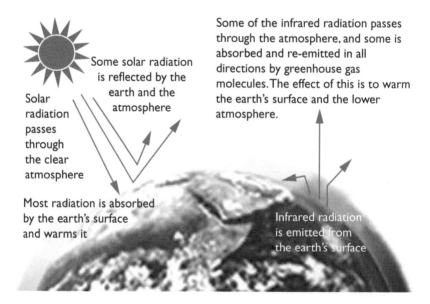

Some solar radiation is reflected by the earth and the atmosphere

Solar radiation passes through the clear atmosphere

Some of the infrared radiation passes through the atmosphere, and some is absorbed and re-emitted in all directions by greenhouse gas molecules. The effect of this is to warm the earth's surface and the lower atmosphere.

Most radiation is absorbed by the earth's surface and warms it

Infrared radiation is emitted from the earth's surface

Figure 3.4 The greenhouse effect.[13]

(Source: http://www.epa.gov/globalwarming/).[13]

3.6.1 Greenhouse impacts

Table 3.10 Details of greenhouse gases and their impacts. (drawn from various sources).[3, 7, 12, 23]

| Greenhouse gases (GHGs) | Concentration[3, 7] | | % Change | Sources | % Share & important features |
	Year 1750	Present			
CO_2	280 ppm	360 ppm	29	Fossil fuel combustion, organic decay, deforestation, burning wood and wood-products.	Human induced CO_2 is pollutant. Its share is 65%
Methane	0.7 ppm	1.7 ppm	143	Biomass, organic decay, rice field, landfills. Production, transportation of coal, oil and natural gas.	It traps over 21 times more heat per molecule than CO_2. Its share is 19%
Nitrous oxide	280 ppb	310 ppb	11	Fossil fuel, nitrogen fertilizers, biomass burning,	It absorbs 270 times more heat per molecule than CO_2. Its share is 6%
Ozone	Unknown	varies	–	Occurs naturally	
Very powerful GHGs not naturally occurring include *hydrofluorocarbons* (HFCs), *perfluorocarbons* (PFCs), and *sulfur hexafluoride* (SF6)	0	990 ppt	–	Refrigerators, cleaning solvent, aerosol sprays, foams	HFCs and PFCs are the most heat-absorbent. CFC's share is 8%. Halocarbons' share is 2%. They are generated in a variety of industrial processes.
Remarks	• Naturally occurring greenhouse gases include water vapor, carbon dioxide, methane, nitrous oxide, and ozone. Certain human activities, however, add to the levels of most of these naturally occurring gases.				
	• The GHGs are generated due to fossil fuel use, certain agricultural and industrial activities, and deforestation. These gases absorb infrared radiations.				
	• Each greenhouse gas differs in its ability to absorb heat in the atmosphere.				
	• Adverse impacts of global warming are not felt equally across the earth. They may differ from one region to another.				
	• The earth temperature has been increased by 1° Fahrenheit since the last two decades (Figure 3.5).				
	• Thus, GHGs are responsible for the 'Climate Change'. Increasing concentrations of GHGs in the atmosphere has the potential, over the next few generations, to significantly alter global climate.				
	• This would result in large changes in ecosystems, leading to possibly catastrophic disruptions of livelihoods, economic activity, living conditions, and human health.				
	• On the other hand, abatement of GHGs, would involve significant economic costs.				

(Continued)

Table 3.10 (Continued)

Greenhouse gases (GHGs)	Concentration[3, 7]				% Share & important features
	Year 1750	Present	% Change	Sources	

• While climate change is a global environmental issue, different countries bear different levels of responsibility for the increase in atmospheric GHG concentrations. Further, the adverse impacts of climate change will fall disproportionately on those who have the least responsibility for causing the problem, in particular, developing and undeveloped countries.[13]

Table 3.11 Impacts of green house gases to the ecosystem.[2, 3, 7, 13, 20, 23]

Eco system	Changing climate is expected to:
Physical	1 GHGs are likely to accelerate the rate of climate change. 2 Scientists expect that the average global surface temperature could rise 1–4.5°F (0.6–2.5°C) in the next fifty years, and 2.2–10°F (1.4–5.8°C) in the next century, with significant regional variation.[7, 20]
Sea level[13]	Sea level has risen worldwide approximately 15–20 cm (6–8 inches) in the last century. Approximately 2–5 cm (1–2 inches) of the rise has resulted from the melting of mountain glaciers. Another 2–7 cm has resulted from the expansion of ocean water that resulted from warmer ocean temperatures.
Precipitation[13]	Precipitation has increased by about 1 percent over the world's continents in the last century. High latitude areas are tending to see more significant increases in rainfall, while precipitation has actually declined in many tropical areas.
Biotic 1 Human health	• Increase the risk of some infectious diseases such as malaria, dengue fever & yellow fever. • Higher air temperatures also increase the concentration of ozone at ground level, which can cause chest pains & pulmonary congestion.
2 Birds	• Life cycles could be influenced. Some regions are less hospitable to birds.
3 Fisheries	• Some bodies of water may become too warm for the fish that currently inhabit them. • Global warming may also change the chemical composition of the water. • Loss of wetlands could diminish habitat and alter the availability of food for some fish species.
4 Wildlife[23]	• Researchers reported in 1999 a decline in the health of polar bears. Decline by 50% in the number of penguins in Antarctica.
5 Forests	• Adverse impacts include: • Warmer temperatures may enable trees to colonize into areas that are currently cold. • Some forests may tend to have a less diverse mix of tree species. • Poor soil. • On the negative side, forest fires are likely to become more frequent and severe. • On the positive side, CO_2 has a beneficial fertilization effect on plants.

(Continued)

Table 3.11 (Continued)

Eco system	Changing climate is expected to:
Abiotic 1 Water resources	• Evaporation will increase as the climate warms, which will increase average global precipitation. Soil moisture is likely to decline in many regions, and intense rainstorms are likely to become more frequent. • It will cause expansion of the oceans and the gradual melting of polar ice caps with a consequent rise in sea level. This would in turn cause flooding of low-lying land where many capital cities lie. For example:[35] • The snow cover in the Northern Hemisphere and floating ice in the Arctic Ocean has decreased. • Globally, sea level has risen 4–8 inches over the past century. • Worldwide precipitation over land has increased by about one percent. The frequency of extreme rainfall events has increased throughout much of the United States.
2 Deserts	• Relatively little research has been conducted on the possible impact of climate change on deserts. Deserts tend to be found at latitudes where soils are extremely dry.
3 Mountains Glaciers	• Changes in climate are already affecting many mountain glaciers around the world. Also refer Figure 2.9.
Overall impact	Figure 3.7; illustrates impact of 2–3°C change in Australia.
Remarks	The effect of global warming is being felt the world over including these continents: Africa, Asia, Europe & Russia, Central America, North America, South America and Oceania. Thus, this would result in large changes in ecosystems, leading to possibly catastrophic disruptions of livelihoods, economic activity, living conditions, and human health.

3.6.2 Changing climate[2, 3, 7, 8, 13, 23]

Global mean surface temperatures have increased 0.5–1.0°F since the late 19th century.[35] The 20th century's 10 warmest years all occurred in the last 15 years of the century. Of these, 1999 was the warmest year on record. The snow cover in the Northern Hemisphere and floating ice in the Arctic Ocean has decreased. Globally, sea level has risen 4–8 inches over the past century. Worldwide precipitation over land has increased by about one percent. The frequency of extreme rainfall events has increased throughout much of the United States. Temperature rise impacts not only on the ecosystem (biotic and abiotic components) but also on trade and industries as shown in Figure 3.7.

Increasing concentrations of greenhouse gases are likely to accelerate the rate of climate change. Scientists expect that the average global surface temperature could rise 1–4.5°F (0.6–2.5°C) in the next fifty years, and 2.2–10°F (1.4–5.8°C) in the next century, with significant regional variation.[7]

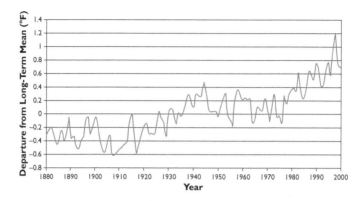

Figure 3.5 Global temperature change during period 1880–2000.[35]

Source: US National Climate Data Center.

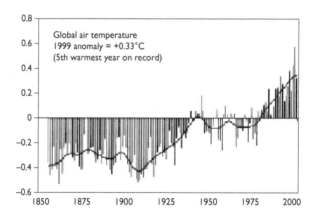

Figure 3.6 Global temperature and atmospheric CO_2.

Source: Climate Research unit, University of East Anglia.[8]

Figure 3.5 presents global temperature change during period 1880–2000. Relation between Global temperature and atmospheric CO_2 has been shown in Figure 3.6. Examples 3.3 to 3.5 are the working examples to compute carbon and CO_2 emissions.

3.6.3   Calculation of CO_2 emission from hydrocarbons

Example 3.3: Calculation of CO_2 emission from hydrocarbons.

Find out CO_2 emission by 100 gms of butane.
1   Write the reaction equation.
   $C_4H_{10} + O_2 = CO_2 + H_2O$
   (Butane) (Oxygen)
2   Balance it
   $2C_4H_{10} + 13O_2 = 8CO_2 + 6H_2O$

3 Find out molecular mass (or weight) of the concerned reactant, as well as that of the product.

Molecular mass of butane $= 2(4 \times 12 + 1 \times 10) = 116$

Molecular mass of CO_2 produced $= 8(12 + 32) = 352$

4 Calculate amount of emission of the desired gas.

Thus, 116 gms of butane produces 352 gms of CO_2

100 gms of butane will produce $352 \times 100/116 = 303$ gms of CO_2

Example 3.4: Calculation of CO_2 emission from fossil fuels.

Example: Calculation of CO_2 Emission from Fossil Fuels

If fossil fuels consumption per year is 3×10^{20} Joules, and all the energy is given by CH_4 gas. Energy content of CH_4 is 3.9×10^7 J/m^3 at STP. Calculate at what rate CO_2 would be emitted to the atmosphere. Also express emission of C in the Air per year.

1 Calculating amount of fuel required to generate the required amount of energy.

Amount of CH_4 required/year $= 3 \times 10^{20}/3.9 \times 10^7 = 769 \times 10^{10}$ m^3

2 Write the reaction equation: $CH_4 + O_2 = CO_2 + H_2O$
Balance it: $2CH_4 + 4O_2 = 2CO_2 + 4H_2O$
$[2 \times (12 + 4) = 32]$ $[2(12 + 32) = 88]$
1 mol of CH_4 produces 1 mol of CO_2

3 Convert Vol. of Methane into Mols. $= 769 \times 10^{10}$ m^3 / $0.0224 = 3.43 \times 10^{14}$

4 Calculate amount of **emission of the desired gas.**

Since 1 mol of CH_4 produces 1 mol of CO_2 i.e. 44 gms of CO_2
3.43×10^{14} mols of methane will produce $= (44) \times 3.43 \times 10^{14}$ gms of CO_2.
$= 1.5 \times 10^{16}$ gms/yr
$= 1.5 \times 10^{10}$ tonnes/yr.
$= 15$ Giga tonnes/yr.

Calculations for Carbon: $CO_2 = C + O_2$

Proportion of C in $CO_2 = (12/44) \times 1.5 \times 10^{10}$ tonnes/yr $= 4.09 \times 10^9$ tonnes
$= 4.09$ giga tonnes/yr.

Example 3.5: Estimation C in the atmosphere corresponding to a concentration of CO_2.

Example: Estimation C in the atmosphere corresponding to a concentration of CO_2.

Estimate the tonnes. of carbon in the atmosphere corresponding to a concentration of 360 ppm of CO_2. Assuming total mass of air to be 5.1×10^{18} kg. The density of air at the standard temperature and pressure (0°C and 1 atmosphere pressure) is 1.29 kg/m^3.

CO_2 in mg/m^3 = (ppm $\times$ mol.wt/22.4) $\times$ $(T_0/T_1) \times (P_1/P_0)$
$= (360 \times 44/22.4)(273/273) \times (1/1) = 707.14$ mg/m^3 $= 0.7071$ g/m^3

Proportion of C in CO_2; Equation $C + O_2 = CO_2$; $12 + 32 = 44$

Thus, C concentration $= (12/44) \times 0.7071$ g/m^3 $= 0.19284$ g/m^3

Total vol. of air = (V = M/D) = 5.1 × 10^{18}/1.29 m^3 = 3.953 × 10^{18} m^3
C weight in air = (0.19284 g/m^3) × (03.953 × 10^{18} m^3)/1000 × 1000
= 762.390 × 10^9 = 762 Giga-Tonnes (GT).

If the concentration of CO_2 in atmosphere is 1 ppm, then amount of C in air will be 762.390/360 = 2.12 Giga-Tonnes (GT).

Or 2.12 GT of C gives the concentration of CO_2 to be 1 ppm.

Reserves	Fossil fuels		
	Gas	Oil	Coal
Quads	3600	4100	22000
GT Carbon	52	85	550
Resources			
Quads	10000	11000	280000
GT Carbon	145	230	7100

Energy and Carbon contents of the world resources.

For natural gas the total carbon content is 145 GT in the resource Table. If all is released and its half remains in the atmosphere; calculate concentration of CO_2 in ppm?

= (145/2.12) × 50% = 34 ppm of CO_2;
Similarly for Oil: it would be 54 ppm of CO_2 and
For Coal: 1675 ppm of CO_2

Australian public can identify with impacts from 2–3° change		
Tourism	**Primary**	**Insurance**
• 97% of the Barrier Reef bleached • 80% of Kakadu wetlands lost • Snow area shrinks by 55%	• Macquarie River Basin flows fall by 20% • Livestock carrying capacity falls by 40% • Australian net primary production falls by 6%	• Temparature related deaths over 65 rise by 100% • Wind speed increases 7.5% • Dengue fever transmission zone reaches Brisbane

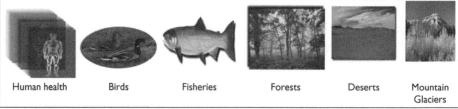

Human health   Birds   Fisheries   Forests   Deserts   Mountain Glaciers

Figure 3.7 Temperature rise affects not only ecosystem (biotic and abiotics components) but also on trade and industries.[2, 18] Upper illustration after Bart[2] (Temperature given in degree centigrade.)

Source: http://www.maltaweather.info/acid.jpg.[18]

3.7   ACID RAIN[4, 6, 9, 20, 24, 25, 32]

What goes up must come down – somewhere, sometime, in some form. A published reference was made in 1872, about acid rain. In 1970's it became evident as lakes without any known source of acid in US, Canada, Scandinavian countries were becoming increasingly acidic and that fish were disappearing from them (Figure 3.10).

Acid or acid forming materials may be deposited from the air in the form of snow, fog, or even as gases or dry particulate matters – as well as in the form of rain, as shown in Figure 3.8 (left).

It is not surprising as from the last 100 years or more we are burning coal and oil, smelting ores of many metals (Cu, Pb, Zn, etc.) containing sulfur. Combustion in automobiles, chemical and power plants, smelters also creates oxides of nitrogen, mostly as nitrogen combines with oxygen.

Gravity keeps the things in the atmosphere from moving off into space, and this leaves only two possibilities, either sulfur or nitrogen come as acid or they are still up there somewhere. It is most probable that almost all have returned back to earth as dry deposition or acid precipitation somewhere down wind or where it was formed.

3.7.1   How acid rain is formed?[4, 6, 9, 20, 24, 25, 32]

It has been discovered that air pollution from the burning of fossil fuels is the major cause of acid rain. The main chemicals in air pollution that create acid rain are sulfur dioxide (SO_2) and nitrogen oxides (NO_x). Acid rain usually forms high in the clouds where sulfur dioxide and nitrogen oxides react with water, oxygen, and oxidants. This mixture forms a mild solution of sulfuric acid and nitric acid. Sunlight increases the rate of most of these reactions. Rainwater, snow, fog, and other forms of precipitation contain those mild solutions of sulfuric and nitric acids which fall to earth as *acid rain*.

The chemistry: due to Nitrogen Dioxide[27, 32]

Bacteria action in soil releases Nitrous Oxides (N_2O) to the atmosphere. In the upper troposphere and stratosphere, atomic oxygen reacts with nitrous oxide to form nitric oxide (NO).

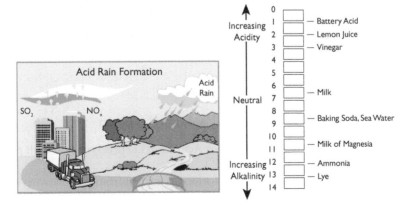

Figure 3.8 Acid rain formation.[18]

Source: http://www.maltaweather.info/acid.jpg.

$N_2O + O = 2NO$ (Production of NO_2 is about 0.45×10^{15} gms/year globally).   (3.1)

$NO + O_3 = NO_2 + O_2$ (3.2)

Ultimately, the NO_2 is converted to either NO_2^- or NO_3^- in particulate form. These particulates are then washed out by precipitation. The dissolution of nitrate in the water droplet allows for the formation of HNO_3 Nitric acid. This is partly accountable for the acid rain in the industrialized areas.

$$[2NO_{x(g)} + H_2O_{(1)} \longrightarrow 2H_{(aq)}^+ + 2NO_{3(aq)}^-$$ (3.3)

Oxides of nitrogen react with water to form nitric acid (NO_3^-)]

Due to Sulfur Oxides (SO_x)[27, 32]

Sulfur oxides may be both primary and secondary pollutants. Power plants, industry, oceans emit SO_2, SO_3, SO_4^- directly as primary pollutants. In addition, biological decays and some industries emit H_2S, which is oxidized to form secondary SO_2.

$H_2S + O_3 = H_2O + SO_2$ (3.4)

Combustion of fossil fuels containing sulfur yield sulfur dioxide in direct proportion to the sulfur content in the fuel

$S + O_2 = SO_2$ (3.5)

This reaction implies that for every gram of sulfur two grams, just double, of SO_2, are emitted to the atmosphere. The ultimate fate of SO_2 is the formation of sulfate salts, which are removed by sedimentation or precipitation. The conversion into sulfates is either of two ways: Catalytic Oxidation or Photochemical Oxidation:

The first method is most effective if water droplets contain Fe, Mn or NH_3

$$2SO_2 + 2H_2O + O_2 \xrightarrow{\text{catalyst}} = H_2SO_4$$ (3.6)

With low relative humidity by photochemical oxidation:

$SO_2 + hv = SO_2;$ $SO_2 + O_2 = SO_3 + O;$ hv – Photon of light energy (3.7)

Sulfates are very much hygroscopic and they absorb water resulting into:

$SO_3 + H_2O = H_2SO_4$ (3.8)

$$[SO_{2(g)} + 2H_2O_{(1)} \longrightarrow 2H_{(aq)}^+ + SO_{4(aq)}^{2-} + H_{2(g)}$$ (3.9)

Sulfur dioxide reacts with water to form sulfuric acid (SO_4^{2-})]

This reaction is a partly responsible for Acid Rain i.e. precipitation with pH Value less than 5.6.

In addition carbolic acid could also be formed as evident from the following equation:

$$CO_{2(g)} + H_2O_{(1)} \longrightarrow H_{(aq)}^+ + HCO_{3(aq)}^-$$ (3.10)

Carbon dioxide (CO_2) reacts with water to form carbonic acid (HCO_3^-)

Figure 3.9 Effects of Acid Rain to materials/rocks. Deterioration in the shape of Sandstone figure over the portal of a castle in Westphalia, Germany due to acid rain.[18]

Source: Malta Weather Information. www.maltaweather.info/pollution.html.

The hydraulic acid could also be formed as evident by the following equation

$$2Cl_{2(g)} + 2H_2O_{(l)} \longrightarrow 4H_{(aq)} + 4Cl^-_{(aq)} + O_{2(g)}$$ (3.11)

Chlorine (Cl_2) reacts with water to form Hydrochloric acid (Cl^-)

Acid Rain: Acid rain is rain that has a larger amount of acid in it than what is normal. Acid rain is one of the most dangerous and widespread forms of pollution.

3.7.2 pH

What is pH? A pH scale is used to measure the amount of acid in liquid-like water, a shown in Figure 3.8 (right). Because acids release hydrogen ions, the acid content of a solution is based on the concentration of hydrogen ions and is expressed as "pH."

Adverse impacts of acid rain are shown in Figure 3.9 and Table 3.9. Example 3.6 illustrates determination of sulfur quantity and its ultimate contribution in the acid rain formation.

3.7.3 Calculation related to acid rain

Example 3.6: Determination of sulfur quantity and its ultimate contribution in the formation of acid rain.

Example: How sulfur present in coal ultimately results in acid rain

Coal is burnt @ 1 kg/sec. It contains 3% sulfur. Calculate the amount of SO_2 that would be produced by it annually. Sulfur when burnt produces 5% ash by weight.

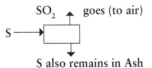

S in = S out = SO_2 (to air) + S with Ash

S produced by coal/sec. = 1 × 3% = 1 × 3/100 = 0.003 kg

S produced by coal/year = 0.003 × 60 × 24 × 365 = 9.46 × 10⁵ kg   (1)

Amount of S in ash = 5% × (1) = 0.005 × 9.46 × 10⁵ per year
= 4.73 × 10⁴ kg/year   (2)

Amount of S available for formation of SO_2 = (1)–(2)
= 8.99 × 10⁵ kg/yr.

The Reaction: $S + O_2 = SO_2$

(Thus, 32 gms of S yields 64 gms of SO_2; which is double the amount of sulfur)

Hence, 8.99 × 10⁵ kg of S would yield = (64/32) × 8.99 × 10⁵ kg of SO_2
= 1.8 × 10⁶ kg of SO_2/Year.

The ultimate fate of SO_2 is the formation of sulfate salts, which are removed by sedimentation or precipitation. The conversion into sulfates is by either of two ways: Catalytic Oxidation or Photochemical Oxidation:
The first method is most effective if water droplets contain Fe, Mn or NH_3

$$2SO_2 + 2H_2O + O_2 \xrightarrow{\text{catalyst}} = H_2SO_4$$

With low relative humidity by photochemical oxidation:

$$SO_2 + hv = SO_2; \quad SO_2 + O_2 = SO_3 + O; \quad hv - \text{Photon of light energy}$$

Sulfates are very much hygroscopic and they absorb water resulting in:

$$SO_3 + H_2O = H_2SO_4$$

These reactions illustrate how sulfur contributes in causing the acid rain having pH less than 5.6.

3.7.4   Adverse impacts of acid rain

Before the Industrial Revolution, the pH of rain was generally between 5 and 6, so the term acid rain is now used to describe rain with a pH below 5.

But over recent decades, rain in North America and Europe close to industrial areas has had a pH close to 4.5 and sometimes as low as 2.1 (equivalent to lemon juice – This is typical around volcanoes, where the sulfur dioxide and hydrogen sulfide form sulfuric acid in the rain).

Acid rain is one of the most dangerous and widespread forms of pollution. Sometimes called "the unseen plague", acid rain can go undetected in an area for years. Since rain travels over long distances in clouds, acid rain is a global problem, and needs a global solution. Figure 3.10 illustrates how acid rain is affecting European countries. Table 3.12 describes its adverse impacts.

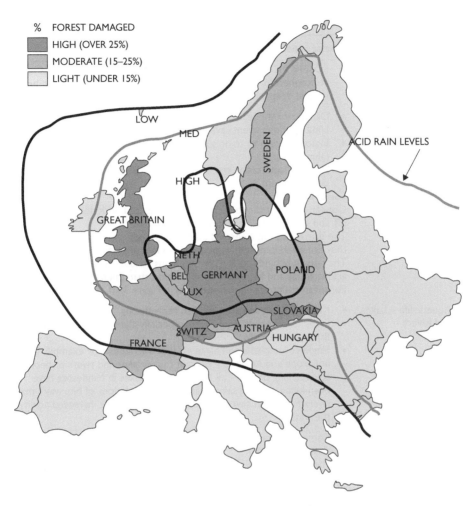

Figure 3.10 Impacts of acid rain in Europe.[18]

Source: www.maltaweather.info/pollution.html.

Table 3.12 Adverse impacts of acid rain. (Taken from various sources).[4, 6, 9, 20, 24, 25, 27, 32]

Receptors	Adverse impacts are
Nature – Physical	• 50–70% Visibility impairment.
Living 1 Human (Health)	The harm to people from acid rain is not direct. Many toxic metals are held in the ground in compounds. Some of these compounds can, however, break down by acid rain, freeing the metals and washing them into water sources such as rivers. In Sweden, nearly 10,000 lakes now have such high mercury concentrations that people are advised not to eat fish caught in them. In Sweden, the drinking water reached a stage where it contained high amounts of copper, which can cause diarrhea in young children, and can damage livers and kidneys.

(Continued)

Table 3.12 (Continued)

Receptors	Adverse impacts are
	• SO$_2$ and NO$_x$ also damage human health. These gases interact in the atmosphere to form fine sulfate and nitrate particles that can be inhaled deep into people's lungs.
2 Flora (Trees & Forests)	• Damage to the leaves of trees • Limiting the nutrients available to trees • Exposing them to toxic substances that are released from the soil (aluminum) • Flora – Plants, crops, trees and forests – a slow attack that ultimately destroys them. For example: Conifer trees appear to be particularly affected, with needles dropping off, and seedlings failing to produce new trees. The damage to the trees will be higher in high mountain regions because they are often exposed to greater amounts of acid than other forests. The reason behind this is that mountains tend to be surrounded by acidic clouds and fog that are more acidic than rainfall. Forest damage in European countries has been shown in Figure 3.10.
3 Aquatic life – Lakes and Rivers	• Kill lots of fish and stop the fish reproducing • Frees toxic metals from rocks, especially aluminum, which prevents fish from breathing • Single-celled plants and algae in lakes also suffer. For example, impacts in Europe and America: Acid rain causes lakes and rivers to become acidic, killing off fish – all the fish in 140 lakes in Minnesota have been killed, and the salmon and trout populations of Norway's major rivers have been severely reduced because of the increased acidity of the water.
Non-living systems Buildings & aesthetic values of materials	• Damage to certain materials, particularly limestone and marble, as shown in Figure 3.9 • The acid dissolves the calcium carbonate in the stone, forming crystals within the stone. As these crystals grow, they break apart the stone, and the structure crumbles. • Materials – aesthetic value of cars, buildings, bridges and cultural statues, monuments tombs etc.). It corrodes them and makes them dirty, as shown in Figure 3.9.
Remarks	pH 0 = maximum acidity; pH 7 = neutral point in the middle of the scale; pH 14 = maximum alkalinity (the opposite of acidity). Refer to Figure 3.8 (right). Thus, to treat an acidic effluent use alkaline such as lime before it is discharged to the predetermined source/destination.

3.8   OZONE GAS & PHOTOCHEMICAL SMOG (PCS)[3, 4, 11, 15, 16, 26]

Ozone (Table 3.3) is a gas composed of three atoms of oxygen. Ozone occurs both in the Earth's upper atmosphere and at ground level. Ozone can be good or bad, depending on where it is found.

Referring Figure 3.11, Top: this is the 'Good' ozone layer in the stratosphere (6–30 miles or 10 to 50 km above the earth's surface) that protects life on Earth from the Sun's harmful ultraviolet (UV) rays. Manmade chemicals are gradually destroying this beneficial ozone. An area where the protective "ozone layer" has been significantly depleted-for example, over the North or South Pole-is sometimes called "the ozone hole".

Middle: Antarctic Ozone Thinning shown in blue and purple, extended out over 16 million square miles or about the same size as North America (2001 NASA Satellite image)

Bottom: 'Bad' ozone at ground level is harmful to breathe, and it damages crops, trees and vegetation.

Photochemical smog is a mixture of pollutants, which includes particulates, nitrogen oxides, ozone, aldehydes, peroxyethanoyl nitrate (PAN), unreacted hydrocarbons, etc. The smog often has a brown haze due to the presence of nitrogen dioxide, which is the major component of photochemical smog. PCS causes pain to the eyes. Sunlight, hydrocarbons, nitrogen oxides and particulates are the major ingredient for the formation of PCS.

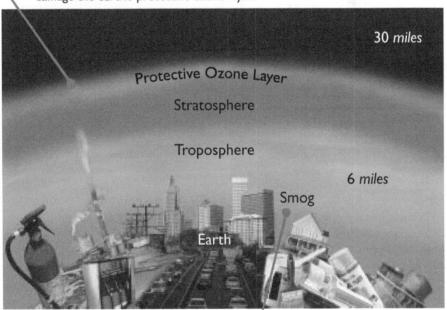

Too little there... Many popular consumer products like air conditioners and refrigerators involve CFCs or halons during either manufacture or use. Over time, these chemicals damage the earth's protective ozone layer.

Even though we have reduced or eliminated the use of many ODSs, their use in the past can still affect the ozone layer.

30 miles

Protective Ozone Layer

Stratosphere

Troposphere

6 miles

Smog

Earth

Too much here... Cars, trucks, power plants and factories all emit air pollution that forms groundlevel ozone, a primary component of smog.

Figure 3.11 Ozone layer and formation of smog (ozone) at the ground level.

Source: EPA, USA.[11, 15]

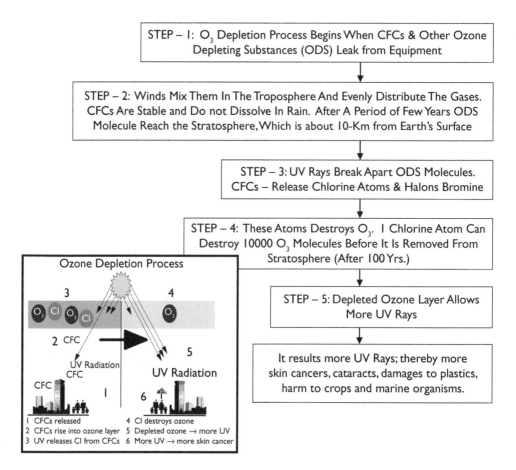

STEP – 1: O_3 Depletion Process Begins When CFCs & Other Ozone Depleting Substances (ODS) Leak from Equipment

STEP – 2: Winds Mix Them In The Troposphere And Evenly Distribute The Gases. CFCs Are Stable and Do not Dissolve In Rain. After A Period of Few Years ODS Molecule Reach the Stratosphere, Which is about 10-Km from Earth's Surface

STEP – 3: UV Rays Break Apart ODS Molecules. CFCs – Release Chlorine Atoms & Halons Bromine

STEP – 4: These Atoms Destroys O_3. 1 Chlorine Atom Can Destroy 10000 O_3 Molecules Before It Is Removed From Stratosphere (After 100 Yrs.)

STEP – 5: Depleted Ozone Layer Allows More UV Rays

It results more UV Rays; thereby more skin cancers, cataracts, damages to plastics, harm to crops and marine organisms.

Ozone Depletion Process

3   4

O_3 Cl O_3 Cl O_3

2 CFC

UV Radiation
CFC   5

CFC   UV Radiation

1   6

1 CFCs released	4 Cl destroys ozone
2 CFCs rise into ozone layer	5 Depleted ozone → more UV
3 UV releases Cl from CFCs	6 More UV → more skin cancer

Figure 3.12 Steps involved in the ozone depletion process.[15]

Source: http://www.epa.gov/ozone/defns.html#ozone.

Photochemical smog causes headache, eye, nose and throat irritations, impaired lung function, coughing and wheezing. Some people feel the effect of O_3 when present at only 0.001 ppm. At 0.12 ppm and above it produces chest discomfort, irritation of respiratory tract and reduction in pulmonary (lung) function. With more concentration asthma and chronic bronchitis are aggravated.

3.8.1 Ozone depletion process

Figure 3.12 depicts the steps involved in the ozone depletion process. Figure 3.13 (Left) shows the mechanism/process involved in destroying ozone. Figure 3.14 (Left) shows the worsening ozone hole and Figure 3.14 (right) the decline in the ozone shield.

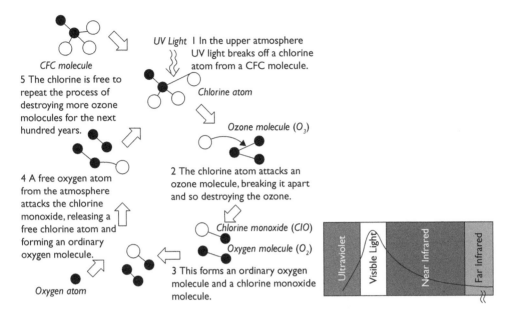

Figure 3.13 (Left) Mechanism/process involved in destroying ozone.

Source: U.S. EPA.

(Right) Solar radiation spectrum.

Source: University of Wisconsin-Stevens – Lecture on radiation energy.[15]

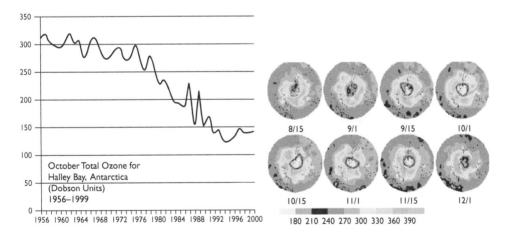

Figure 3.14 (Left) Worsening ozone hole.

Source: British Antarctic Survey;[19] http://www.nerc-bas.ac.uk.

(Right) Decline in ozone shield.

Source: U.S. EPA. http://www.epa.gov/ozone/science/hole/holecomp.html.[16]

3.9 NOISE POLLUTION[22, 24, 25, 30, 31, 34, 36]

Persistent exposure to elevated noise levels has been established to result in significant adverse health impacts. When level of sound becomes objectionable it is called Noise. Sound becomes noise when:

- Sound level is so high that hearing can be permanently impaired.
- It interferes normal work.
- It disturbs or prevents sleep.
- The third party raises objection.

3.9.1 Noise sources

- Unbalanced rotating or reciprocating machines. High speed of rotation and moving parts.
- Repeated Impacts (particularly from heights), Friction, Alternating electromagnetic effects.
- Air turbulence from fast rotating parts.
- Fluid movement within ducts and through orifices (high flow rate & pressure).
- Explosion or sudden release of gas under pressure.
- Loudspeakers, automobile horns, fireworks and continuous traffic by rail, road and air.

3.9.2 Industrial noise

Industry creates serious noise problems. It is responsible for high noise emissions indoors as well as outdoors of plants. In industrial vicinities it has been estimated that 30–49% or more of the working population is affected by sound pressure.

This noise is due to machinery of all kinds and often increases with the power of the machines. The characteristics of industrial noise vary considerably, depending on specific equipment. Rotating and reciprocating machines generate sound that is dominated by tonal and harmonic components; air-moving equipment tends to generate sounds with a wide frequency range.

Machinery that moves air is of special interest because it usually creates noise with a large component of low frequencies. Unlike noise containing mainly higher frequencies, walls or other structures less attenuate low frequency noise and it can cross great distances with little energy loss due to atmospheric and ground attenuation.

The major sources of noise in conventional power stations are turbo-generators, feed pumps, forced draught (FD) and induced draught (ID) fans, generator transformers, circuit breakers, cooling towers and mobile equipment. Added to these are transient sources such as steam or air releases from safety valves, drains or soot-blowers, fuel handling and delivery plant, alarms and auxiliary generators. The sources of noise from gas turbine plant include air inlet and exhaust, gas generator and oil cooling fans and pumps.

3.9.3   Important relations for sound/noise measurement[36]

The sound pressure level, L_p, is expressed in dimensionless form as the ratio between two sound energies. These are proportional to the squares of sound pressures, p. The basic unit is called bel (B)

$$L_p = \log\left\langle \frac{P}{P_r} \right\rangle^2 \tag{3.12}$$

L_p – Sound pressure level; P – Actual pressure level of sound in Pa;
P_r – Reference pressure level in, Pa.
A more convenient unit is decibel (dB)

$$L_p = 10\log\left\langle \frac{P}{P_r} \right\rangle^2 = 20\log\left(\frac{P}{P_r} \right) \tag{3.13}$$

Threshold of hearing, represents the lowest sound that can be heard by the human ear and it is taken as reference pressure level P_r having the value equal to 2×10^{-5} or 20 µPa. This reference pressure of 20 µPa represents the sound level of 0 dB.

$$L_p = 20\log\left\langle \frac{P}{20 \times 10^{-6}} \right\rangle, dB \tag{3.14}$$

Sound Intensity and Power
Amount of sound energy flowing through a unit area in a unit time is known as its intensity and this can be represented by the following equation.

$$I = \frac{P^2}{WC} \tag{3.15}$$

I – Intensity; w/m² – (watts per square meter); P – Sound pressure in Pa.
W – Density of medium = 1.2 kg/m³; C – Velocity or speed of sound in m/sec.

Sound pressure level and intensity level are related to each other as follows.
I_r is the reference intensity level in watts/m²; Acoustic power, N, of a sound source is given by:

$$L_p = 10\log\left\langle \frac{P}{P_r} \right\rangle^2 = 10\log\left(\frac{I}{I_r} \right), dB \tag{3.16}$$

$N = I \times S$ in watts.

Table 3.13(a) Noise threshold limits (American Standard).

Duration per day, hours	Sound level dB(A)
16	80
8	85
4	90
2	95
1	100
½	105
¼	110
1/8	110

S is surface area of the sphere = $4\pi r^2$ in m^2 but for a source placed on a level ground it is taken as $2\pi r^2$. r is the distance at which sound pressure intensity is measured.

Relation: $(C_1/T_1) + (C_2/T_2) + (C_3/T_3) + (C_4/T_4) + \cdots + (C_n/T_n) = 1$
Whereas C_1 – Total duration of exposure at the site;
T_1 – permissible exposure duration at that level.

If the sum of these fractions exceeds 1, noise exposure exceeds permissible limits and up to 1 it is within the allowable limits.

Summation of Sound Pressure Levels
If the pressure levels of two sources are P_1 and P_2; then combined pressure level in dB can be calculated using the following relation. If more than two sources are there, then calculations proceed in the same manner in a group of two values.

$$L_p = 10\log\left\langle\frac{P_1^2 + P_2^2}{20 \times 10^{-6}}\right\rangle, dB \tag{3.17}$$

Alternately, use the following relation and the Table 3.13(b).

VALUE (1) + Difference of two values, f

3.9.4 Noise control techniques

Figure 3.15 illustrates as how noise could be controlled by the application of various techniques that are in practice at the various industrial setups.

3.9.5 Noise related calculations

Example 3.6: Calculation of Noise levels.

Table 3.13(b) Threshold limits value of the impulsive or impact noise. No exposure exceeding 140 dB(A) peak sound pressure is allowed. Impact noise included those variations that involve maximum interval greater than 1 second. Where intervals are less than 1 second, it should be considered continuous. Noise levels are measured in decibels (dB), and the noise standards listed in this Specification refer to the 'A' frequency rating, dB(A), which covers the range audible to the human ear.

Difference between dB values of two sources	Add this value to the higher dB value
0	3
1	2.5
2	2
3	2
4	1.5
5	1
6	1
7	1
8	0.5
9	0.5
10	0
>10	0

Sound level dB(A)	Permitted numbers of impulses or impacts/day
140	100
130	1000
120	10000

3.6(1). In a factory noise level noted are as under: 95 dB (A) – 2 hrs. 90 dB(A) – 4 hrs.; 80 dB(A) – 2 hrs.; Find out noise exposure status, using standard table.

Noise level (dB(A))	95	90	80
Measured, hrs.	2	4	2
Permissible, hrs.	2	4	16

Using formula: $(C_1/T_1) + (C_2/T_2) + (C_3/T_3) + (C_4/T_4) + \cdots + (C_n/T_n) = 1$
$2/2 + 4/4 + 2/16 = 17/8 = 2.1$; It is exceeding 1.
This is exceeding unity, hence noise level exceed the permissible limits.

3.6(2). In a factory noise level noted are as under: 95 dB (A) – 1 hrs. 90 dB (A) – 1 hrs.; 70 dB(A) – 2 hrs.; Find out noise exposure status, using standard table.

Solution: Using Table 3.13(a): $1/2 + 1/4 + 0 < 1$. Remark: Noise level Within Limits.

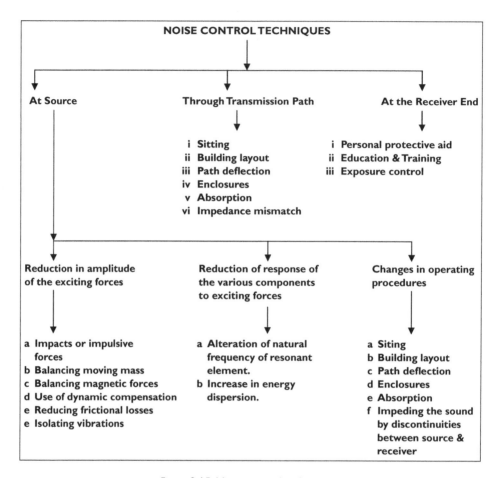

Figure 3.15 Noise control techniques.

3.6(3). Values of sound level are as follows: 85, 81, 83. Calculate combined sound level.
Solution: using standard Table 3.13(b)
So, the combine value for two readings is $85 + f(4) = 85 + 1.5 = 86.5$ next value is 83.
Calculation: $86.5 + f(86.5 - 83) = 86.5 + 1.75 = 88.25$.

Standards for environmental noise are a measure of sound levels imme-diately "outside the boundary fence", taken at the affected area. Standards for environmental noise are based on the premise that the environment can tolerate a defined level of noise without presenting a hazard to human health or harming the environment.

Noise levels are measured in decibels (dB), and the noise standards listed in this Specification refer to the 'A' frequency rating, dB(A), which covers the range audible to the human ear. "Leq" is a measurement unit applied to an average number of decibels over a specified period of time.

The noise limits which apply specifically to "Industrial plants and public works" and "Airports" in a country in the Gulf region are presented in Table 3.14 for the purpose of illustration. These norms are prescribed in every country and may differ based on the local conditions. Table 3.15 depicts the adverse health effects of noise to an industrial worker and level of responsibility lies to the concerned.

3.9.6   Noise threshold limits

3.10   VIBRATIONS

Vibration can be both a safety and a health problem, causing:

- Fatigue and failure in metal parts
- Fixtures to work loose
- Objects resting on high surfaces to move and fall
- Inability to read instruments and instructions
- Ill health such as 'white finger' among persons whose hands or bodies are exposed to excessive vibrations.

Like noise, much unwanted vibration can be avoided by careful design and initial choice of machines and by improved mounting. Also refer sec. 7.4.2.4.

Table 3.14 Noise threshold limits in a Gulf country.[28, 34]

Type of district	Leq, T, dB(A)		
	Workdays 7.00 am–6.00 pm	Workdays 6.00 pm–11.00 pm	All Holidays & Workdays 11.00 pm–7.00 am
Rural, residential, recreational	45	40	35
Suburban residential	50	45	40
Urban residential	55	50	45
Urban residential (with some workshops and business)	60	55	50
Industrial and commercial	70	70	70

Table 3.15 Effect of noise pollution to the industrial worker.[22]

Noise induced hearing loss							
Policy		To practice					Results
Risk Assessment Noise	System Controls, Planning, Design, Medical Surveillance	Operational Control	Unsafe Acts – Failure to attenuate	Unsafe Conditions – Noise Environment	Incident: Exposure to Noise +85 dBA	Threshold Limit 85 dBA	Effects Noise Induced Hearing Loss
Snr. Management		Middle Management					Individual

3.11   DUST[1, 4, 6, 22, 25, 31, 34, 36]

Dust can be defined as finely divided solid matter and can be considered from two aspects:

- Explosive properties and
- Harmful physiological effects.

3.11.1   Conditions for dust to become nuisance

Dust becomes a nuisance:

- When it is in the form of clouds, reducing visibility, creating an uncomfortable environment (irritation of the eyes, ears, nose, throat and skin).
- Increase equipment maintenance costs due to excessive wear and premature failure of components.
- Dust also causes higher production costs in the industry by increasing the accident frequency and undue delays on account of dust-prone air at work.
- *Explosion hazard.* The possibility of a dust explosion is always present in coalmines. To minimize this risk water spraying and use of noncombustible stone dust barriers becomes mandatory.
- *Health risk.* Due to the inhalation of fine dust particles and their retention in the alveoli of the lungs, there is a health risk, depending upon exposure time and the nature of the dust, particularly its concentration and physicochemical properties. The term 'physicochemical' is used to describe all lung diseases caused by accumulation of insoluble dust in the lungs.

3.11.2   Factors affecting the degree of health risk

Physiological Properties of Dust:[36]
   The properties of dust affecting the development and severity of lung diseases are:

- *Composition.* Free silica is the most dangerous component of dust affecting the behavior or alveolar macrophages.
- *Size.* By conducting post mortem examinations of the lungs of numerous workers, it has been determined that pneumoconiosis is caused by dust particles below 5 μm. There is some evidence that the greatest danger is from particles ranging from 1 to 2 μm in size. In general, irregular particles fall more slowly than spheres of the same mass and for this reason irregular particles having a mass greater than that of 10 μm diameter unit density sphere can be found in the lungs.
- *Concentration.* Concentration of dust can be expressed as follows:
    - Mass of dust per unit volume of air. The mass concentration of respirable dust is the best single parameter to measure for assessment of the risk of pneumoconiosis from coal dust
    - Number of particles per unit volume, and

- Surface area of particles per unit volume. For quartz dust, the surface area of the respirable particles is probably the best parameter to measure although it is usually measured with a gravimetric dust-sampling instrument.

- Exposure duration: The human respiratory system has a certain capacity for disposing of inhaled dust. Under overloading conditions, larger lung dosages produce faster development of pneumoconiosis. Thus, the time of exposure to a certain dust concentration is an important factor in the development of pneumoconiosis.

3.11.3   Physiological effects of dusts

A classification of dusts with respect to potential hazard to the health and safety of workers may be divided into five categories.

1   *Toxic dusts.* They are poisonous to body tissue or to specific organs. Some metalliferous ores fall into this category. The most hazardous include compounds of arsenic, lead, and uranium and other radioactive minerals, mercury, cadmium, selenium, manganese, tungsten, silver and nickel.
2   *Carcinogenic dusts.* The cell mutations that can be caused by alpha, beta and gamma radiation from decay of the uranium series make radon daughters the most hazardous of the carcinogenic particulates. A combination of abrasion of lung tissue and surface chemical action can result in tumor formation from asbestos fibers and, to a lesser extent, freshly produced quartz particles.
3   *Fibrous dust.* The scouring action of many dusts causes microscopic scarring of lung tissue. If continued over long periods this can produce a fibrous growth of tissue resulting in loss of lung elasticity and a greatly reduced area for gas exchange.
4   *Explosive dusts.* These are a concern of safety rather than health. Many organic materials, including coals other than anthracite, become explosive when finely divided at high concentrations in air. Sulfide ores and many metallic dusts are also explosive.
5   *Nuisance dusts.* Quite apart from adverse effects on the health of personnel, all dusts can be irritating to the eyes, nose and throat.

Table 3.16 gives adverse impacts of dust to the industrial worker and the level of responsibility involved to the concerned.

Table 3.16 Effect of dust to the industrial worker.[22]

Policy		To practice					Results
Occupational lung disease due to dust							
Risk Assessment Dust	System Controls, Planning, Design, Medical Surveillance	Operational Control	Unsafe Acts – Failure to Filter	Unsafe Conditions – Dusty atmosphere	Incident: Exposure to Dust + 0.1 mg/m³	Threshold Limit + 0.1 mg/m³	Effects Occupational lung disease
Snr. Management		Middle Management					Individual

3.11.4   Sources of dust

The main sources of dust are mining, civil, construction, chemical and many other industries where bulk material handling takes place. Material handling consists of loading, transportation and unloading operations in any industrial setup. It is obvious that large quantities of dust are produced when the mineral is broken down and reduced to a size convenient for handling, as shown in Figure 3.16. The main mining operations responsible for the dust in mine air, i.e. airborne dust, are: drilling, cutting, blasting, loading, continuous mining, dumping mine-cars and drawing chutes.

3.11.5   Control of dust

There are four rules regarding dust control:

- Avoid it
- Produce a minimum amount of dust and prevent it from becoming airborne at its source
- Dilute it as soon as possible by ventilation and
- Separate it.

Every effort should be made to prevent the formation of dust as well as to prevent it from becoming airborne at its source. Water is mainly used for this purpose. In order

Figure 3.16 Dust at mining site at Sultanate of Oman.

to suppress airborne dust by spraying, it is necessary to spray the material before it is broken up, or to confine any dust by a curtain of water sprays. An adequate quantity of good quality water supplied through a well-designed spray system is essential. The pump selected should be powerful enough to give the quantity and pressure required. An appropriate pipe size is also important. A non-clogging filter system is also required to maintain the quality of water.

Excessive use of water should be avoided as it can result in the deterioration of climatic conditions as well as in problems in the preparation of the coal, for example it can cause sticking to the screens and bunkers.

Dust separators can be used to remove airborne dust from the air. The methods available for dust separation are:[36]

- Gravity methods;
- Centrifugal methods;
- Filtration;
- Wet scrubbers;
- Electrostatic precipitators.

3.12   PARTICULATE MATTER (PM)[3, 4, 10]

Apart from dust, there is a complex mixture of very small-sized particles and liquid droplets, which are known as 'Particulate Matter (PM)'. PM is made up of a number of components, including acids (such as nitrates and sulfates), organic chemicals, metals, and soil or dust particles.

Particles of 10 micrometers or smaller in diameter when they enter the lungs through the throat or nose can cause health problems.

3.12.1   Grouping particulate matter

EPA groups particle pollution into two categories:[4] EPA Q & A. [Accessed Aug. 2009]

1   "Inhalable coarse particles," such as those found near roadways and dusty industries, are larger than 2.5 micrometers and smaller than 10 micrometers in diameter.
2   "Fine particles," such as those found in smoke and haze, are 2.5 micrometers in diameter and smaller. These particles can be directly emitted from sources such as forest fires, or they can form when gases emitted from power plants, industries and automobiles react in the air. Graphs given below shows types and sources of PM.

Figure 3.17 gives a graphical presentation of types and sources of PM in the year 2002 in USA.

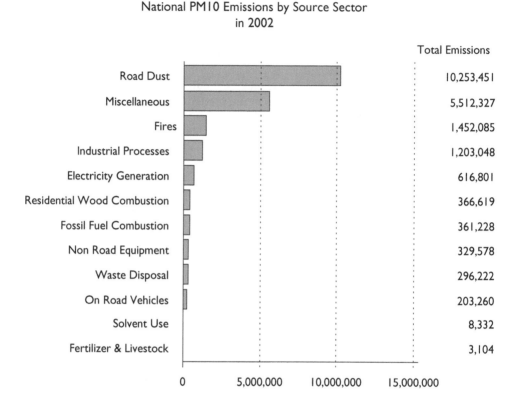

Figure 3.17 Graphs showing types and source of PM year 2002, USA.[10]

As stated above, particle pollution includes "inhalable coarse particles," with diameters larger than 2.5 micrometers and smaller than 10 micrometers and "fine particles," with diameters that are 2.5 micrometers and smaller. How small is 2.5 micrometers? Think about a single hair from your head. The average human hair is about 70 micrometers in diameter – making it 30 times larger than the largest fine particle. These particles can penetrate into the lungs and have been associated with a number of serious health problems such as asthma, bronchitis, heart problems, and even premature death.

Health: Particle pollution contains microscopic solids or liquid droplets that are so small that they can get deep into the lungs and cause serious health problems. The size of particles is directly linked to their potential for causing health problems. Small particles less than 10 micrometers in diameter pose the greatest problems, because they can get deep into your lungs, and some may even get into your bloodstream.

Visibility: Fine particles ($PM_{2.5}$) are the major cause of reduced visibility (haze) in many parts of the world including the United States where many of its treasured national parks and wilderness areas have been affected.

3.13   AIR SAMPLES

Air samples can be collected by the following methods:

- Water displacement methods: their drawback is that some gases are soluble in water.
- Air displacement methods: these use dry tubes and air is pumped into them.
- Evacuation: A sampling bottle fitted with a stopcock is evacuated using a suction pump and the sample is taken by opening the stopcock.

Instant analyzing methods:

1   Flame safety lamp: for CH_4 and CO_2 detection. Lamp will show the flame for 1.25 to 5% CH_4. Lamp will extinguish when O_2 is in the range of 14–17%. It will also extinguish at about 3% CO_2.
2   Canaries – are still used in coalmines to detect CO.
3   Detection Tubes – for O_2; CO_2; CH_4; CO; H_2S; SO_2; Oxides of nitrogen (NO_x)

In addition to this there are some filament instruments that are used for this purpose as shown in Table 3.17.

3.13.1   Gas detection techniques

Table 3.17 Gas detection techniques (prepared from various sources).[26, 29–31, 34, 36]

Instrument/Method	Gases	Lab./Portable	Remark
Flame safety lamp	CH_4; O_2 Deficiency	Portable	Coal mines
Canaries	CO		Mines
Detection Tubes	O_2; CO_2; CH_4; CO; H_2S; SO_2 Oxides of nitrogen (NO_x)	Portable	
Filament instruments	Combustible gases such as CH_4	Portable as well as lab.	
Optical instruments Infrared radiation	CO	Portable as well as lab.	Digital readout and alarming system
Chemiluminescence	NO/NO_2;O_3	Portable	
Interference fringes	Combustible gases such as CO_2		
Gas chromatography	Up to 8 gases (O_2; N; CO_2; CH_4; CO.	Laboratory	
Gas chromatography	Non-methane volatile hydrocarbons	Laboratory	
Electron capture instrument	NO_2; CH_4; Cl; HCl	Portable	
Electrochemical instruments	O_2; CO; H_2S; SO_2; Oxides of nitrogen (NO_x)	Portable as well as on line	
Catalytic oxidation instruments	CO		Can read directly in ppm.

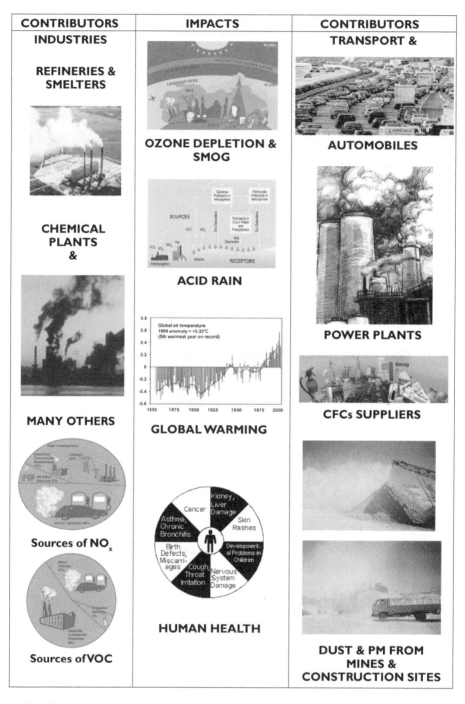

CONTRIBUTORS	IMPACTS	CONTRIBUTORS
INDUSTRIES		TRANSPORT &
REFINERIES & SMELTERS		
	OZONE DEPLETION & SMOG	AUTOMOBILES
CHEMICAL PLANTS &	ACID RAIN	
	GLOBAL WARMING	POWER PLANTS
MANY OTHERS		CFCs SUPPLIERS
Sources of NO$_x$		
Sources of VOC	HUMAN HEALTH	DUST & PM FROM MINES & CONSTRUCTION SITES

Figure 3.18 The culprits for air pollution results in: acid rain, global warming, ozone depletion and photochemical smog. Ultimately all these are causing damage to human beings, flora and fauna. (Taken from various sources).[2, 6–9, 13, 15]

3.14 REMEDIAL MEASURES

Figure 3.18 attempts to summarize the main culprits for air pollution that results in: acid rain, global warming, ozone depletion and photochemical smog. Ultimately all these are causing damage to human beings, flora and fauna. The remedial measures lie in adopting the following strategy:

- Reduction at the source by way of

    - Choosing eco-friendly technology
    - Minimizing wastes of all kinds (sec. 9.3; sec. 5.4)
    - Adopting best practices and applying them through adequate training to those concerned. (Ref. Table 2.3; sec. 9.10)
    - Abide by the norms, rules, regulations and law of the land (sec. 9.11).

- Fossil fuels are the main culprits (as described in sec. 2.8.5), the remedy is making us aware and educating us for their minimum use. Encouraging the use of conventional non-fossil fuel energy such as solar, wind etc.
- Putting restrictions to those materials which are the cause of environment-related problems and degradation of the environment
- Bringing change in policy at the international level, for example it has been advocated to adopt the following guidelines in the case of GHGs to address climate change issues:

    1 Adherence to the principle of common but differentiated responsibilities, and the respective capabilities of different countries, in respect of both mitigation of GHGs, and adaptation measures.
    2 Reliance on multilateral approaches, as opposed to bilateral or plurilateral or unilateral measures.
    3 Equal per-capita entitlements of global environmental resources to all countries.
    4 Over-riding priority of the right to development.

3.15 CONCLUDING REMARKS

Repeating once again: 'Fresh air, safe drinking water, unadulterated food and clean environment are our basic needs.' Should we not strive hard to achieve it by educating ourselves? We keep mother-earth neat and clean. We participate to raise our awareness for a clean environment.

QUESTIONS

1 Give the composition of dry and clean air. What is a pollutant? What are the main sources of air pollution? Classify them using a line diagram.
2 How EPA has grouped particle pollution? Write it down.

3 The air quality standard for CO (used at 8 hrs. measurement basis) is 10 ppm. Express this standard as % by volume as well as in mg/m^3 at 1 atmospheric pressure and 25°C.

4 The air quality standard for nitrogen dioxide NO_2 is 475 μg/m^3 (at temperature of 25°C and 1 atm. Pressure). Express the concentration in ppm.

5 Calculate the density of oxygen at a temperature of 273.0 K and at a pressure of 98.0 kPa.

6 Coal is burnt @ 1.2 kg/sec. If an analysis of coal indicates 3% S, what amount of SO_2 will be produced annually? Sulfur when burnt produces 5% ash by weight.

7 Find out the CO_2 emission from 110 gms of butane (C_4H_{10}). Hint: $(C_4H_{10}) + O_2 = CO_2 + H_2O$

8 Given the emission factor for CO from fuel oil = 70 g/1000 liters and sales of oil/day are 40,000 liters, find out the CO produced/day.

9 If fossil fuel energy consumption per year is 3.2×10^{20} Joules, and if all the energy is given by CH_4 gas having an energy content 3.9×10^7 J/m^3 at STP, at what rate would CO_2 be emitted to the atmosphere? Also express the emission of C in the air per year (not of CO_2).

10 Categorise dusts with respect to potential health hazard to workers.

11 List the main sources of sound in industry. When does sound becomes noise? What techniques can be applied to control noise pollution? How is it harmful to human beings?

12 List major air pollutants and write briefly their effects on human beings and also on materials. Give the toxic rank of major air pollutants as adopted by NAAQS of USA. What is PSI Value? Name major sinks of CO, CO_2, NO, NO_2, SO_2, Hydrocarbons and H_2S. List air sample collecting methods.

13 What is the maximum noise level allowed in the commercial and industrial factories in your own country?

14 Suggest remedial measures for minimizing air pollution.

15 The combined noise level for two machines running in a factory is 96(dB). If noise level of one of them is 90(dB), find the noise level of the other, using the standard Table given in section 3.9.

16 What is bad Ozone? How it is formed? What is good Ozone how it can be preserved? Give necessary illustrations and reactions.

17 What is the greenhouse effect and what is its global impact? What are the green house gases and give an account of their relative contribution (share). What is the CO_2 generation trend? Which waves are mainly responsible for causing the 'Greenhouse' effect?

18 What is photochemical smog and how it is formed? Illustrate the mechanism by giving a figure. What are its adverse effects?

19 What are the culprits for air pollution; what are its results?

20 Write the pollutants that are responsible for acid rain. Give the necessary reactions for this purpose. List the adverse impacts of acid rain. In a factory the effluent is acidic, what treatment will you suggest to neutralize it?

Hint.: For solving questions 3 to 9, and 15, please refer to solved examples.

REFERENCES

1 Abdullah Al-Furqani (2002) *Term project – Studies on Dusts*, Sultan Qaboos University, Oman.

2 Bart de Haan (2007) Corporate Climate Change Policies. Presentation in: *Mining 2020, The AIMEX International Mining Conference, Sydney, Australia, Sept. 5–6.*

3 Environment Health Center, National Safety Council. Reporting on climate change: Understanding the Science, 2nd. Edt. June 2000.

4 EPA (Environmental Protection Agency, USA) *Alphabetic list of all topics.* EPA Q & A. [Accessed Aug. 2009].

5 Glover, Thomas J. (2006) *Limits for air contaminants.* Pocket Ref.; Pub. Sequoia Pub. Inc. Littleton, USA. pp. 22.

6 http://cfpub.epa.gov/airnow/index.cfm?action=health_prof.main [Accessed Aug. 2009].

7 http://climatechange.gc.ca/

8 http://www.cru.uea.ac.uk/cru/press/pj9601/index.htm (Greenhouse gases) Source: Climatic Research Unit, University of East Anglia.

9 http://www.epa.gov/airmarkets/acidrain/effects [Accessed 2009].

10 http://www.epa.gov/cgi-bin/broker

11 http://www.epa.gov/docs/ozone/

12 http://www.epa.gov/ebtpages/air.html (data and information for the period 1980–2000.)

13 http://www.epa.gov/globalwarming/

14 http://www.epa.gov/oar/oaqps/takingtoxics/ (Air toxics)

15 http://www.epa.gov/ozone/defns.html#ozone

16 http://www.epa.gov/ozone/science/hole/holecomp.html (Decline in ozone shield)

17 http://www.epa.gov/ttn/uatw/hlthef/heffref.html (Air toxics)

18 http://www.maltaweather.info/acid.jpg

19 http://www.nerc-bas.ac.uk (Worsening ozone hole. British Academy Survey)

20 http://www.psr.org/breathe.htm (Health effects)

21 http://www.unep.ch/

22 Jager, K.D. (2006) Wellness in the Workplace. *International conference focusing on safety and health on 14–16 November 2006, Johannesburg, South Africa.* London ICMM. (Permission: AngloGold Ashanti)

23 Jokha al-Busaidy (2002) *Term-Project: Global warming*, Sultan Qaboos University, College of Engineering, Department of Petroleum & Mineral Resources.

24 Khaled Al-Kalbany (2003) Term project, Air pollution, Sultan Qaboos university, Oman. http:// images.google.com; http:// hk:geocities.com

25 Kupchella, C.E. & Hyland, M.C. (1993) *Environmental Science.* Prentice-Hall International Inc. pp. 255–280, 297–325.

26 Malcolm, J. McPherson. (1993). *Subsurface Ventilation and Environmental Engineering.* London, Chapman and Hall.

27 Misra, S.G. & Tiwari S.D. (1990) *Air and atmospheric pollutants.* New Delhi, Venus Publishing House.
28 Petroleum Development of Oman (2000–04) – HSE management guidelines.
29 Rao, C.S. (1991) *Environment Pollution Control Engineering.* New Delhi, Wiley Eastern Ltd.
30 Rao, M.N. & Rao, H.V.N. (1989) *Air pollution.* New Delhi, Tata McGraw Hill Publishing Ltd.
31 Saxena, N.C.; Singh, G. & Ghosh, R. (2002) *Environment management in mining areas,* Scientific Pub. India. pp. 250–260.
32 Stern, A.C.; Bouble, R.W.; Turner, D.B. & Fox, D.L. (1984) *Fundamentals of air pollution* Academic press, Inc. 2nd ed.
33 Tatiya, R.R. (2000). Impacts of Mining on environment in Oman. *International Conference on Geo-environment 2000, Sultan Qaboos university, Muscat, Oman, 4–7, March.*
34 Tatiya, R.R. (2000–04). *Course material for Petroleum and Natural Resources Engg; Chemical Engg. And Petroleum and Mineral Resources Engg.* Sultan Qaboos University, Oman.
35 US National Climate Data Center. Global temperature change during period 1880–2000.
36 Vutukuri, V.S. & Lama, R.D. (1986). *Environmental Engineering in Mines.* Cambridge University Press.

Chapter 4

Water pollution

'Besides losing millions of lives, a huge sum of money is spent the world over every year on curing diseases caused by industrial pollution. Prevention is always far better than cure.' There is an old saying in industry: 'Dilution is the solution to pollution'.[6]

Keywords: Water Cycle, Resources, Quality Standards, Water Uses and Pollutants. Receiving Environment, Groundwater, Groundwater Contamination, Point and Non-Point Pollution Sources, Eutrophication, Biological Oxygen Demand, Dissolved Oxygen, Acidity, Red Tide Sewage, Suspended Solids, Marine Pollution, Acid Mine Drainage, Yellow Boy, Bottled Water, Subsurface Water.

4.1 INTRODUCTION[6, 26, 30]

There is a famous proverb in Indian Culture to express the importance of water – '*Bin Pani Sab Soon*' without water every thing would come to a halt. Yes, together with sun and air water is our basic need. Water occupies about two-thirds of the earth's area (294,000,000 cubic miles of water)[6] but out of this how much is potable water? The safe and clean drinking water is the one that does not have any harmful effect on human beings, and it is within the prescribed allowable limits (as prescribed under the prevalent regulations of any country) of the various constituents that are usually present in water. The freshwater resources are comprised of the river systems, groundwater, and wetlands. Each of these has a unique role, and characteristic linkages to other environmental entities.

Water is the Universal Solvent and it's this property/characteristic prevents it to remain pure. Its excellent sweeping, cleaning and cooling properties makes its use indispensable; but when it is used for these purposes it gets dirty and polluted. The water cycle described in Figure 4.1 reveals that it cannot remain pure as it holds a number of minerals, suspended solids, and many more items as described in Table 4.2. The presence of these ingredients to a certain limit, which is known as the tolerance or prescribed limit, makes it suitable for various purposes including drinking, but beyond that it needs treatment.

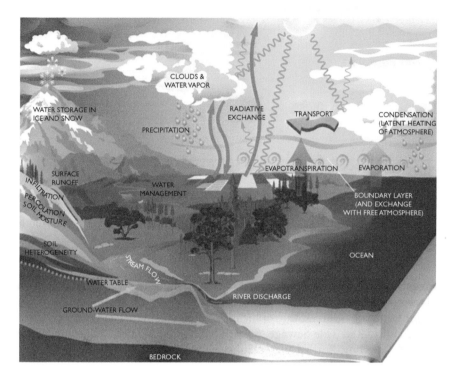

Figure 4.1 Water cycle and water bodies.

Source: http://www.usgcrp.gov/usgcrp/images/
ocp2003/WaterCycle-optimized.jpg).[11]

4.1.1   Water cycle[9, 11]

As illustrated in Figure 4.1,[11] this consists of the following 4 stages

- Evaporation (and transpiration)
- Condensation
- Precipitation
- Collection.

Evaporation: this is caused by the sun, which heats up water in the various water-bodies, ultimately converting it into vapor or steam. The water vapor or steam leaves these water-bodies (river, lake, stream or ocean) and goes into the air. Like people/human-beings, plants also sweat i.e. plants loose water through their leaves. This process is known as transpiration. Transpiration helps in getting water vapor back into the air.

Condensation: In the air water vapor gets cold and changes back into liquid, forming clouds. This process is known as condensation.

Precipitation: It occurs when so much water has condensed that the air cannot hold it anymore. The clouds get heavy and water falls back to the earth in the form of rain, hail, sleet or snow.

Collection: Water falls back as precipitation, at any place on earth. From the land it could soak into the earth and becomes the part of 'groundwater', which is used by us, plants and animals, else it may run-over the earth and go to water-bodies: rivers, streams, ponds, lakes and oceans from where the water cycle restarts.[9]

4.2 WORLDWIDE WATER RESOURCES – SOME FACTS[5, 6, 30]

This refers to the water resources that are useful, particularly, for agricultural, industrial, domestic, recreational and environmental activities of humanity. All these activities require fresh water. Table 4.1 depicts an account of water resources and their distribution. About 4×10^{20} gallons (or 1.5×10^{21} liters); of water is above, on and in the earth; 97.5% of water on the earth is salt water, leaving only 2.5% as fresh water of which over two thirds is frozen in glaciers and polar ice caps. The remaining unfrozen freshwater is mainly found as groundwater, with only a small fraction present above ground (surface water that is in lakes, rivers, streams, and in soils) or in the air.[5, 30]

Some facts:

1   Water demand already exceeds supply in many parts of the world, and as the world population continues to rise at an unprecedented rate (1.8% annually and with a doubling period of about 39 years), many more areas are expected to experience this imbalance in the near future. This would be mainly influenced by the growth in population, the increase in their living standards and their lifestyle. This aspect has been illustrated in Figures 2.11 and 2.12. In fact the amount of fresh and clean water required per man per day for his biological needs – drinking & food; hardly exceeds 0.5 gallons or 2 liters, whereas, per capita water consumption presents astonishing data/figures, for example in Europe per capita water consumtion is from 76 to 270 liters per person per day; and in some of the developing countries it is as low as 2.5 liters.[16] In USA it is about 285–300 liters/man/day.**

**Residential demands account for about three-fourths of the total urban water demand. Indoor use accounts for roughly 60 percent of all residential use. Low-flow plumbing fixtures and retrofit programs are permanent, one-time conservation measures that add little or no additional cost over their life. The City of Corpus Christi, for example, has estimated that an average three-member household can reduce its water use by 54,000 gallons annually and can lower water bills by about $60 per year if water-efficient plumbing fixtures are used.[3] EPA Q & A. [Accessed Aug. 2009]

2   Surface water is naturally replenished by precipitation and naturally lost through discharge to the oceans, evaporation, and sub-surface seepage.
3   Methods of increasing water supply include construction of dams, desalination of seawater, ice mining and cloud seeding. All these have environmental impacts as a result of required energy for these options.
4   With increased water scarcity, conflicts are likely between agriculture and industry, towns in the same country and between nations.
5   Buying water-efficient plumbing fixtures, turning off the tap when we brush our teeth, using a hotel towel two or more days, watering the lawn in the early morning. There are probably as many ways to save water, at any time of year, as there are people.

Table 4.1 Global water resources (information from various sources).[5, 6, 16, 26, 30]

Water resources	Quantity $10^{15}\,m^3$	%	Remarks
Oceans	1350.0	97.20	It is **Saline**. Besides numurous advantages, and being one of the most important natural resources, sea water provides major transportation routes. Salinity makes it unsuitable to use as fresh-water for drinking, industrial and agricultural use. Howver, it could be used as cooling water in power plants. Desalination is costly and has limitations on account of the energy required and adverse environmental impacts, and so is confined to arid regions where other water resources are limted. However, in excess of 3 billion gallons of seawater is converted into potable water everyday.[16]
Ice caps & Glaciers	29.0	2.0	Out of avaialble fresh water, **over two thirds is frozen as ice-caps and glaciers**. Its use as fresh water is questionable.
Groundwater >1 km depth	4.2	0.30	Partly saline. And imagine!!!, the rest is the water which is mostly used as fresh water for drinking, agriculture and industries. Its
Groundwater <1 km depth	4.2	0.30	proportion is very small. **Thus, Only very small percentage (0.16%)[6] of the total water in, on and above the earth is accessible to humans at any time as fresh water**. 13–20 times the surface-water is available from aquifers.
Freshwater lakes (surface water)	0.135	Traces	Fresh water is confined in freshwater-lakes, rivers and undergroundwater sources (aquifers – confined and unconfined). In 1 barrel (55 gallons) in lakes and rivers, less than 1 ounce of water is available.[16] In USA for example:[3] The five Great Lakes – Erie, Huron, Michigan, Ontario, and Superior – are the largest surface freshwater source on the Earth. They contain more than 90 percent of the U.S.'s freshwater supply and more than 20 percent of the world's. Despite their considerable size, the Great Lakes are sensitive to the effects of a wide range of pollutants. To address this issue the Great Lakes Interagency Task Force was created in 2004 to provide strategic direction on Great Lakes policy, priorities, and programs.
Saline lakes & inland seas	0.104	0.007	Saline
Soils' moisture	0.067	0.005	In the soils
Atmospheric vapor	0.013	0.0009	In the air at any time only 5 gallons out of 100,000 gallons is in motion as precipitation – recharging reservoirs and going to salty bodies including oceans.
Water in biomass	0.003	0.0002	In the flora
Streams, wetlands etc.	0.001	0.00007	Surface water is naturally replenished by precipitation and naturally lost through discharge to the oceans, evaporation, and sub-surface seepage. Wetlands provide: natural water filtration, habitat for commercial fisheries and protection against flooding. Wetlands are valuable because they replenish and clean water supplies and reduce flood risks.

4.3 WATER QUALITY STANDARDS (WQS)

Every country has its own standards and norms for water, which is used for various purposes. The World's citizens should not be deprived of wholesome safe drinking water (with contaminants well below allowable limits).

Chronic effects occur after people consume a contaminant at levels over EPA safety standards (or safety standards as applicable to the concerned region) for many years. The drinking water contaminants that can have chronic effects are chemicals (such as disinfection by-products, solvents, and pesticides), radio-nuclides (such as

Table 4.2 Drinking water standards as adopted by EPA,[26] Safe Drinking Water Act, and US Public Health Service in (1974 as amended in 1986 and 1996) and adopted by American Water Works Association (AWWA)[4] have been summarized in column 2.

Parameters/variable	Standard EPA USA**	AWWA
Alpha particle activity (gross)	15 pCi/L	
Arsenic	0.01 mg/l	0.01 mg/l
Bacteria	4/100 ml	
Barium	2.0 mg/l	
Benzene (organic)	0.005 mg/l	
Beta particle & photon radioactivity	4 merm/yr	
Cadmium	0.005 mg/l	0.01 mg/l
Calcium (as Ca^{2+})	75 mg/l	
Carbon tetra chloride (organic)	0.005 mg/l	
Chloride	250.0 mg/l	250 mg/l
Coliform	5%	1
Color (platinum – cobalt scale)	15 units	10 Hazen units
Chromium (hexavalent)	0.01 mg/l	0.05 mg/l
Copper	1.3 mg/l	0.05
Cyanide	0.05 mg/l	0.05 mg/l
1,1 Dichloroethylene (organic)	0.007 mg/l	
1,2 Dichloroethylene (inorganic)	0.005 mg/l	
Dioxin (2,3,7,8,-TCDD)	0.00000003	
Endrin (organic)	0.002 mg/l	
Foaming agent	0.05 mg/l	
Fluoride	4.0 mg/l	0.6 to 1.2 mg/l
Iron	0.3 mg/l	0.3 mg/l
Lead	0.015 mg/l	0.1 mg/l
Lindanium (organic)	0.0002 mg/l	
Magnesium		30 mg/l
Manganese	0.05 mg/l	0.1 mg/l
Mercury	0.002 mg/l	0.001 mg/l
Methoxychlor (organic)	0.04 mg/l	
Mineral oil		0.01 mg/l
Nitrogen – nitrate	10.0 mg/l	45 mg/l
Nitrogen – organic	3	
Odor (threshold odor)		Unobjectionable

(Continued)

Table 4.2 (Continued)

Parameters/variable	Standard EPA USA**	AWWA
p-Dichlorobenzene (organic)	0.075 mg/l	
pH	6.5–8.5	6.5–8.5
Phenolic compounds		0.001 mg/l
Radium –226 and –228	5pCi/L	
Radioactive – Alpha emitter Beta		10^{-8} uc/ml 10^{-7} uc/ml
Selenium	0.05 mg/l	0.01 mg/l
Silver	0.1 mg/l	
Sulfate (SO_4)(> has a laxative effect)	250.0 mg/l	150 mg/l
Taste		Acceptable
Total dissolved solids	500.0 mg/l	500 mg/l
Total hardness as $CaCO_3$		300 mg/l
Toxaphene (organic)	0.003 mg/l	
1,1,1 Trichloroethylene (organic)	0.2 mg/l	
Trichloroethylene (organic)	0.1 mg/l	
Turbidity (silica scale)	1 to 5 TU	10(NTU)
Uranium	30 μg/L	
Vinayal chloride (organic)	0.02 mg/l	
Zinc	5.0 mg/l	5.0 mg/l
2, 4 – D (Organic)	0.07	

**EPA has set standards for more than 80 contaminants, as shown in Table 4.2, that may occur in drinking water and pose a risk to human health. EPA sets these standards to protect the health of everybody, including vulnerable groups like children. The contaminants fall into two groups according to the health effects that they cause. As per practice the water suppliers alert those concerned through the media, mail, or other means if there is a potential acute or chronic health effect from compounds in the drinking water. However, for additional information the supplier could be contacted. Acute effects occur within hours or days of the time that a person consumes a contaminant. People can suffer acute health effects from almost any contaminant if they are exposed to extraordinarily high levels (as in the case of a spill). In drinking water, microbes, such as bacteria and viruses, are the contaminants with the greatest chance of reaching levels high enough to cause acute health effects. Most people's bodies can fight off these microbial contaminants the way they fight off germs, and these acute contaminants typically don't have permanent effects. Nonetheless, when high enough levels occur, they can make people ill, and can be dangerous or deadly for a person whose immune system is already weak due to HIV/AIDS, chemotherapy, steroid use, or another reason. EPA Q & A. [Accessed Aug. 2009]

radium), and minerals (such as arsenic). Examples of the chronic effects of drinking water contaminants are cancer, liver or kidney problems, or reproductive difficulties.[3] EPA Q & A. [Accessed Aug. 2009]

4.3.1   Water quality standards based on receiving environment[19, 20, 23, 24, 26]

In the event that there are no local regulations, the principle of best practice should be applied to ensure that the effluent discharges do not adversely affect the receiving environment. For this the water quality standards (WQS) could be adopted. The WQS are a measure of the concentration of specific substances in the receiving water following an initial period of dispersion and dilution. A key feature of WQS for water is that they are derived in relation to specific water quality objectives (WQO). WQO include the protection of the receiving environment/end use, which could be:[24]

- Freshwater
- Saltwater (seawater)
- Drinking water
- Irrigation of crops
- Watering of livestock.

Table 4.3 Water quality standards based on the receiving environment/end user.[19, 20, 24]

Component	Receiving environment/end use	Maximum allowable concentration
BOD*	Freshwater	3000 µg O_2/l (T 95)[1]
	Saltwater (seawater)	–
	Drinking water	<1500 µg O_2/l (T)
	Irrigation of crops	–
	Watering of livestock	300 µg O_2/l (T)
COD*	Freshwater	–
	Saltwater (seawater)	–
	Drinking water	30,000 µg O_2/l (T 90)
	Irrigation of crops	–
	Watering of livestock	–
pH*	Freshwater	6.0–9.0
	Saltwater (seawater)	6.5–8.5
	Drinking water	6.5–8.5
	Irrigation of crops	5.5–8.5
	Watering of livestock	–
Free Cl_2*	Freshwater	860,000 µg/l (T)
	Saltwater (seawater)	–
	Drinking water	200,000 µg/l (T)
	Irrigation of crops	30,000 µg/l (T)
	Watering of livestock	1,200 µg/l (T)
Ammoniacal nitrogen*	Freshwater	–
	Saltwater (seawater)	–
	Drinking water	50,000 µg/l (T)
	Irrigation of crops	–
	Watering of livestock	90,000 µg/l (T)
Oil in water**	Freshwater	–
	Saltwater (seawater)	–
	Drinking water	10 µg/l (T)
	Irrigation of crops	–
	Watering of livestock	–
Total coliforms*	Recreational water use	10000 per 100 ml

* Mandatory for sewage effluent only.
** Mandatory for production water only.
T Total concentration (i.e., without filtration).
90 90-percentile.
95 95-percentile.
[1] Suitable for the protection of freshwater fish only.

Table 4.4(a) Water quality with permissible limits for various crops and vegetation.[19, 20, 24]

Parameters	Standard A	Standard B
BOD (5 days @ 20°C)	15.0 mg/l	20.0 mg/l
COD	150.0 mg/l	200.0 mg/l
Suspended solids	15.0 mg/l	30.0 mg/l
Total dissolved solids	1500.0 mg/l	2000.0 mg/l
Electrical conductivity	2000.0 micro S./cm	2700.0 micro S./cm
Sodium absorption ratio	10.0	10.0
pH	6–9	6–9
Aluminium	5.0 mg/l	5.0 mg/l
Arsenic	0.1 mg/l	0.1 mg/l
Barium	1.0 mg/l	2.0 mg/l
Beryllium	0.1 mg/l	0.3 mg/l
Boron	0.5 mg/l	1.0 mg/l
Cadmium	0.01 mg/l	0.01 mg/l
Chloride	650.0 mg/l	650.0 mg/l
Chromium	0.05 mg/l	0.05 mg/l
Cobalt	0.05 mg/l	0.05 mg/l
Copper	0.5 mg/l	1.0 mg/l
Cyanide	0.05 mg/l	0.10 mg/l
Fluoride	1.0 mg/l	2.0 mg/l
Iron	1.0 mg/l	5.0 mg/l
Lead	0.1 mg/l	0.2 mg/l
Lithium	0.07 mg/l	0.07 mg/l
Magnesium	150.0 mg/l	150.0 mg/l
Manganese	0.1 mg/l	0.5 mg/l
Mercury	0.001 mg/l	0.001 mg/l
Molybdenum	0.01 mg/l	0.05 mg/l
Nickel	0.10 mg/l	0.10 mg/l
Nitrogen – ammoniacal	5.0 mg/l	10.0 mg/l
Nitrogen – nitrate	50.0 mg/l	50.0 mg/l
Nitrogen – organic	5.0 mg/l	10.0 mg/l
Oil and grease	0.5 mg/l	0.5 mg/l
Phenols	0.001 mg/l	0.002 mg/l
Phosphorous	30.0 mg/l	30.0 mg/l
Selenium	0.02 mg/l	0.002 mg/l
Silver	0.01 mg/l	0.01 mg/l
Sodium	200.0 mg/l	300.0 mg/l
Sulfate	400.0 mg/l	400.0 mg/l
Sulfide	0.1 mg/l	0.1 mg/l
Vanadium	0.1 mg/l	0.1 mg/l
Zinc	5.0 mg/l	5.0 mg/l
Faecal coliform bacteria	*200.0 per 100 ml*	*1000.0 per 100 ml*
Viable nematode ova	<1.0 per litre	<1.0 per litre

* The definitions relating to Standards A and B are detailed below.

Table 4.4(b) Definitions relating to Standards A and B.[19, 20, 24]

	Standard A	*Standard B*
Crops	Vegetables likely to be eaten raw. Fruit likely to be eaten raw and within 2 weeks of any irrigation.	Vegetables to be cooked or processed. Fruit if no irrigation within weeks of cropping. Fodder, cereal and seed crops
Grasses and Ornamental Areas	Public parks, hotel lawns, recreational areas. Areas with public access. Lakes with public contact (except places which may be used for praying and hand washing).	Pastures. Areas with no public access
Aquifer Recharge	All controlled aquifer recharge.	
Method of Irrigation	Spray or any other method of aerial irrigation not permitted in areas with public access unless with timing control.	
Any other Reuse Application	Subject to the approval by the regulatory authorities	

As a standard practice, an assessment of (eco) toxicological test results, including information on persistence in the aquatic environment and potential for bioaccumulation, are used to derive water quality standard (WQS), which is intended to protect the identified WQO. The permissible levels, which are applicable in one Gulf country, are detailed in Table 4.3. And also Tables 4.4(a) and (b) depict WQS for water to be used for agriculture purposes. Also refer sec. 7.5.4.

4.4 GROUNDWATER[6, 10, 13, 30]

Groundwater is present in underground aquifers (Figures 4.1 and 4.2). 13–20 times more water is available from aquifers than surface water-bodies excluding glaciers and oceans. Aquifers near the surface are subject to annual recharge from precipitation, but the rate of recharge is impacted by human interference.

Deep aquifers, on the other hand, occur below a substratum of hard rock. The deep aquifers generally contain very pure water, but since they are recharged only over many millennia, must be conserved for use only in periods of calamitous drought such as may happen only once in several hundred years.

The water table has been falling rapidly in many areas in recent decades. This is largely due to withdrawal for agricultural, industrial, and urban use, in excess of annual recharge.

In urban areas, apart from withdrawals for domestic and industrial use, housing and infrastructure such as roads prevent sufficient recharge.

The result is that inefficient withdrawals of groundwater by the users in some countries are leading to the situation of falling water tables. Falling water tables have

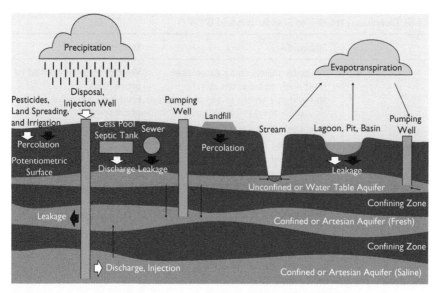

Figure 4.2 Groundwater contamination – sources and routes. Adapted from Environmental Protection Agency, Office of Water Supply and Solid Waste Management Programs, Waste Disposal Practices and their Effects on Groundwater (Courtesy: U.S. Government, Washington D.C).[10]

several adverse social impacts, apart from the likelihood of mining deep aquifers, *"the drinking water source of last resort"*.

The efficient use of groundwater would, accordingly, require that the practice of non-metering of electric supply, which is prevalent in some parts of world, to farmers be discontinued in their own enlightened self-interest. It would also be essential to progressively ensure that the environmental impacts are taken into account in setting electricity tariffs, and diesel pricing.

Increased run-off of precipitation in urban areas is due to impermeable structures and infrastructure preventing groundwater recharge. This is an additional cause of falling water tables in urban areas.

Pollution of groundwater from agricultural chemicals is also linked to their improper use, once again due to pricing policies (in some countries), especially for chemical pesticides, as well as agronomic practices, which do not take the potential environmental impacts into account.

The following action points could help in preventing groundwater depletion at the faster rates:

a   At some locations electricity and diesel prices for farmers are subsidized, and that often leads to misuse of groundwater. Measure to take proper accountability would promote efficient use of this resource.

b   Promoting efficient water use techniques, such as sprinkler or drip irrigation for irrigation purposes. Incentives for efficient use of water should be considered.

c   The practices of contour bunding and revival of traditional methods for enhancing groundwater recharge, which has been an established practice in countries like India, should be encouraged. Those concerned should make requisite financial support available.

d   Rainwater harvesting in all new constructions in relevant urban areas could be made mandatory. This feature should be incorporated in the designs for road surfaces and infrastructures, which would enhance groundwater recharge.

4.4.1   Sources and routes for groundwater contamination

Some pollution of groundwater occurs due to leaching of stored hazardous waste and the use of agricultural chemicals, in particular, pesticides.

Contamination of groundwater is also due to gynogenic causes, such as leaching of arsenic from natural deposits. Since groundwater is frequently a source of drinking water, its pollution leads to serious health impacts.

4.5   WATER USE[12, 30]

4.5.1   Industry[8, 12, 13]

Industry accounts for 22% of all water withdrawals worldwide, but this figure ranges from 89% in developed countries compared to a mere 8% in less-developed countries. In the coming decades, industry's share of water withdrawals is set to grow.[12] However,

Table 4.5 Sources and pollutants for groundwater contamination.[6, 10, 13, 20, 29, 30]

Sources	Contaminants
Underground and above-ground petroleum storage tanks	Gasoline, diesel, waste oil, kerosene, hydraulic fluid and few others.
Industries and land disposal facilities including hazardous waste facilities and unauthorized wastewater discharges	Trichloroethylene, arsenic, PCBs, VOCs, solvents, creosote, chromium, lead, heavy metals, phenols, chlorides, pesticides, sulfates, cadmium and many others.[8]
Mining activities: Mine structures such as tunnels, shafts, drives/drifts etc. when driven through the water-table	The water bearing formations or strata releases water, which could be acidic or contain heavy metals.
Various types of underground injection wells	Sodium chloride, chlorides, refined oil, crude oil, hydrochloric acid, total dissolved solids etc.
Existing and abandoned municipal and industrial waste facilities	Total dissolved solids, chlorides, trichloroethylene, vinyl chloride and few others.
Septic tanks, CAFOs, Class V wells, sludge disposal, wastewater treatment facilities	Nitrates, chlorides, salt water, copper, selenium, iron, ammonia and few others.
Agricultural activities including the application of pesticides	Arsenic, prometon, Atrazine, dicamba, metolachlor, propazine, bromacil, DDT dieldrin, DDE and many others.

it is estimated that 15%–25% of worldwide water use is industrial (Figure 4.3). The nature of industrial wastewater can differ markedly within the same industry or from one industry to another. The impact of industrial discharges depends not only on their collective characteristics, such as biochemical oxygen demand and the amount of suspended solids, but also on their contents, which could be inorganic or organic.

Three options (which are not mutually exclusive) are available in controlling industrial wastewater:

1   Control at source – at the point of generation within the plant/factory;
2   Pretreatment for discharge into the municipal treatment systems; or
3   Treating completely at the plant and either reused or discharged directly into receiving waters.

Major industrial users include:[12]

- Power plants for power generation and cooling
- Mining and minerals processing – washing, concentration, smelting, refining,. oil refineries, which use water in chemical processes.
- Manufacturing plants (which are numerous), use water as a solvent, and for cleaning, sweeping, washing/bleaching, coloring etc. etc. and to carry out wastes of many kinds.

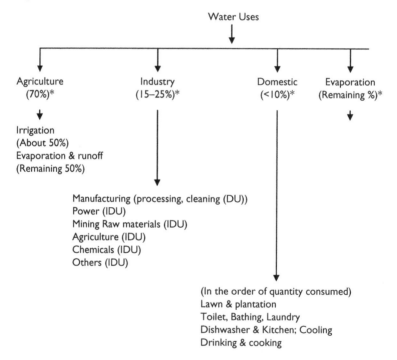

Figure 4.3 Classification – Water uses.

* Percentage may vary from one nation to another, this illustration pertains to USA. DU – Direct use. IDU – Indirect use.

In USA industry is the greatest source of pollution, accounting for more than half the volume of all water pollution and for the most deadly pollutants.

In USA for example, some 370,000 manufacturing facilities use huge quantities of freshwater to carry away wastes of many kinds.[8] The waste-bearing water, or effluent, is discharged into streams, lakes, or oceans, which in turn disperse the polluting substances.

- In its *National Water Quality Inventory*, reported to Congress in 1996, the U.S. EPA concluded that approximately 40% of the nation's surveyed lakes, rivers, and estuaries were too polluted for such basic uses as drinking supply, fishing, and swimming.[8]
- The pollutants include grit, asbestos, phosphates and nitrates, mercury, lead, caustic soda and other sodium compounds, sulfur and sulfuric acid, oils, and petrochemicals.
- In addition, numerous manufacturing plants pour off undiluted corrosives, poisons, and other noxious byproducts.
- The construction industry discharges slurries of gypsum, cement, abrasives, metals, and poisonous solvents.
- Another pervasive group of contaminants entering food chains is the polychlorinated biphenyl (PCB) compounds, components of lubricants, plastic wrappers, and adhesives.
- In yet another instance of pollution, hot water discharged by factories and power plants causes so-called thermal pollution by increasing water temperatures. Such increases change the level of oxygen dissolved in a water-body, thereby disrupting the water's ecological balance, and killing off some plant and animal species while encouraging the overgrowth of others.
- Clearing of land can lead to erosion of soil into the river.

4.5.2   Mining[12, 13]

Mining operations could influence water quality and quantity in the area of operation, if suitable measures are not taken. The excavation made requires proper diversion of the waterways and catchments areas.

- If the underground mine openings such as tunnels, shafts, drives/drifts etc. are driven through the water bearing formations or strata; water is bound to be encountered. Water is encountered while working below the water table. This phenomenon results in taking away fresh-water from the underground aquifers, if effective sealing does not prevent it from doing so. Of course, effective sealing is an established practice in such cases. The mine-water (the quality of which has been described in sec. 4.9) can pollute it.
- At surface mines also, when mine workings cross the water table, it takes away the useful freshwater.
- The rainwater carries the silt from the waste rock dumps, if the construction of effective bunds around the waste rock-dumps does not prevent it.

- Water can become contaminated with toxic or radioactive materials from mine sites and abandoned hazardous waste sites.
- If mine water is not treated before its discharge, it could result in adverse impacts as illustrated through the case study in section 4.11.

4.5.3 Agriculture[6, 12, 13, 30]

Agriculture, including commercial livestock and poultry farming, is the source of many organic and inorganic pollutants in surface waters and groundwater. Aquaculture, whether carried out in ponds or holding tanks on land or in pens in seawater, can also be a cause of pollution.

- Farms often use large amounts of herbicides and pesticides, both of which are toxic pollutants. These substances are particularly dangerous to life in rivers, streams and lakes, where toxic substances can build up over a period of time.
- Farms also frequently use large amounts of chemical fertilizers that are washed into the waterways and damage the water supply and the life within it. Fertilizers can increase the amounts of nitrates and phosphates in the water, which can lead to the process of eutrophication.[6]
- Allowing livestock to graze near water sources often results in organic waste products being washed into the waterways. This sudden introduction of organic material increases the amount of nitrogen in the water, and can also lead to eutrophication.
- Four hundred million tonnes of soils are carried by the Mississippi River to the Gulf of Mexico each year.[6] A great deal of this siltation is due to runoff from the exposed soil of agricultural fields. Excessive amounts of sediment in waterways can block sunlight, preventing aquatic plants from photosynthesizing, and can suffocate fish by clogging their gills.
- Agricultural contaminants include both sediment from the erosion of cropland and compounds of phosphorus and nitrogen that partly originate in animal wastes and commercial fertilizers.
- Aquaculture – animal wastes tend to be particularly high in oxygen-demanding material, nitrogen, and phosphorus, and they often harbor pathogenic organisms. Wastes from commercial feeders are typically contained and disposed of on land; the main threat to natural waters, therefore, is via runoff and leaching.

Control may involve settling basins for liquids, limited biological treatment in aerobic or anaerobic lagoons, and a variety of other methods. Among other antipollution measures, erosion control techniques are helpful in reducing nutrient pollution. Other measures include:

- Manage animal waste to minimize contamination of surface water and ground water.
- Protect drinking water by using less pesticides and fertilizers.
- Reduce soil erosion by using conservation practices and other applicable best management practices.
- Use planned grazing systems on pasture and rangeland.
- Dispose of pesticide containers, and reinstate tanks in an approved manner.

4.6 WATER POLLUTION (INFORMATION TAKEN FROM VARIOUS SOURCES)[6, 8, 12, 13, 30]

Water pollution occurs when a water-body is adversely affected due to the addition of large amounts of materials to the water as shown in Figure 4.4. The sources of water pollution are categorized as:[3]

1   Point source
2   Non-point source (NPS)

4.6.1   Point sources of pollution

Point sources of pollution occur when the polluting substance is emitted directly into the waterway. Thus point sources are localized. Examples: oil spill from a ship; pipe discharging toxic chemicals directly into a river, etc. Point sources are generally much easier to monitor and control.

4.6.2   Non-point sources of pollution

A non-point source occurs where the localized source cannot be identified. There is runoff of pollutants into a waterway, for instance, when fertilizer from a field is carried into a stream by surface runoff. In many industrialized countries most of the pollutants in streams and lakes today derive from non-point sources.

Non-point source pollution is caused by rainfall or snowmelt "runoff" moving over and through the ground. As the runoff moves, it picks up and carries away natural

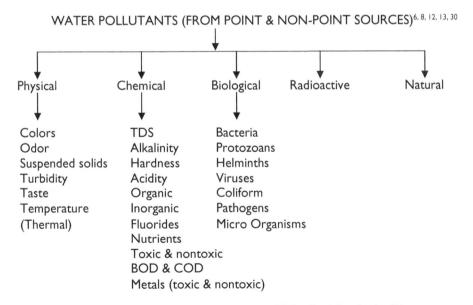

WATER POLLUTANTS (FROM POINT & NON-POINT SOURCES)[6, 8, 12, 13, 30]

Physical	Chemical	Biological	Radioactive	Natural
Colors	TDS	Bacteria		
Odor	Alkalinity	Protozoans		
Suspended solids	Hardness	Helminths		
Turbidity	Acidity	Viruses		
Taste	Organic	Coliform		
Temperature	Inorganic	Pathogens		
(Thermal)	Fluorides	Micro Organisms		
	Nutrients			
	Toxic & nontoxic			
	BOD & COD			
	Metals (toxic & nontoxic)			

Figure 4.4 Classification – water pollutants. TDS – Total dissolved solids.

and human-made pollutants, finally depositing them into lakes, rivers, wetlands, coastal waters, and even our underground sources of drinking water.

In a survey to investigate the leading cause of water quality problems in the U.S. today, it was found to be the non-point sources (NPS).

Measures to minimize NPS pollution; the below mentioned guidelines given by EPA are helpful in addressing this problem; you can modify them to suit your local conditions and requirements.

- Urban Storm-water** Runoff

  - Keep litter, pet wastes, leaves, and debris out of street gutters and storm drains, as these outlets drain directly to lake, streams, rivers, and wetlands.
  - Apply lawn and garden chemicals sparingly and according to directions.
  - Dispose of used oil, antifreeze, paints, and other household chemicals properly, not in storm sewers or drains. If your community does not already have a program for collecting household hazardous wastes, ask your local government to establish one.
  - Clean up spilled brake fluid, oil, grease, and antifreeze. Do not throw them into the street where they can eventually reach local streams and lakes.
  - Control soil erosion on your property by planting ground cover and stabilizing erosion-prone areas.
  - Encourage local government officials to develop construction erosion/sediment control ordinances in your community.
  - Have your septic system inspected and pumped, at a minimum, every 3–5 years so that it operates properly.
  - Purchase household detergents and cleaners that are low in phosphorous to reduce the amount of nutrients discharged into our lakes, streams and coastal waters.

**Storm-water runoff is generated when precipitation from rain and snowmelt events flows over land or impervious surfaces and does not percolate into the ground. As the runoff flows over the land or impervious surfaces (paved streets, parking lots, and building rooftops), it accumulates debris, chemicals, sediment or other pollutants that could adversely affect water quality if the runoff is discharged untreated. The primary method to control storm-water discharges is the use of 'Best Management Practices (BMPs)'. In addition, most storm-water discharges are considered point sources and require special measures to tackle them.[3]

- Mining

  - Become involved in local mining issues by voicing your concerns about acid mine drainage, disposal of the waste rocks, mine water and reclamation projects in your area.

- Forestry

  - Use proper logging and erosion control practices on your forestlands by ensuring proper construction, maintenance, and closure of logging roads and skid trails.

- Report questionable logging practices to state and federal forestry and state water quality agencies.

- Agriculture

  - As given in sec. 4.5.3

4.6.3   Types of water pollutants[6, 13, 16, 30]

4.6.3.1 Biological agents

Throughout civilization the most troubling thing to human society is the presence of biological agents present in the drinking water. It is the main cause of many diseases including the epidemics that could spread any time anywhere throughout the globe. The types of organisms that are responsible for different types of dieses are shown in Table 4.6:[16]

They are measured/assessed as 'Coliform Count', which means the number of bacteria present in a water sample. In normal drinking water it should not exceed 1/100 ml. Good sanitation and hygiene is the key to avoid diseases listed in the Table 4.6. Disinfections become mandatory to control these pathogens.

Table 4.6 Presence of biological agents in drinking water and their impacts on human health.[6, 13, 16, 30]

Pathogens (Organisms that can cause disease)	Diseases (Contamination of humans' drinking water with human diseases and organism-bearing human waste).	Remarks
Bacteria	Cholera, Typhoid, Enteritis, Dysentery, Pneumonia-like pulmonary disease (caused by inhaling the organism)	Epidemics: Cholera epidemic outbreaks in London in mid 19th century. Typhoid – Pennsylvania (1885), New York & Massachusetts states (1890) are few examples.
Protozoan	Amoebic dysentery, Giardiasis	It is found in surface water and proper filtration and disinfection is mandatory to prevent their access to the human body.
Helminthes	Schistosomiasis, Ascariasis	
Viruses	Fever, headache, nausea, diarrhea, muscular pain, upper and lower respiratory tract infections, inflammation of the eyes, common cold, hepatitis (specially in children), nausea and few others.	Microorganisms found other than in human-waste; such as those found in fruits, vegetables and slaughterhouses can also find their way into water, but most of them are relatively harmless.

4.6.3.2   Toxic substances

A toxic substance is a chemical pollutant that is not a naturally occurring substance in aquatic ecosystems. The greatest contributors to toxic pollution are herbicides, pesticides and industrial compounds.

4.6.3.3   Organic substances

Organic pollution occurs when an excess of organic matter, such as manure or sewage, enters into the water. When excessive quantity of compounds discharged into a water-body or its stretch; it makes life miserable for the species that need oxygen to survive and are known as 'aerobic' for example fish, zooplankton. This has an oxygen depleting impact (Table 4.7). Amongst bacteria there are species that can live without oxygen, known as 'anaerobic species'. When the organic compounds with elements carbon, sulfur and nitrogen are discharged into the water-body, where oxygen is present, or absent, the following products are resulted:

A type of organic pollution can occur when inorganic pollutants such as nitrogen and phosphates accumulate in aquatic ecosystems. High levels of these nutrients cause an overgrowth of plants and algae. As the plants and algae die, they become organic material in the water. The enormous decay of this plant matter in turn lowers the oxygen level. The process of rapid plant growth followed by increased activity by decomposers and a depletion of the oxygen level is called '**eutrophication**'. Thus, organic elements have adverse impact on water-bodies.

4.6.3.4   Thermal pollution[13, 30] (increase in temperature)

Thermal pollution can occur when water is used as a coolant near a power or industrial plant and then is returned to the aquatic environment at a higher temperature than it was originally. Thermal pollution can lead to a depletion of the dissolved oxygen level in the water; also increases the biological demand of aquatic organisms for oxygen.

4.6.4   Natural pollution

Ecological pollution takes place when chemical pollution, organic pollution or thermal pollution is caused by nature rather than by human activity. This includes pollution caused by natural calamities such as floods (carrying animals and organic, inorganic materials), volcanoes eruption, landslide (increase in silt in waterways), volcanoes, storms.

Table 4.7 Impact of discharging organic matters into water-bodies.[16]

Chemical compound with element	Where O_2 present	Where O_2 absent
C	CO_2	Methane CH_4
S	SO_4 (Sulfate salts)	Hydrogen Sulfide H_2S – it is known as rotten gas, it stinks and is poisonous
N	NO_3 (Nitrate salts)	NH_3 Ammonia; it stinks and is poisonous

4.6.4.1   Dissolved Oxygen (DO)

Flora and fauna present in the surface water-bodies are dependent on the dissolved oxygen, which they get from water. The solubility of atmospheric oxygen is 14.6 mg/l at 0°C to about 7 mg/l at 35°C at 1 atm pressure.[30] The solubility of oxygen decreases in salty water and also in polluted water.

4.6.4.2   Biological Oxygen Demand (BOD)

Organic matter received by water-bodies serves as food for bacteria present and also its oxidation releases energy. The higher the BOD in water-body, the less will be the DO. It is the quantative expression of oxygen depleting impact (via the action of decomposers) of a given amount of organic matter. It shows how much oxygen is needed by the microbes to oxidize the organic matter. Organic matter will undergo chemical oxidation even in the absence of decomposers; as there is also a straight Chemical Oxygen Demnd (COD)). COD is thus the total quantity of oxygen required for oxidation of waste to CO_2.

4.6.4.3   Hardness[4]

This is a function of dissolved calcium salts, magnesium salts, iron and aluminum. They occur as calcium and magnesium bicarbonates (referred to as temporary hardness) and sulfates and chlorides (referred to as permanent hardness).

The most obvious effect of hardness is preventing soap from lathering. Most people cannot tolerate drinking water that exceeds 300 ppm carbonates; 1500 ppm chlorides, or 2000 ppm sulfates, as more than 500 ppm sulfates can produce a laxative effect in the body. Livestock can tolerate higher level of sulfates but when TDS >10000; it could cause problems.

The formula used to calculate hardness is as follows:[4]

Total hardness in ppm Carbonates = (ppm Calcium × 2.497)
+ (ppm Magnesium × 4.115) + (ppm Iron
× 1.792) + (ppm Manganese × 1.822)

Hardness is treated either by a Zeolite process (home water softeners or by lime-soda ash process (large operations). Hardness is also measured in grains/gallon (1 ppm = 0.058 grains/US gallon).[4]

4.6.4.4   Acidity

Acid precipitation is caused when the burning of fossil fuels emits sulfur dioxide into the atmosphere. The sulfur dioxide reacts with the water in the atmosphere, creating rainfall, which contains sulfuric acid. As acid precipitation falls into lakes, streams and ponds, it can lower the overall pH of the waterway, killing vital plant life, thereby affecting the whole food chain. It can also leach heavy metals from the soil into the water, killing fish and other aquatic organisms. Because of this, air pollution is potentially one of the most threatening forms of pollution to aquatic ecosystems. pH7 is neutral, below it shows acidity and above it shows alkalinity.

4.6.4.5 Alkalinity[6]

Bio-carbonates present represent the alkalinity. For example salts of borates, silicates and phosphates etc. They are produced in the soap, leather and textile industries as byproducts.

4.6.4.6 Colors[6]

Some of the pollutants can change watercolor. Coloring agents are the most objectionable ingredients to human beings.

4.6.4.7 Radioactive[6, 13]

Nuclear waste can have detrimental effects on marine habitats. Nuclear waste is generated from the sources that include:

- Nuclear power stations.
- Nuclear-fuel reprocessing plants in Northern Europe are the biggest sources of man-made nuclear waste in the surrounding ocean. Radioactive traces from these plants have been found as far away as Greenland.
- Mining and refining of uranium and thorium are also causes of marine nuclear waste.
- Waste is also produced in the nuclear fuel cycle, which is used in many industrial, medical and scientific processes.

4.6.4.8 Oils & Petrochemicals[3, 6, 8, 13]

Oil does not dissolve in water; instead it forms a thick layer on the water surface. This can stop marine plants receiving enough light for photosynthesis. It is also harmful for fish and marine birds. Petrochemicals are formed from gas or petrol and can be toxic to marine life. Used oil from one oil-change can contaminates 1 million gallons of fresh water – a year's supply for 50 people (US Citizens)!.[3]

4.6.4.9 Red tide[3]

A "red tide" or "harmful algal bloom" is a natural event caused by rapid growth of microscopic, single-celled algae that makes the ocean look red or brown, especially in summertime, and especially when rains wash more nutrients into rivers and the sea. Most kinds of algae are harmless plants but some types produce a natural toxin that can contaminate shellfish that eat the algae. Animals or humans who eat the contaminated shellfish can be poisoned. Local fishing areas are usually closed during major red tides to prevent harvest and sales of contaminated shellfish. Local fish advisors tell you when to avoid buying or eating shellfish.[3]

4.7 SEWAGE[13, 23]

Sewage is the term used for wastewater that often contains faeces, urine and laundry waste. Thus, production of sewage is huge. It is treated in the water treatment plant and the waste is often disposed of in the sea or any water-body that has been permitted by the legal authorities. A network of pipelines carries it away. Untreated sewage can cause diseases as it carries harmful viruses and bacteria. The disease such as diarrhea is common. Sewage generated by houses or runoff from septic tanks into nearby waterways, introduces organic pollutants that can cause eutrophication.

4.7.1   Suspended or sedimentary solids[6, 8, 13]

Solids that are not dissolved affect water physically. Dumping of litter in a water-body can cause huge problems. Litter items such as a 6-pack ring packaging can get caught in marine animals and may result in death. Different items take different lengths of time to degrade in water, for example:[6]

Cardboard – Takes 2 weeks to degrade.
Newspaper – Takes 6 weeks to degrade.
Photodegradable packaging – Takes 6 weeks to degrade.
Foam – Takes 50 years to degrade.
Styrofoam – Takes 80 years to degrade.
Aluminum – Takes 200 years to degrade.
Plastic packaging – Takes 400 years to degrade.
Glass – It takes so long to degrade that we don't know how much it is?

Suspended solids could be organic or inorganic. Water treatment processes are influenced by their presence – quantity as well as quality. Control measures to minimize suspended solids, or their separation mechanisms, include: Mechanical separation, Centrifuges, Cyclones, Screening, Filtering, Thickening, including Plate Settlers, Flocculation, Floatation, Floto-Flocculation and Settling Ponds.[6] Measures to control/minimize dissolved solids include: Neutralization, Adsorption, Ion-Excahange, Reverse-Osmosis and Freezing.

Figure 4.5 Sewage.

4.7.2   Polluted municipality water[13]

The purpose of treating wastewater from homes and commercial establishments is to reduce its content of suspended solids, oxygen-demanding materials, dissolved inorganic compounds (particularly compounds of phosphorus and nitrogen), and harmful bacteria. It involves three steps:

1   Primary treatment, including grit removal, screening, grinding, flocculation, and sedimentation;
2   Secondary treatment, which entails oxidation of dissolved organic matter, using biologically active sludge, which is then filtered off; and
3   Tertiary treatment, using advanced biological methods of nitrogen removal and chemical and physical methods such as granular filtration and activated carbon adsorption.

More recently, stress has also been placed on improving the means of disposal of the solid residues from municipal treatment processes.

4.8   MARINE POLLUTION[13, 23]

Marine water is polluted by:

• Dredging – in some countries where mining by dredging is in vogue, the waste is directly discharged into the marine environment.
• Industrial waste, discharges from oil and gas drill rigs, and sewage sludge.
• Runoff containing nutrient materials, such as from fertilizers, is a problem particularly in coastal waters.
• Ships constitute another noteworthy cause of marine pollution, releasing into the ocean bilge water, ballast water, and sewage and other wastes, along with unintentional spills of substances such as petroleum products.
• Atmospheric deposition is a key avenue for the entry of nitrogen, metals such as lead and mercury, and chemicals such as DDT and PCBs into coastal and ocean environments.
• Oil Spills – Large-scale accidental discharges of liquid petroleum products along shorelines.

    • The supertankers used for oil transport and also many other ships also spill oil, and also offshore drilling operations contribute a large share of the pollution.
    • The largest spill in the U.S. (240,000 barrels) was that of the tanker *Exxon Valdez* in Prince William Sound, Gulf of Alaska, in March 1989. Within a week, under high winds, this spill had become a 6700-sq-km (2600-sq-mi) slick that endangered wildlife and fisheries in the entire Gulf area.
    • The oil spills in the Persian Gulf in 1983, during the Iran-Iraq conflict, and in 1991, during the Persian Gulf War, resulted in enormous damage to the entire area, especially to the marine life.

Table 4.8 Water Quality Discharge limits for discharges to the marine environment in a Gulf country.[19, 20, 24]

Variables	Discharge limit
Arsenic	0.05 mg/l
Cadmium	0.05 mg/l
Chromium	0.50 mg/l
Copper	0.50 mg/l
Cyanide	0.10 mg/l
Iron	2.00 mg/l
Lead	0.10 mg/l
Mercury	0.001 mg/l
Nickel	0.10 mg/l
Selenium	0.02 mg/l
Silver	0.005 mg/l
Zinc	0.10 mg/l
Chlorine (salt)	2.50 mg/l (minimum)
Hydrogen ions	6–9 units
Sulfide salts	0.10 mg/l
Sticking solid particles	30.0 mg/l
Sludge	75.0 Jackson sight unit
BOD	30.0 mg/l
Oil & grease	5.0 mg/l
Carbolic acids (phenols)	0.10 mg/l
Ammonium nitrates	40.0 mg/l
Phosphates	0.10 mg/l
Faeces	100 mpn/100 mm (80% samples)
Faeces – samples of floating bacteria	100 mkn/100 mm (100% units)
Sal-ammoniac	mbn/l (invisible)
Internal viruses	pfu 10 litres

Effluent Discharges – Marine Disposal
In addition to the regulatory requirements, effluent discharges should not result in:

- Visible oil or grease on the surface of receiving waters
- A change in color of receiving waters
- Emission of foul smells
- Any harmful effect, or change, which may lead to a harmful effect on marine life or the marine environment.

Table 4.8 depicts water quality discharge limits for disposal into the marine environment in a Gulf country.

4.9 WATER IN SUBSURFACE (UNDERGROUND) AREAS[26, 27]

Water enters underground workings for various reasons. Since geological and hydrological conditions of different formations vary, the inflow of water differs from one location to other. This inflow is often described as the make of water. Underground water is variable in chemical properties; it is often unsuitable for drinking and industrial use. Sometimes it contains free sulfuric acid, and then it is called acidic water. This water is very harmful and needs special care to deal with it. Subsurface drainage includes: prevention of entry of surface water into mines/tunnels and their protection from a sudden inrush of water, and pumping of water to the surface.

The ability of rocks to contain water as a consequence of their porosity is called moisture retention. The coefficient of permeability defines the degree of permeability of different rocks. Coefficient of permeability is the rate (speed) of flow of water through it under a hydraulic gradient equal to 1. The presence of water in rock can change its physical and mechanical properties considerably. The action of water in a rock can be to leach some of its ingredients, making it porous and permeable to water.

4.9.1   The main sources of water

- Water is bound to be encountered if openings such as tunnels, shafts, drives/drifts etc. are driven through the water bearing formations or strata. This is also the case when working below the water table.
- A sudden inrush of water from the water bodies such as lakes, rivers, sea, etc. when these are penetrated as a result of ground subsidence. During the rainy season the direction of flow of the surface water should be checked, and diverted from large cracks, subsided area, old workings etc. through which it could pass underground.
- From underlying or overlying strata if it gets through by joints or fractures.
- The water that has been brought down into the mines and tunnels to carry out the unit operations such drilling and dust suppression etc.
- High water make and inflow (flooding) – While approaching water-bearing zones, probing holes are often drilled 10–30 m ahead of tunneling face. Sometimes ground treatment ahead of the tunnel face becomes essential. Special measures are taken to deal with these situations to avoid ground collapse due to excessive water seepage.
- During operational phase due to the failure of linings (support work), which otherwise were sealing off the fissures, could also bring flooding. Inadequate barrier/parting between the subsurface excavation and water bodies, or waterlogged areas; and puncturing these barriers have been the causes of inundation in mines that have taken the lives of hundreds of miners.
- Sometimes it becomes essential to lower the water table using the appropriate techniques.

4.9.2   Effects of subsurface water

Direct effects:

- It adds pumping cost.
- If the make of water is abnormal, special techniques including treating the ground by grouting etc. are applied to seal off the source. This costs additionally.
- A sudden inrush of water can cause loss of men, machines, equipment and production (progress).

- Working under watery conditions makes mining and tunneling operations slow, tedious, less productive and risky.

Indirect effects:

- Damage to stability of openings (mine workings such as tunnels, shafts, drives etc.) in the presence of moisture if the rocks forming them are sensitive to it. Floor heaving and roof falls are common problems if they are of shale or mud-stone, as these rocks are very sensitive to moisture and the presence of even a very small amount of water.
- Increasing the problem of humidity, particularly when mines and tunnels are deep seated. Extra ventilation is required to improve the working conditions.
- High maintenance costs of the sets of equipment that are subjected to water.
- Watery holes during blasting require special types of explosives.
- Large scale dewatering can cause subsidence.

Presence of water with sulfide formations adds to the problems. This water could be corrosive, resulting in damage to service lines such as tracks, pipes, equipment and support systems (corroding concrete and steel support work). It can also corrode boots and clothes of the working crews.

4.10 ACID MINE DRAINAGE[5, 18, 22, 24, 29]

This phenomenon, which is known as Acid Mine Drainage (AMD), or, Acid Rock Drainage (ARD) is usually confined to coal as well as metal mines from where outflow of acidic water takes place. This is due to the presence of sulfur as pyrite (FeS_2), or sulfide ores of copper, zinc, nickle and a few others including sulfide minerals. The mines could be active as well as abondnaed. Copper mines usually have this problem. However, other areas where the earth has been disturbed (e.g. construction sites, subdivisions, transportation corridors, etc.) may also contribute acid rock drainage to the environment.

As mining progresses from the surface to greater depth once it enters the watertable, water starts accumulating and it needs pumping simultaneousely. When mines are abondoned this water get accumulated and this is usually the initiation of AMD particulaly at the mines descibed above. Tailing ponds have also been a source for surface as well as ground water contamination. There are thousands of sites located in different parts of the world including the mines discharges which are suffering from AMD; a few mining sites are bulleted below:[29]

- Buller coalfield in the north-west of the South Island, New Zealand
- Iron Mountain Mine, Shasta County, California, USA
- Various Coal Mines in the anthracite and bituminous coal regions of Pennsylvania, USA
- Davis Pyrite Mine in NW Massachusetts
- Whittle Colliery, Northumberland, England
- Woolley Colliery, Yorkshire, England
- Aznalcollar mine on the Agrio River, Spain
- Potosi, Bolivia, metal mines in and around Cerro Rico
- Cerro de Pasco, Peru, metal mine in the Central plain of the Peruvian Andes

4.10.1 Chemistry[29]

The chemistry of the process is complex but to illustrate; pyrites oxidation could be descibed by considering a general equation for this process as follows:

$$2FeS_2(s) + 7O_2(g) + 2H_2O(l) \rightarrow 2Fe^{2+}(aq) + 4SO_4^{2-}(aq) + 4H^+(aq)$$

The oxidation of the sulfide to sulfate solubilizes the ferrous iron (*iron II*), which is subsequently oxidized to ferric iron (iron III):

$$4Fe^{2+}(aq) + O_2(g) + 4H^+(aq) \rightarrow 4Fe^{3+}(aq) + 2H_2O(l)$$

Either of these reactions can occur spontaneously or can be catalyzed by micro-organisms that derive energy from the oxidation reaction. The ferric irons produced can also oxidize additional pyrite:

$$FeS_2(s) + 14Fe^{3+}(aq) + 8H_2O(l) \rightarrow 15Fe^{2+}(aq) + 2SO_4^{2-}(aq) + 16H^+(aq)$$

The net effect of these reactions is to release H$^+$, which lowers the pH and maintains the solubility of the ferric ion.

4.10.2 Yellow boy

When the pH of ARD exceeds 3, either through contact with fresh water or neutralizing minerals, previously soluble iron ions precipitate as iron hydroxide, a yellow-orange solid colloquially known as Yellow boy. Figure 4.6 illustrates this phenomenon. Yellow boy discolors water and smothers plant and animal life on the streambed, disrupting

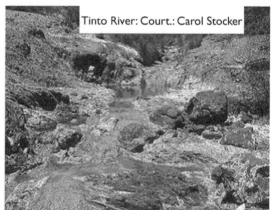

Tinto River: Court.: Carol Stocker

Figure 4.6 (Left) Yellow boy in a stream receiving acid drainage from a surface coal mine shown.[29] (see back cover page for colour picture)

stream ecosystems. The process also produces additional hydrogen ions, which can further decrease pH. Research is currently being conducted as to the feasibility of using Yellow boy as a commercial pigment.

4.11   CASE STUDY: WATER POLLUTION DUE TO MINING, PETROLEUM PRODUCTS HANDLING AND INDUSTRIAL ACTIVITIES[1, 14, 15, 17, 19, 25]

The case study presented is concerned with mining and petroleum products tanker loading & unloading facilities, and industrial activities in a developing country in the Gulf region. All these activities are not very old, covering a period of about 15 years.

4.11.1 Study areas

4.11.1.1   Copper mining

The copper mining and smelting operations began in 1982. Out of three mines two were underground and the third one a surface mine. Mining and consequently processing of copper ore continued until 1994 when farmers reported that the groundwater salinity near the copper mines area has advanced to 85% of the village wells in Sagha and Wali (two adjacent villages). 50% of former agricultural wells in the area were thus become unsuitable for irrigation. It has been established that salinity of water (the pollutant plume) has been progressing with increasing desertification hazards, and the productivity of the soil has been seriously affected. An unpublished draft report cited the following:

- Water is already salty (in Faraj A'Souq, Kheshishet, Misial A'Sidr, and Aarja).
- Symptoms such as hair falling off goats are observed.
- Trees in gardens and farms have died.
- Some of the children and women are suffering from coughing, asthma, and allergy.
- In Sagha (a village), honeybees have disappeared.
- Due to high salinity, groundwater is not potable; most of the householders are buying water.

Figure 4.7 (top) shows the location of a tailing dam and mining area, whereas Figure 4.7 (Bottom) displays the polluted area. Trenches that were dug to prevent further pollution of the groundwater have been also shown.

Based on government decisions and orders, mining operations were ceased in 1994 (MWR, 1996). During this period, some 11 million tonnes of sulfide-rich tailings and five million cubic meters of seawater have been disposed of at the plant's unlined tailings dam.

4.11.1.2   Areas adjacent to the petroleum refinery, PDO area and Shell marketing area

The second area chosen was the Petroleum Refinery and PDO (Petroleum Development of Oman) area are located in the coastal region of Muscat. The area extends

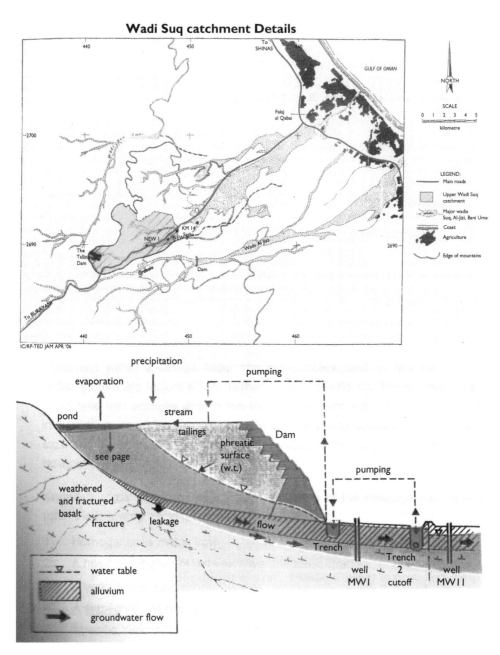

Wadi Suq catchment Details

Figure 4.7 (Top) Location of tailing dam and mining area. (Bottom) Layout of the polluted area, Trenches to safeguard from further pollution of the groundwater shown.[13, 14, 23]

offshore to a distance of approximately four km; within which tanker loading/unloading activities occur at three designated areas.

4.11.1.3 Industrial area

The third area chosen is an industrial area (Rusail Industrial Estate), which became operational in 1985 and by 1999 it had 119 factories. Rock types are characterized by three major rock formations from the bottom to the top: Seeb Formation (limestone), Rusail Formation (shale), and Jafnayn Formation (limestone). Some faults occur in the area that may act as conduits for the fluids to a great subsurface depth.

4.11.2 Sample collection and preparation

Twenty-one soil samples (seven from each site) of approximately 1 kg in weight were collected on judgmental-basis from selected sites for physical and chemical analysis. The samples were dried at room temperature and sieved to extract 2 mm-fractions.

Thirty-nine groundwater samples were used for this purpose; out of which 22 samples were from the Mining area; six from the Shell marketing Company, and 11 from Rusail Industrial Area. Samples were collected from boreholes and dug-wells and then stored in 250-ml glass bottles. Electrical conductivity and pH measurements were recorded after collection of samples.

4.11.2.1 Analysis

This study was confined to assessing copper (Cu), zinc (Zn), lead (Pb), Chromium (Cr) iron (Fe), and manganese (Mn) concentrations in soil and groundwater. An Inductively Coupled Plasma-Optical Emission Spectrometer (ICP) was used to measure heavy metal concentrations. In addition, X-ray Diffraction Spectrometer (XRD) and Scanning Electron Microscopy (SEM) were used to characterize soil mineralogy at those sites where heavy metal concentrations were measured.

4.11.3 Observations

- Results of analysis for the mining area are shown in Tables 4.9 and 4.10. Table 4.9 presents Chemical analysis & heavy metals concentrations of the groundwater samples. Table 4.10 details the physical and chemical properties & heavy metals concentrations of the soil samples.
- Table 4.11 depicts metal concentration of selected groundwater and irrigation water samples (in mg/l) collected from the Industrial Area and the Shell Marketing Company Area.
- Table 4.12 details chemical analysis & heavy metals concentrations of soil samples collected from the Industrial Area, PDO, Refinery and Shell Area.

4.11.4 Inference drawn – Physical and chemical properties

It is obvious from Table 4.9(a) that the continuous disposal of the sulfide-rich tailing (some 11 million tonnes) and seawater (some 5 million cubic meters) from the

Table 4.9(a) Chemical analysis & heavy metals concentrations of selected groundwater samples collected from copper mining areas (see table 4.8(b) for details).[15]

Sample I.D	D.F.S [*] (m)	Chemical analysis [+]										Heavy metals concentration [+]				
		pH	EC [*]	TDS [♦]	SAR	Na	Ca	Mg	Cl	SO$_4$	NO$_3$	Fe	Zn	Cu	Pb	Mn
Trench1	1.15	6.90	42600	28800	21.40	5570	4130	595	16148	2175	10.00	**0.32**	0.03	0.03	0.01	**1.1**
Trench2	1.75	6.90	38900	28321	18.57	4970	4040	823	16207	2124	16.00	**0.23**	0.03	0.03	0.01	**0.06**
MW1	1.30	6.20	49980	35573	26.91	7400	4690	616	20749	1980	1.10	**0.35**	0.05	0.03	0.01	**1.07**
MW2	1.25	6.90	53600	39813	27.79	8110	5030	847	23620	2014	3.2	**0.4**	0.05	0.03	0.01	**1.24**
MW3	1.70	6.60	47800	33415	21.25	6060	4450	1020	19625	1997	23.80	**0.34**	0.05	0.01	**0.03**	**0.72**
MW5	1.85	7.40	15000	9839	14.49	2200	1120	374	5305	712	14.0	0.09	4.47	0.03	0.01	**0.16**
MW9	1.70	7.60	12245	9437	6.95	1010	994	361	6345	617	1.1	0.03	**6.10**	0.03	**0.03**	**0.61**
MW11	1.85	6.90	26100	37541	25.31	2670	775	40.40	10194	2065	1.1	**0.18**	**9.55**	0.03	0.01	**0.17**
MW12	1.95	7.10	6730	8644	7.29	687	443	137	2166	292	1.10	0.05	0.37	0.01	0.01	0.03
WS 5	8.50	7.70	11520	6868	10.37	1310	459	448	602	17.10	0.06	0.03	0.03	0.01	0.03	0.03
WS 6	10.60	8.30	1430	980	2.90	149	39	96	326	147	2.2	0.04	0.03	0.03	0.01	0.05
WS 7	10.65	8.20	3520	2148	11.63	581	122	40	723	550	14.8	0.03	0.03	0.03	0.01	0.03
WS 9	10.45	7.40	4127	5269	9.55	1622	1286	536	5269	484	20.20	0.02	0.03	0.03	0.01	0.03
GSA 2	3.65	7.00	17300	11673	6.77	1310	1800	616	7162	681	21.40	0.11	**5.92**	0.03	0.01	0.03
JDU	5.400	8.10	1066	1066	7.70	207	123	27	262	323	16.6	0.03	0.03	0.03	0.01	0.03
JDD	5.70	8.00	2010	1256	5.56	269	112	39	361	369	14.9	0.03	0.03	0.03	0.01	0.03
Arja A4	3.45	8.00	3380	1935	5.67	387	115	142	824	301	13	0.03	0.24	0.03	0.01	0.03
Sagah 1	7.25	7.50	18800	10962	9.96	1740	907	841	6715	609	11.40	0.1	0.03	0.03	**0.08**	0.03
Sagah 2	6.55	7.60	11868	6774	9.58	1210	654	332	3985	464	10.5	0.05	0.45	0.03	0.01	0.03
Sagah 3	7.35	7.60	4315	2786	8.31	579	215	91	948	750	53.30	0.03	0.03	0.01	**0.05**	0.03
Km 14	8.25	7.90	8125	5094	10.70	1100	318	288	2583	630	20.00	0.03	0.03	0.03	0.01	0.03
F.Q	>11.00	8.70	1255	693	3.16	125	47	43	222	131	15.40	0.03	0.03	0.03	0.01	0.03

[*] D.F.S: Distance from plantsite

[*] EC: Electrical Conductivity (μS/cm)

[♦] TDS: Total dissolved solid

[+] Concentration in mg/l (unless otherwise indicated)

Table 4.9(b) Details of the monitoring points that has been surveyed and located at Copper Mines Area.[15]

Monitoring point	Type	*Depth (m)	General information
Trench 1	Open trench	3–4	Located Behind the tailing dam, water drained from the dam is pumped back to the dam.
Trench 2	Large dugged borehole	>10	Used to stop any further deep-water seepage from trench 1.
MW 1	Borehole	4–6	First monitoring points after trench1, and mainly used to check the EC level.
MW 2	Borehole	4–6	Immediately after MW1, used also to check EC level.
MW 3	Borehole	1.5–2	Located further dawn from MW2 in the main wadi channel
MW4	Borehole	1.2	Used to check for any seepage to the north of the tailing dam
MW5	Borehole	7–10	Used to check for any seepage to the north of the tailing dam
MW 9	Borehole	7–10	Located to the north-east of the tailing dam
MW 11	Borehole	4–6	Located further dawn from MW3 in the main wadi channel
MW 12	Borehole	4–6	Located further dawn from MW11 in the main wadi channel
WS 5	Dug well	5	Located in the Wadi Souq main channel, to the left side of Buriami-Sohar road.
WS 6	Dug well	15	Located in the Wadi Souq main channel, to the right side of Buriami-Sohar road.
WS 7	Dug well	10	Located in the end of Wadi Souq channel
GSA 2	Borehole	-	This is one of the most recent boreholes in Wadi Souq, further dawn of GSA 1.
JDU	Dug well	5	Located in a small local farm (water is not suitable for irrigation and for human animal consumption)
JDD	Dug well	4	Located in a small local farm (water is not suitable for irrigation and for human animal consumption).
KM 14	Dug well	3	Located in the main wadi channel, just after Sagah 1.
Sagah 1	Dug well	2.5	Located further dawn of MW12, and it is in the right side of Buraimi-Sohar road.
Sagah 2	Dug well	4	Located in a small local farm (water is not suitable for irrigation and for human animal consumption).
Sagah 3	Dug well	-	Located in the main wadi channel, and to the left side of Buraimi-Sohar road.
Falag Al Qabial	Narrow stream or Falaj	<0.5	Fresh groundwater, used for drinking and for irrigation purposes

* The depth was taken from the ground level to the water surface.

mining area during period 1982–1994; have resulted the groundwater highly saline with appreciable sulfate concentrations. Trench 1 is an open ditch located behind the tailing dam securing water dried from the dam, whereas trench 2 is used to stop any further seepage from trench 1. The electrical conductivity of water samples collected from the trench is as high as that of seawater, being saline and sodic with high sulfate concentrations.

• Water samples collected from wells located between 1.25 and 14 km from the plant site and 2–10 m deep, were then analyzed. The electrical conductivity of

Table 4.10 Physical and chemical properties & heavy metals concentrations of soil samples collected from the mining area.[15]

Sample I.D	D.F.S (m)*	Direction from site	Physical properties %				Chemical properties						Heavy metals concentration (mg/kg⁻¹) air-dry soil			
			Sand	Silt	Clay	Texture•	pH	EC* (µS/cm)	ESP	Na (mg/l)	Ca (mg/l)	Mg (mg/l)	Cu	Zn	Mn	Fe
MIN 1	100	East	69.7	20.2	10.1	SL	8.7	142	0.79	85	144	1.3	1250	1500	370	675
MIN 2	150	North	79.8	14.1	6.1	LS	8.4	168	1.02	30	700	2.5	1000	750	1000	395
MIN 3	100	North	85.8	5.0	9.1	LS	8.3	102	1.02	8.5	533	1.3	1000	170	1500	425
MIN 4	50	North-west	86.8	9.1	4.0	LS	8.2	169	1.01	35	700	2.5	12000	100	600	48000
MIN 5	125	West	64.6	28.3	7.1	SL	7.8	566	1.02	28	925	6.3	27950	2000	600	72700
MIN 6	150	South	76.7	16.2	7.1	LS	8.1	171	1.01	35	800	2.5	5850	500	610	72700
MIN 7	70	South-west	65.6	24.2	10.1	SL	8.5	65	1.02	9.6	144	0.5	2300	150	1000	47000

◆ D.F.S: Distance from plantsite (approximate measurements).

∗ EC: Electrical conductivity.

• Soil Texture: Sandy Loam (SL) and Loamy Sand (LS).

Table 4.11 Metal concentration of selected groundwater and irrigation water samples (in mg/l) collected from the Industrial Area and the Shell Marketing Company Area.[15]

Sample I.D	Zn	Pb	Cu	Cd	Cr
Rusail Industrial Area					
RIE/MAIN	0.09	0.01	0.05	0.005	0.04
MAIN/SOURCE	0.31	0.01	0.05	0.005	0.02
RUSAIL/MAIN	0.07	0.01	0.05	0.005	0.02
ROAD/10/KIMJI	0.13	0.01	0.05	0.005	0.05
ROAD/15	0.12	0.01	0.05	0.005	0.02
AL-KHOD WELL1	0.02	0.01	0.05	0.005	0.02
AL-KHOD WELL2	0.03	0.01	0.05	0.005	0.02
WADI AL-KHOD	0.02	0.01	0.05	0.005	0.02
ROAD/3/10/A	0.13	0.01	0.05	0.005	0.05
ROAD/3/10/B	0.15	0.01	0.17	0.005	0.05
Shell Marketing Company					
BW-1	0.05	0.04	0.05	0.005	0.02
BW-2	0.05	0.06	0.05	0.005	0.02
BW-3	0.26	0.04	0.05	1.570	0.02
BW-4	0.63	0.14	0.17	0.005	0.02
BW-5	0.47	0.03	0.34	0.005	0.02
BW-6	0.45	0.28	0.27	0.005	0.02

• For Maximum Omani Standard (MOD) for element concerned (in mg l⁻¹) are: Zn, 5.0; Cu, 0.05; Pb, 0.05; Cd, 0.01; Cr, 0.01; Fe, 0.1 and Mn, 0.05.

these samples were directly proportional to the distance from the plant site ranging between 53600 to 1430 μs/cm with the Sodium Adsorption Ratios (SAR) ranging from 27 to 3.0.

• The quality of groundwater around the plant side for irrigation-purposes is classified as having very high salinity, very high sodicity hazard within 10-km range, and with high sulfate concentration within 2 km. About 85% of the wells in Sagha and Wali villages have been seriously affected and more than 50% of them have been abandoned.

• Falaj Al-Qabail (another stream) water has not been affected, originating from a different area.

• The time-series (Figure 4.8) of sodium concentration in groundwater samples collected during the years 1984–1997 from wells mw1, mw2 and km 14 located 1.3, 1.2, and 8.2 km from the monitoring plant site are shown. Sodium concentrations have increased from about 500 ppm in 1984 to 8000 ppm in 1997.

• The soils around Sohar Mining Area range from sandy loam to loamy sand with clay contents of less than 10%. Low salinity and low exchangeable sodium percentage (ESP) characterize soil samples taken from the surface (0–15 cm).

• Soils of the Industrial Area were similar to those of Copper Mines in their physico-chemical characteristics, whereas those of the Refinery area were sandy (being close to the beach).

4.11.5 Heavy metal concentration

The heavy metal concentrations of selected groundwater samples from the Copper Mines (Table 4.9(a)) show that these samples were not contaminated with Cu. Even through Fe and Mn concentrations were exceeding the allowable limits, as shown in Table 4.9(a), within 2 km range from the plant site; they were not excessively high. Some isolated wells showed slightly higher Zn and Pb concentrations exceeding the allowable limits by the prevalent regulations.

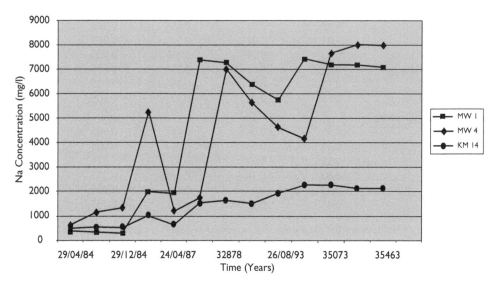

Figure 4.8 Time series of sodium (Na) concentration in groundwater samples from the mining area.

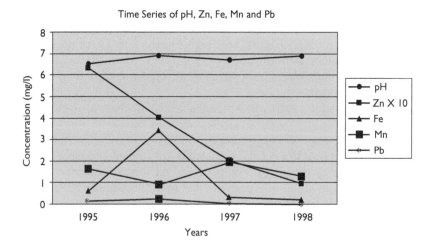

Figure 4.9 Time series for Ph, Zn, Fe, Pb concentration in the groundwater samples from the mining area.

- In general, the problem of the Copper Mining Area was that of groundwater salinity and sulfates and not of copper contamination. The time series for the years 1995–1998 of pH Mn, Fe, Zn and Pb from well-11 located 1.85 kilo-meters from the plant site is given in Figure 4.7. No profound increases in concentrations were noticed over those years. Zinc in fact decreased from 63 to 9.5 ppm. Fe and Pb concentrations increased with pH, whereas Mn decreased with pH.

- The metal concentrations of selected groundwater and irrigation water samples collected from the Rusail Industrial and Shell Marketing Company areas is given in Table 4.12. The analysis revealed contamination of chromium exceeding the allowable limits by as high as 5 times in groundwater and irrigation water of the Industrial Area.
- Lead (Pb) concentrations in groundwater around the Shell Marketing area were twice the allowable limits on the average with higher concentrations in wells near to storage tanks. Copper (Cu) concentration also exceeded the limits set by regulations in half of the wells investigated, being as high as six times in one of the boreholes.
- Generally, water around the Industrial area were Cr contaminated whereas those around the Shell Marketing Company were Pb and Cu contaminated.
- Soil sample analysis for metals are given in Table 4.10 for the Copper Mines Area and Table 4.12 for the Industrial and Shell refinery areas. High concentrations of iron and copper were detected in the soil samples collected from the Copper Mines area within 100 m distance; Fe was as high as 72,700 ppm; and Cu as high as 27950 ppm.
- Soil contaminated with metals was not evident around the Industrial area. Fe concentrations in the soil were excessive around the Shell Marketing Area, about 22,000 ppm on the average whereas Cu and manganese Mn were noticeable in the Refinery Area being as high as 1500 and 2000 ppm, respectively.

Another study was conducted by the petroleum company in the same country to assess whether any groundwater contamination occurred through leakage of diesel from a tank associated with a terminal building. The findings of the study concluded that there is significant concentration of hydrocarbons (up to 3800 mg/kg) to a depth of 4 to 6 meters. Hydrocarbon contamination was also identified to a depth of 10 meters with concentrations of 500 mg/kg. The groundwater which was encountered at depth between 4 to 5.5 m below that ground level indicated that the contamination had reached the underlying groundwater.[24]

4.11.6 Conclusion

- Mining activities employing seawater were responsible for groundwater salinity, and copper and lead contaminants for the soils.
- Industrial activities elevated chromium content in groundwater and irrigation water.
- Lead and copper contamination was associated with the petroleum activities.

Table 4.12 Physical and chemical properties & heavy metals concentrations of soil samples collected from the Industrial Area and PDO, Refinery, Shell Area.[15]

| | Physical properties | | | | | | Heavy metals concentration (mg/kg⁻¹) air-dry soil | | | | | | |
Sample I.D	% Sand	% Silt	% Clay	Text-ure	pH	EC* (µS/cm)	Cu	Zn	Mn	Fe	Al	Br	Cr
Industrial Area													
WA/C	69.09	21.95	8.18	SL	9.00	83.3	5.0	5.0	*	*	5.0	2.95	0.10
WA/C1	*	*	*	*	1.60	32.3	5.0	5.0	*	*	125.0	74.0	2.00
WA/DP	*	*	*	*	8.17	619	5.0	5.0	*	*	40.0	1.0	2.00
WA/3	*	*	*	*	7.55	80.2	5.0	5.0	*	*	1.34	53.0	2.00
WA/5	81.84	10.89	9.06	LS	9.00	80.3	5.0	5.0	*	*	210.0	46.0	2.00
WA/7	77.86	16.02	8.56	LS	7.55	2.0	5.0	5.0	*	*	81.0	40.0	2.00
WA/9	82.79	10.81	8.10	LS	9.16	80.2	5.0	5.0	*	*	138.0	46.0	2.00
WA/10	61.94	25.38	10.83	LS	9.14	110.7	5.0	5.0	*	*	86.0	34.0	2.00
Refinery, PDO and Shell Area													
REF-1	93.93	2.02	4.04	Sand	7.4	130	25.0	60.0	310.0	210	*	*	*
REF-2	94.94	2.02	3.03	Sand	8.5	98	105.0	105.0	320.0	265.0	*	*	*
REF-3	91.91	4.04	4.04	Sand	8.5	94	120.0	125.0	315.0	265.0	*	*	*
REF-4	92.92	3.03	4.04	Sand	8.5	106	20.0	40.0	310.0	20.5	*	*	*
REF-5	94.95	1.01	4.04	Sand	8.7	85	370.0	500.0	335.0	295.0	*	*	*
REF-6	93.93	3.03	3.03	Sand	8.7	73	155.0	500.0	340.0	230.0	*	*	*
REF-7	94.94	2.02	3.03	Sand	8.6	283	1500.0	165.0	2000.0	545.0	*	*	*
R.GAS	91.92	3.03	5.05	Sand	8.8	0.6	30.0	55.0	365.0	22900.0	*	*	*
R.GAS	85.85	7.07	7.07	LS	9.3	0.6	30.0	55.0	405.0	26250.0	*	*	*
R.GAS	93.95	1.00	5.05	Sand	8.8	0.43	21.5	50.0	335.0	21950.0	*	*	*
R.GAS	89.90	6.06	4.04	LS	8.6	0.38	19.0	35.0	355.0	20900.0	*	*	*
R.GAS	93.94	3.03	3.03	Sand	8.3	0.59	29.5	75.0	340.0	22350.0	*	*	*
R.GAS	92.93	3.03	4.04	Sand	9.0	0.42	21.0	50.0	315.0	20950.0	*	*	*

Based on these facts and figures and reports from local inhabitants, government agencies, petroleum and mining companies, consultants engaged for this purpose, the government has formulated a strict policy and enforced regulations. Research is on the way to model the plume of Wadi Suq enabling government to take appropriate measures.

The purpose of presenting this case study is to demonstrate a typical case of alkalinity and salinity that could pollute surface as well as groundwater due to mining activities, to show how mine water if not treated well could change pH, and how heavy metals could pollute the groundwater.

It also illustrates as how important it is to protect any industrial setup from the adverse impacts of pollution right from Conceptual to execution and regular production phases, and shows how land could be polluted by salinity of the seawater.

The figures and tables illustrate the type of sampling, monitoring and record-keeping which is essential while undertaking such studies. The author happened to be involved as one of the investigators in the research work that was undertaken.

4.12   BOTTLED WATER[2]

Based on the ideas and views expressed through various website on this subject matter; below outlined text briefly describes what use of bottled-water may result in. It gives a sensible idea for our consideration based on data from USA.

4.12.1   Bottled water – Do we need it? Some facts[2]

- In USA it costs more than $ 1.5/bottle, which is 1900 times more than tap water, which means bad news for our wallet, as average US citizen spends more than $400 on bottled water/year.
- Health can be damaged by chemicals like Bisphenol-A (BPA), leached from bottled water. BPA intake can generate cancerous cells in our body. The environment is tainted by it's the production, transportation, packaging and disposal of plastic bottled water.
- In the year 2004 bottled water usage was marked as 26,000,000,000 liters; which is nearly 28,000,000,000 plastic bottles in a year. Of which 1500 bottles end up in garbage/second.
- 28,000,000,000 plastic bottles means 17,000,000 barrels of oil was used to manufacture those plastic bottles. This could have been enough to fuel 100,000 cars that year. It also means an addition of 2,500,000 tonnes of carbon dioxide that was produced in manufacturing the plastic bottles.
- It means $100,000,000,000 (one hundred billion dollars) spent every year by consumers on the bottled water. Research shows that for a fraction of this amount every one on this planet could have safe drinking water!!!!
- If we drink bottled water then do we know all the facts about it? If we don't then lets say we are in for a great surprise:

    - We spend money
    - We pollute the earth

- We risk polluting the aquifer and other water-bodies
- All for bottled water??? A survey shows that 35% bottled water drinkers think that it is safer than tap water?
- Do you keep bottled water in your car? And think that it is safe; but do you know that heat in the car and the plastic of bottle can leach and produce substances, which could cause breast cancer and other types of cancer. And even if not kept it in your car, do you know extreme temperatures it has been through before you bought–

    1   Stored in warehouse temperature Variation: 26 to 85°F
    2   Transportation in trucks, temperature Variation: 100 to 150°F
    3   Loading and unloading temperature Variation: 45 to 100°F
    4   Taken to local suppliers, temperature Variation: 55 to 100°F

No matter which option we choose when thinking of Going Green; ditching bottled water keeps Mother Earth and our wallet Green.

4.13   CONCLUDING REMARKS

Lets us make it a point: 'A Drop of Water' matters us and we make sure it is utilized judiciously.

QUESTIONS

1   'Bottled water' - should we limit its use? Why?
2   Descibe phenomenon of 'Yellow boy'.
3   Do you think out of a barrel of water only a bucketful is suitable for our use? Name the sources of fresh water suitable for drinking.
4   Do you think some pollutants are actually beneficial to the aquatic eco-system? But there are others, which are harmful at almost any level; what are they?
5   How is marine water polluted?
6   How much water is there throughout the globe and how much of it (%) can be utilized for human needs? What are the primary sources of water for human use?
7   How do solids that are not dissolved affect water physically?
8   List the operations involved in completing the 'Water Cycle'? Also draw it.
9   List problems due to presence of water in underground mines. And what special care is required to deal with it?
10   List some 'Biological agents' responsible for human diseases.
11   List the sources of ground water contamination. How can this pollution be minimized?
12   List the types of sampling, monitoring and record keeping which are essential while undertaking studies to assess underground water pollution.
13   Prepare a line diagram to present water pollutants from point and non-point sources.
14   What does 'Coliform Count' signify?
15   What are surface water pollutants?

16 What do you mean by potable water? What would the quality of water be for 3 samples having their pH value of 1, 7, 10 respectively?

17 What is Red Tide?

18 What makes water a major cause of health problems? What diseases could be caused due to polluted water?

19 What would you look for in a water sample to check whether it is potable or not? For how many contaminants (in water) has EPA set standards?

20 Where is Acid Mine Drainage usually confined? List its impacts to the surrounding environmment.

21 Which biological agents in drinking water could cause problems to human health?

22 Which substances are the greatest contributors to toxic pollution?

REFERENCES

1 Al-Maktoumi, Ali (2000) *Modeling of Groundwater Contamination at Wadi Suq*, Master Degree Thesis, Department of Soil and Water, College of Agriculture, Sultan Qaboos University, Oman.

2 Bottled water – Information source: http://www.earth911.com/; http://www.earthpolicy.org/; http://www.commondream.org/; http://www.filterforgood/; http://us.oneword.net. [Accessed Aug. 2009]

3 EPA Q & A. [Accessed Aug. 2009]

4 Glover, T.J. (2006) Water hardness. Pocket Ref.; Pub. Sequoia Pub. Inc. Littleton, USA; pp. 637.

5 GreenFacts Website: "Scientific Facts on Water: State of the Resource". Retrieved on 2008–01–31. [Accessed June 2009]

6 http://in.answers.yahoo.com/question/index?qi; http://www.google.co.in/imgres. [Accessed June 2009]

7 http://www.eea.europa.eu/themes/water/water-pollution/figures-and-maps/sources-of-pollution/image_preview. [Accessed June 2009]

8 http://www.infoplease.com/ce6/sci/A0861892.html; Dangers of Water Pollution. [Accessed June 2009]

9 http://www.kidzone.ws/water. [Accessed June 2009]

10 http://www.texasep.org/html/wql/wql_3grw.html [Accessed June 2009]

11 http://www.usgcrp.gov/usgcrp/images/ocp2003/WaterCycle-optimized.jpg [Accessed Aug. 2009]

12 http://www.wateryear2003.org/en/ev.php-[Accessed Aug. 2009]

13 http://www.worldalmanacforkids.com/ [Accessed Aug. 2009]

14 Japan International Corporation Agency (JICA) and MCI, Ministry of Commerce and Industry (MCI) – Sultanate of Oman (2001). The Feasibility Study on Mine Pollution Control in Sohar Mine Area, Sultanate of Oman. Unpublished Summary Report, Mitsubishi Materials Natural Resources Development Corp. and E and E Solutions Incorporation, Japan.

15 Juma Khalfan Al-Handhaly, Hayder AbdelRahman & Said Salim Al-Ismaily. (2000) Impact of industrialization and mining wastes on heavy metal deposition in the environment, sultanate of Oman. *International Conference on*

Geo-environment 2000, Sultan Qaboos University, Muscat, Oman, 4–7, March.

16 Kupchella, C.E. & Hyland, M.C. (1993) *Environment Science*. Living within the system of nature, Pub.: Prentice-Hall Inter. Ltd. pp. 193–205; 337–379.

17 Master Plan For Groundwater Pollution Protection in the Sultanate of Oman (1995) (Contract B8/95). Study Conducted by Mott MacDonald Company.

18 Mielke, R.E., Pace, D.L., Porter, T. & Southam, G. (2003) A critical stage in the formation of acid mine drainage: Colonization of pyrite by Acidithiobacillus ferrooxidans under pH-neutral conditions. Geobiology 1 (1): 81–90. doi:10.1046/j.1472 4669.2003.00005.x. [Accessed 2009]

19 Ministry of Regional Municipality and Environment. Groundwater monitoring in the vicinity of Oman mining companies. Changes and trends during 1995. MRM & E, Sultanate of Oman.

20 Ministry of Water Resources (1996) Ground Water Pollution and Remediation in Wadi Souq. MWR, Sultanate of Oman.

21 National environment policy – Draft for comments. (2004) Ministry of environment and forests; Govt. of India, New Delhi. 30th Nov.

22 Nordstrom, D.K. Alpers, C.N. Ptacek, C.J. & Blowes, D.W. (2000) Negative pH and Extremely Acidic Mine Waters from Iron Mountain, California. Environmental Science & Technology 34 (2): 254–258. doi:10.1021/es990646v. http://ca.water.usgs.gov/water_quality/acid/.[Accessed 2008]

23 Ocean dumping of sewage sludge is prohibited in the United States by the Marine Protection, Research, and Sanctuaries Act (MPRSA).

24 Petroleum Development of Oman – HSE management, course material, reports and documents and interaction with those concerned during period: 1996–2004.

25 Satti, O., Al-Rawahy, K. & Tatiya, R.R. (2003) Characterization of Geo-environment problems using geostatistics, the case of Wadi Suq. *International Conference on Soil and groundwater contamination, edited by Mohmad Ahmed, Sultan Qaboos University, Oman, 20–23 Jan.*

26 Saxena, N.C., Singh, G. & Ghosh, R. (2002) *Environment management in mining areas*. Scientific Pub. India. pp. 20, 205, 206–221.

27 Tatiya, R.R. (1996–2004) *Course material for Chemical Engg. Petroleum and Mineral Resources Engg.* Sultan Qaboos University, Oman.

28 Tatiya, R. R. (2000) Impacts of Mining on environment in Oman. *International Conference on Geo-environment 2000, Sultan Qaboos University, Muscat, Oman, 4–7, March.*

29 Wikipedia: acid mine drainage-free encyclopedia. [Accessed July 2009]

30 Wikipedia: Making Life Easier. http://en.wikipedia.org/wiki/Water_resources #Surface_water. [Accessed Aug. 2009]

Chapter 5

Solid industrial waste & land degradation

'The success of any setup whether industrial or non-industrial lies in minimizing wastage of resources, which could be: natural (minerals, land, air, water, flora, fauna) or man-made (man-himself, machines, equipment, materials and energy)'

Keywords: Hazardous Wastes, Waste Disposal, Waste Generation, Waste Accumulation, Waste Management in Petroleum & Mining Industries, Solid Waste, Source Reduction, Recycling/Recovery, Reuse, Treatment, Responsible Disposal, Modern Landfill, Particulate Matter, Composting, Incineration, Land Degradation, Soil Degradation Dumping Site, Gaseous Emissions, Waste Consignment Note, Material Substitution, Equipment Modifications, Proactive Outlook.

5.1 INTRODUCTION

Waste is a material which has no direct value to the producer, and so it must be disposed off. It is usually generated by the non-value-adding activities of any operation/process. It can be a solid, liquid or gaseous material. Every industry produces waste but the quantity and nature varies, and thus it is difficult to quantify and measure. It can easily be noticed with a little care and awareness. Its handling, storage and disposal costs are huge, and *accumulation beyond certain limits* results in adverse impacts to the health, environment and overall economy of the operation, as:

- It is an indication of consuming excessive materials
- It builds up due to excessive losses (in most of the cases)
- It causes accidents and jeopardizes safety
- It necessitates re-work, and ultimately:
- It causes delays.

Waste generation cannot be eliminated, but applying effective management techniques certainly could minimize it.

5.2   CLASSIFICATION[4, 5, 8, 11]

5.2.1   Non-hazardous wastes

In general, it could be defined as: any solid material or semi solid, which does not have any danger to the environment or to the human health, if it is dealt in a safe and scientific way. Thus it is a useless, unwanted and discarded material[4] and in an industrial setup it could be any of the following: kitchen refuse, domestic waste, tree cuttings, office waste, non-hazardous waste chemicals, non-hazardous empty drums, scrap metal, water-based drilling mud (WBM) – in petroleum industry, water-based drilling mud cuttings (WBMC) in petroleum and mining industry and others – specific to the industry it belongs to but non-hazardous.[5] Figure 5.1 classifies wastes.

5.2.2   Hazardous wastes[3, 5, 10, 11]

Hazardous waste is defined as liquid, solid, contained gas, or sludge wastes that contain properties that are dangerous or potentially harmful to human health or the environment. It is thus a useless, unwanted and discarded material that poses threat to

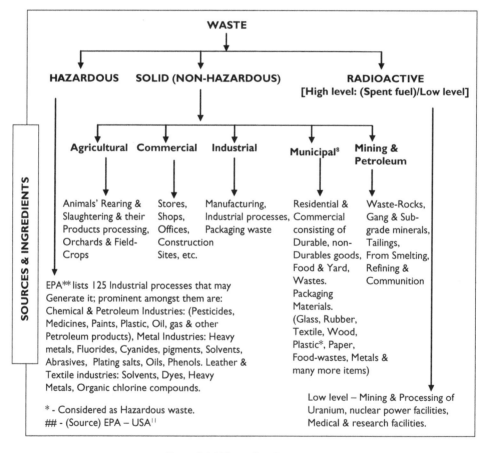

Figure 5.1 Waste classification.

the human health (including flora & fauna) and/or the environment.[3] In an industry, following constitutes hazardous waste:

Batteries, Clinical waste, Hazardous empty drums, Lab waste chemicals, Waste chemicals; Oily sand/soil, Oily sludge, Pigging sludge, Sewage Sludge, Tires, Waste lubricants, Naturally-ccurring radioactive materials (NORM). Oil-based mud (OBM) and OBM cuttings.

This specification does not address aqueous wastes (such as production water, sewage effluent and reverse osmosis plant discharges (in the petroleum industry) etc.) and gaseous wastes. These are also hazardous emissions as detailed in Table 5.1.

Table 5.1 Environmental hazards and possible effects due to solid, aqueous and gaseous wastes.[5, 10, 11]

Hazard	Possible effect	Sources: Type of industry
Gaseous emissions of:		
Methane (CH_4)	Global warming/atmospheric ozone increase	Chemical, Petroleum, Mining and Metallurgical to less hazardous process industries producing goods and services of many kinds.
Sulfur oxides (SO_x)	Acid deposition, water and soil acidification	
Nitrogen oxides (NO_x)	Atmospheric ozone, acid deposition	
Nitrous oxide (N_2O)	Global warming, stratospheric ozone depletion	
Carbon dioxide (CO_2)	Global warming	
Carbon monoxide (CO)	Human health damage	
Hydrogen sulfide (H_2S)	Human health damage, odor nuisance	
Volatile Organic Compounds (VOC)	Atmospheric ozone increase, human health damage	
Organic toxics (PAH, PCB)	Human health damage, ecological damage	
Emissions of:		
Fine particulate matter	Human health damage, soot deposition	Chemical, Petroleum, Mining and metallurgical to less hazardous process industries producing goods and services of many kinds.
Toxic metals	Human health damage, ecological damage	
Odorous compounds	Nuisance	
Radiation	Human health damage, ecological damage	
Heat	Nuisance, ecological damage	
Light	Nuisance	
Noise/vibration	Nuisance	
Chlorofluorocarbons (CFC)	Global warming, stratospheric ozone depletion	Also includes discharge from factories & power plants.
Halons	Global warming, stratospheric ozone depletion	
Spills & leaks of crude oil or distillates	Ecological damage, biological damage	Petroleum
Emissions of:		
Dissolved organic compounds	Ecological damage, biological damage, tainting of fish	Petroleum, Chemical, Mining, Construction and many others.
Soluble heavy metals	Ecological damage, biological damage through accumulation	

(Continued)

Table 5.1 (Continued)

Hazard	Possible effect	Sources: Type of industry
Soluble salts	Increased salinity, biological damage	
Drilling mud/cuttings/ chemicals	Ecological damage, biological damage	Also includes discharge from factories & power plants.
Organic nutrients (NH_4, PO_4)	Eutrophication	
Suspended solids	Ecological damage	
Oil and Grease (O/G)	Ecological damage, biological damage	
Hot/cold effluent	Ecological damage	
Detergents/solvents/ cleaners	Eutrophication, ecological damage, biological damage	
Pathogens	Human health damage	
Anoxic effluent	Ecological damage, biological damage	
Disposal of hazardous wastes on Land	Ecological damage, biological damage, Loss of resources	Applicable to most industries.
Disposal of domestic wastes on Land	Loss of resources	
Land occupied for operations	Habitat loss, ecological damage	
Energy use for operations	Loss of resources	
Volume of water used	Loss of resources	
Volume of raw material used	Loss of resources	
Soil compaction from heavy vehicles	Modification of hydrology	

5.3   THE GROWTH OF WASTE

5.3.1   The waste problem[8, 10]

Waste material can cause lots of problems:

- Disposal of waste is an increasing problem. Traditional disposal sites are rapidly becoming filled and the volume produced continues to grow.
- Due to the huge volume of waste generated and its potential environmental effects, safe and acceptable disposal of waste is becoming increasingly expensive, as evident from the data shown in Table 5.2.
- Environmental impacts of disposal can be high. This pollution can affect soil, water and air, along with the possible loss of land used for disposal.
- Hazardous and toxic materials in the waste stream, such as paints, solvents and dry-cell batteries, may represent a danger to human health and cause significant damage to the environment.

The reasons for this increase are as follows:

There has been phenomenal growth in waste generation due to the increase in population (Figure 5.2) the world over, and the lifestyle and consumption patterns of

Table 5.2 Costs of municipal solid waste disposal using different disposal options.[8]

Waste disposal options	Cost/t (pounds)
Landfill	25–30
Incineration	60–70
Anaerobic composting	40–70
Windrow composting	10–15
In vessel composting	20–30

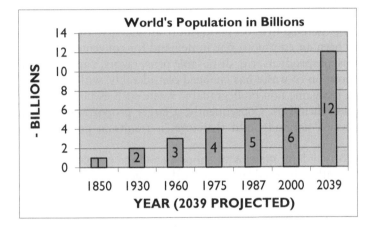

Figure 5.2 Growth in world population, including projection for year 2039 when it would be doubled from its level in the year 2000.

the world's citizens, particularly from the developed countries. It includes huge industrialization in the past years to feed masses of the growing population.

Every year, we produce hundreds of millions of tonnes of new waste, much of which is harmful to humans, plants and animals. Paper/card, metals, plastics, textiles, glass and many others are the major goods and items that constitute solid wastes. Apart from them there are rock fragments, soils, sub-grade minerals, mill tailings, scraps and many more items, which are produced at the mineral based industries.

- This increased quantity of waste is a reflection of the overall growth of a country, for example in the U.S. the consumption of minerals has increased dramatically as shown table 1.2.

5.3.2 Lifestyle

As described in sec. 1.3.2 the world's culture, customs and lifestyles must be considered when dealing with environment-related issues. Large urban areas consume vast quantities of natural resources, which may be transported from hundreds or even thousands of kilometers away. They also produce vast quantities of waste. People living in these urban areas often have little contact with the natural environment and may have little knowledge of, or concern for, their impact on it.

Table 1.3 Illustrates the lifestyle of an average middle-class American citizen. Is it not detrimental to the environment?

- Products with built-in obsolescence: Many products such as toys, electrical appliances, plastic utensils, sporting goods, daily consumables, etc. are not designed to last very long and often the simplest problems are difficult and expensive to repair. This results in a regular demand for the products since their replacement is often cheaper than repair. Many industries rely on this built-in obsolescence to maintain sales.
- There is an increase in the amount of packaging used. Most products have a huge amount of packaging to make goods more attractive to the consumer.
- There has been an enormous rise in demand for convenience products, particularly disposable products, e.g. disposable pens, razors, nappies, cameras, etc.
- The composition of waste has changed considerably, with an increasing proportion being inorganic non-biodegradable waste such as plastics, metal alloys and non-recycled chemicals. Consequently, a larger proportion of the waste stream is not recycled naturally, and therefore, it results in greater accumulation of wastes.
- Buy-now & pay-later systems in shops encourage people to buy new products more frequently. Huge advertisements that consume paints, papers, and other ingredients ultimately results in waste.
- Changes in the models of products in a very short time also encourages people to buy new products and get rid of older models, which is very common in case of electronic goods, automobiles, garments, etc.
- The development of transport refrigeration and rapid transport networks has allowed products to be sent around the world, and most times they require considerable packaging.

5.4   METHODS OF WASTE DISPOSAL (INFORMATION GATHERED FROM VARIOUS SOURCES)[1, 3–5, 10, 11]

The Waste Management Hierarchy is detailed in Figure 5.3. Wastes should be managed in order to reduce their potential to cause harm to health or the environment, and to reduce operating costs and potential future liabilities. The following waste management hierarchy should be considered while developing waste management programs for any industrial set up:

- **Source reduction:** generation of less waste through more efficient processes.
- **Re-use** – the use of materials or products that are reusable in their original form.
- **Recycling/recovery** – the conversion of wastes into usable materials, or the extraction of energy or materials from wastes.
- **Treatment** – the destruction, detoxification and/or neutralization of residues.
- **Responsible disposal** – depositing wastes using methods appropriate for a given situation.

5.4.1   Source reduction

Source reduction means generation of less waste through more efficient processes. It minimizes the amount and/or the toxicity of wastes. This can be achieved by maintaining careful control on chemical inventories, changing operations to minimize losses and leaks, modifying or replacing equipment to generate less wastes, and changing the process to reduce or eliminate the generation of wastes.

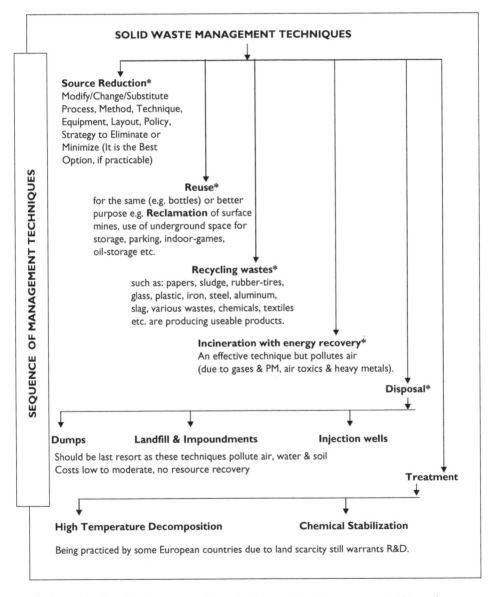

Figure 5.3 Classification – waste disposal techniques based on management hierarchy.

* Details described in following sections/paragraphs.

5.4.2   Reuse

Reuse means the use of materials or products that are reusable in their original form. If the wastes contain valuable components, those can be segregated or separated from the reminder of the waste stream and recovered for reuse.

5.4.3   Recycling of waste

This is the diverting of a waste product from disposal, processing it into a raw material and manufacturing a new product from it.

Recycling can be defined as the collection and separation of materials from the waste stream and their subsequent reuse or processing to produce a marketable product. Some recycling processes reuse materials for the same product, e.g. old bottles are reused, old aluminium cans are used to produce new aluminium cans. Other processes turn the waste into an entirely new product, e.g. old tires become material for road surfacing, and steel cans are turned into cars. Recycling can be divided into four distinct areas:

1   Pre-consumer recycling: this is an industrial recycling as a part of the production process, e.g. selling unwanted material to other industries.
2   Product reuse: This covers a variety of returnable products, such as milk bottles.
3   Primary recovery: is the collection of materials and their subsequent use to replace the use of new raw materials.
4   Secondary recovery: is the use of waste to produce heat and electricity, e.g. heat from incinerators used to produce electricity; collect methane gas from landfill to produce heat or electricity.

Recycling may involve sorting and washing different types of rubbish, using recycling centers like bottle banks, and thinking more about what we buy.

Reuse and recycle must become a way of life. The reused material would diminish the solid waste-handling problem and also it would reduce the problem of a shortage of various raw materials.

5.4.3.1   Merits

Most of the advantages of recycling are of benefit to the environment, although some economic benefits can be found.

1   Recycling some materials, such as paper and aluminium drinks cans, helps to conserve finite resources by reducing demand. This also reduces reliance on raw materials from a single country or group of countries. *Recycling one tonne of aluminium saves four tonnes of bauxite (aluminium ore) and 700 kg of petroleum coke, which has to be extracted from the ground.*[11]
2   Reduction in demand through recycling will reduce production of goods and the associated energy consumption and emissions of greenhouse gases. *Recycling plastic bottles can save 50% to 60% of the energy needed to make new bottles; similarly making new steel from scrap is estimated to save 75% energy.*[11]
3   Recycling results in a reduction of pollution from extraction industries, production processes and waste disposal. *Recycling one tonne of aluminium reduces*

aluminium fluoride emission into the air by 35 kg.[11] In fact, it has been estimated that by doubling aluminium recovery worldwide, one million tonnes of air pollutants would be eliminated annually. (Table 5.4)

4   Recycling may result in reduced waste disposal costs; this is important as the cost of landfill sites is increasing.

5   Recycling brings about job creation in industries such as: recycling, repairing and renovating.

6   Recycling can reduce the problems of litter. Discarded paper, glass, metal, foam and plastics packaging produces an unsightly litter problem, which requires expensive collection and cleaning. By encouraging the use of recyclable containers to reduce packaging, this problem would decline.

7   Participation in recycling helps to raise our awareness of the environment. This means that we have an increasing awareness and responsibility towards waste production.

8   Recycling different materials has different environmental impacts. Table 5.3 summarizes the benefits of using recycled materials in terms of savings in resource consumption and pollution.

5.4.3.2   Limitations

The main disadvantages are economics as at present in terms of profitability, recycling is not able to compete either as a method of waste disposal or as a manufacturing process.

1   Recycling products have to compete with 'virgin' raw materials, which are established in the market and may be subsidised in price. However, comparing costs is difficult because the environmental cost of not recycling is hard to estimate. If disposal of products not recycled is taken into account in the costing of new materials, the economics of recycling become more favorable.

2   The production of recycled materials is not determined by demand but by production of waste. This causes economic problems since the supply of recycled products cannot respond directly to changes in demand. This has occurred significantly in the paper recycling industry, where the amount of recycled paper far exceeded the demand, and resulted in great financial problems to the industry.[8]

3   Recycling may require sponsorship to be able to operate. For example, Pepsi-Cola International is one of several companies providing money for a "collect" scheme in Adur, West Sussex, England.[8]

Table 5.3 Environmental benefits of using recycled materials.[1, 4, 6, 8]

Material	Reduction%				
	Energy wastage	Water usage	Air pollution	Water pollution	Mining waste
Aluminium	90–97	–	95	97	–
Steel	47–74	40	85	76	97
Paper	23–74	58	74	35	–
Glass	4–32	50	20	–	80

Table 5.4 Recycling rates of selected materials in European countries.[8]

Country	Glass%	Aluminium%	Steel can%
France	34	*	24
Italy		29	**
Netherlands	40	*	45
Norway	53	60	**
Sweden	6	82	**
Switzerland	22	26	**
UK	55	10	10
Germany	20	*	15

* Negligible percentage; ** Data not available.

4   Some material recovery may not be environmentally beneficial, since energy and resource consumption for recycling may be greater than simply manufacturing new materials. It has been estimated that at the maximum possible level of paper recycling, emissions of sulfur dioxide, nitrous oxides and carbon dioxide would be greater than if some paper waste was incinerated in waste-to-energy plants.[6]
5   There may be a lack of space, both in the domestic and urban environment, for storing material to be recycled. However, this is not really as important as the preceding points. It is important to note that individuals may have to put up with some personal inconvenience in order to recycle, compared to the overall benefit to the whole community.

Table 5.4 details the recycling rates (in percentage) of selected materials in European countries.[8]

5.4.4   Treatment

The destruction, detoxification and/or neutralization of residues using techniques such as incineration.

5.4.4.1   Incineration

Non-hazardous wastes are burned at extremely high temperature producing ash, which can be used in roadbed, landfill cover or artificial reefs. Incineration is considered a clean process but has limitations as outlined below. An incinerator is a plant that burns a large volume of waste per day at very high temperatures.

5.4.4.1.1   Merits

1   Reduction in volume of waste for final disposal by up to 90%.
2   Sterilization of waste.
3   One tonne of municipal solid waste is equivalent to 9 million BTUs, 65 gallons of no.1 fuel oil, or 90,000 ft^3 of natural gas.[3] Thus, the energy so produced can be used to produce electricity that brings in money to pay for incinerator operation and reduces the burning of fossil fuels which generates greenhouse gases.

4   Occupies less space than landfill.
5   Ash residue for combustion can be recycled, for example, in building roads.

5.4.4.1.2 Limitations

1   Produces air pollution due to burning of some materials, e.g. heavy metals from burning of batteries; nitrous oxide from burning organic waste.
2   Residual ash (solid) may contain toxic materials.
3   Initial cost is high, up to $100 million.

5.4.5   Responsible disposal

Depositing wastes using a method appropriate for a given situation; landfill is an established practice for this purpose.

5.4.5.1   Landfill

Wastes that cannot be reused or recycled must be disposed of to ensure protection of public health and the environment. Landfill techniques are the main method for disposal of waste in many countries in the world. It has some advantages as well as some disadvantages:

5.4.5.1.1   Merits

1   Landfill is a convenient technique and initial cost is low, requires land for the site and arrangements for transportation of waste to the site.
2   Sites produce combustion gas, particularly methane, through the anaerobic decomposition of organic material within the waste. This can be used for heating or to generate electricity.

5.4.5.1.2   Limitations

1   Landfill sites produce hazardous fumes and particles.
2   Possibilities of contamination of ground and surface water.
3   Landfill gases, if not collected, act as significant greenhouse gases within the atmosphere.
4   What's not safe to throw out in the trash? Products such as paints, cleaners, oils, batteries, and pesticides that contain potentially hazardous ingredients require special care when we dispose of them. The dangers of such disposal methods might not be immediately obvious, but improper disposal of these wastes can pollute the environment and pose a threat to human health.

Requirement of Modern Landfill (in some of the regions, these norms have become mandatory, for example in the European Union)[2, 6]

1   Should be far from residential areas.
2   An impermeable lining of natural clay or plastic membrane must be used to prevent leaching.

3   Leaching must be controlled using a network of pipes installed at the bottom of the site and a separate treating plant to collect toxic substances.

4   Landfill gas must be controlled through a pipe network. Methane gas can be used to generate electricity or supply houses.

5   The waste must be compacted and covered every day with a top layer of soil to control airborne pollution and pest populations, e.g. flies and rodents.

6   Trash which can be recycled should be avoided for the landfill option. For example in the U.S. the trash most commonly found in municipal landfills is plain old paper – on average, it accounts for more than 40 percent of a landfill's contents. Newspapers, which can be recycled, can take up as much as 13 percent of the space in U.S. landfills and deteriorates very slowly in a landfill. Research has shown that, when excavated from a landfill, newspapers from the 1960s can be intact and readable.[11]

7   Yard trimmings and food residuals together constitute 23 percent of the U.S. municipal solid waste stream.[11] That's a lot of waste to send to landfills when it could become useful and environmentally beneficial compost instead! Composting offers the obvious benefits of resource efficiency and creates a useful product from organic waste that would otherwise have been landfilled.

Ocean disposal option should be examined carefully. In fact it requires R&D before it becomes an established practice.

5.5   LAND DEGRADATION DUE TO INDUSTRIAL OR DOMESTIC WASTE DISPOSAL

Landfills or dumps are the traditional method of waste disposal. It is the least desirable technique to dispose of the waste as it has ill effects of leaching that can pollute groundwater. It has become a major health problem. Sometimes the waste that has been filled or dumped could be hazardous and radioactive causing very severe health problems.

Soil: Rocks and soils constitute the lithosphere. Thus, soils are important part of the lithosphere with which living things interact. It provides us with nutrients that are helpful in the production of food, clothing and shelter. Soil is formed by the interaction of climate, topography, plants and animals with the bedrock. Its formation is a continuous process in nature but slow and it takes centuries to form soils. Hence, its conservation is an important aspect that should be given due importance.

5.5.1   Land degradation

The degradation of land through soil erosion, alkali salting, water logging, pollution, and reduction in organic matter content has several proximate and underlying causes. The proximate causes include loss of forest and tree cover (leading to erosion by surface water run-off and winds), excessive use of irrigation (in many cases without proper drainage, leading to leaching of sodium and potassium salts), improper use of agricultural chemicals (leading to accumulation of toxic chemicals in the soil), diversion of animal wastes for domestic fuel (leading to a reduction in soil nitrogen and organic matter), and disposal of industrial and domestic wastes on productive land.

These in turn, are driven by implicit and explicit subsidies for water, power, fertilizer and pesticides, and in absence in some countries of proper policies and regulatory systems to enhance people's incentives for afforestation and forest conservation.

It is essential that land degradation be minimized. The following initiatives could be useful:

a   Encouraging the adoption of scientifically planned and traditional sustainable land use practices through research and development. Farmers should be assisted through training.
b   Promoting reclamation of wasteland and degraded forestland through formulation and adoption of multi stockholder partnerships involving the land owning agency, local communities, and investors.
c   Preparing and implementing action plans for arresting and reversing desertification.

5.5.2   Soil degradation/Pollution

Similarly, the immediate and deeper causes of soil degradation/pollution should be analyzed. The following could be elements of an action plan:

a   Developing and implementing schemes for setting up and operating secure landfills and incinerators for non-toxic and nonhazardous wastes.
b   Developing and implementing strategies for the clean up of pre-existing toxic and hazardous waste dumps, in particular in industrial areas, and reclamation of such lands for future sustainable use.
c   Strengthening the capacities of local bodies for segregation, recycling, and reuse of municipal solid wastes, and setting up and operating sanitary landfills, in particular through competitive outsourcing of solid waste management services.
d   Promoting organic farming of traditional crop varieties through research, which is preferred by consumers.
e   Developing and implementing strategies for recycling, reuse, and final eco-friendly disposal of plastics wastes, including promotion of relevant technologies, and the use of incentive-based schemes.

5.6   WASTE GENERATION AND ITS MANAGEMENT IN MINING AND EXCAVATION (CIVIL) INDUSTRIES – SOME BASICS[10]

5.6.1   Surface excavations/mining

Civil as well as mining operations require surface excavations. Formation of slopes and benches over hilly terrain along roadsides is an important civil work. Surface excavations which are essential for surface mining to mine-out the mineral deposits that are outcropping at the surface, lie above surface datum, and extend to shallow depths.

The amount of waste rock would depend from mine to mine, and also from one type of deposit to other. The ore to overburden ratio in case of open cast mines, and stripping ratio in case of open pit mines varies from below 1:1 to as high as 1:20 or even more in some cases. This is mainly governed by the grade and value of the ore, i.e. the useful mineral contents that are ultimately recovered. It also is governed by the cost of mining and the ultimate selling price of the mineral in the market. Waste generation in surface mining amounts to a huge quantity, and therefore the basics of surface mining operations have been dealt briefly in the following paragraphs.

5.6.1.1   Open pit elements

An excavation created to strip a deposit for the purpose of mining is called a pit and since this excavation is exposed to atmosphere; the resultant structure is known as open pit. Sometimes the deposit is outcropping at the surface and is covered by the rocks surrounding it. The rock masses on its hanging and footwall sides are termed *'Hanging and Footwall Wastes (figure 5.5(a)'*. But if the same orebody is located at a certain depth, then the rock-mass covering top of the orebody is known as *'Over-Burden'*. Thus, to strip an orebody suitable for open pit mining, the removal of hanging waste, footwall waste and the over burden is mandatory. But the amount of waste rock enclosed in this envelope is a function of *'Over All Pit Slope Angle'*, which can be defined as the angle formed while joining the *Toe* of the lowest bench to the *Crest* of the top most bench of a pit with the horizontal, when benches reach to their ultimate ends. Refer to Figures 5.5 (a) to (e).

The waste that need to be stripped cannot be taken at a stretch but needs to be divided into convenient steps, which are safe and economical to be mined out; these steps so formed are called 'Benches'.

The deposits are sometimes located near the surface datum but covered by an aqueous body such as lake, tank, river, or even by seawater. Mining of such deposits is also a part of surface mining practices. These are known as aqueous extraction methods. Figure 5.4 classifies the 'Surface Mining Methods'.

5.6.1.2   Stripping ratio

In open pit mines, in order to decide the depth of the pit it is essential to carry out detailed calculations as how much waste rock will be required to remove to strip the orebody. The ratio between the amount of waste rock to be removed to mine out a unit of ore is called the stripping ratio. Since it is a ratio, it should therefore be dimensionless. But in practice different connotations are used; e.g.

> S. R = Total waste rock (tonnes)/(Total ore tonnes); within envelope considered;
>
> OR
>
> S. R = Total waste (m^3) of/Total ore (tonnes); within envelope considered;
>
> OR
>
> S. R = Total waste (m^3)/Total ore (m^3); within envelope considered;

The amount of rock needed to strip the orebody increases as the depth increases, and a situation arises when it becomes uneconomical to go beyond it. This is known as *'Break-Even Depth'*. This is also a function of pit slope angle; the lower the overall pit slope angle, the lower the depth of pit would be and vise-versa.

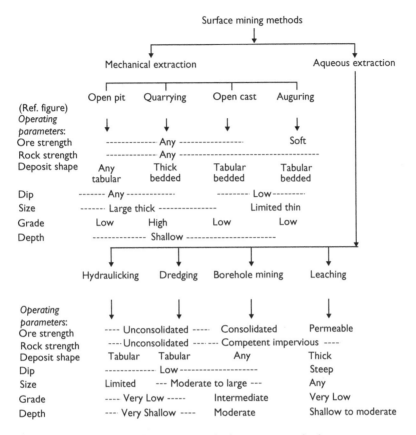

Figure 5.4 Classification – Surface mining methods.

Open cast: as the name implies, it is a surface mining system in which a deposit is opened to the atmosphere, and after removing the orebody, the over burden which was blanketing it is cast back into the worked out area. This is also known as Strip Mining. In Figure 5(c), nomenclatures used have been illustrated.

When any deposit is extending beyond the break-even depth (i.e. the depth at which cost of mining is equal to price fetched) which could be attained by any of the surface mining methods, then underground mining could be applied (Figure (5b)).

The term quarrying of course is very loosely applied to any of the surface mining operations but it should be confined to a surface mining method to mine out the dimensional stones such as slate, marble, granite etc. (Figure 5.5(e)).

Referring Figure 5.5; its details are: (a) Open pit mining – Mining the outcropping or shallow seated deposits by open pit; followed by underground mining beyond the break-even depth. (b) Illustration is typical example of copper mining in Oman. (c) An open cast mine for mining coal/ores, overburden stripping and casting and land reclamation operations shown. (d) Surface mining of soft ore (coal) using Augur (e) Quarry for dimension stones mining.

Refer to Figure 5.6, which shows various schemes of waste rock dumps. As shown in this Figure; the locations of dumps are based on the geometry and suitability of

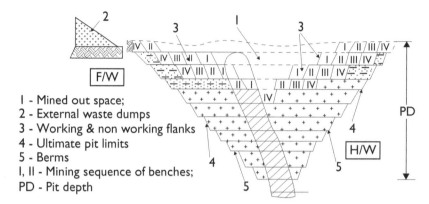

1 - Mined out space;
2 - External waste dumps
3 - Working & non working flanks
4 - Ultimate pit limits
5 - Berms
I, II - Mining sequence of benches;
PD - Pit depth

(a) For steep orebodies a suitable pit slope angle at foot wall side is also essential.

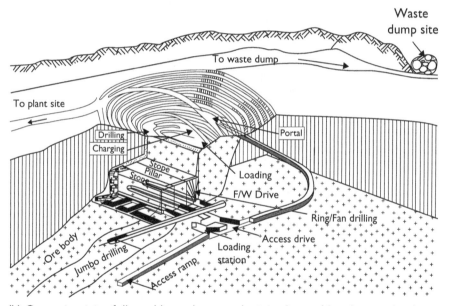

(b) Open pit mining followed by underground mining beyond break-even depth.

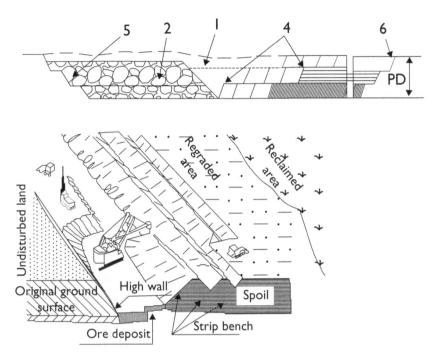

(c) Development and mining by opencast. 1 – Mined-out space; 2 – Internal and external waste dumps;
4–5 – Working and non-working flanks; 6 – Ultimate pit limits; PD – Pit depth.

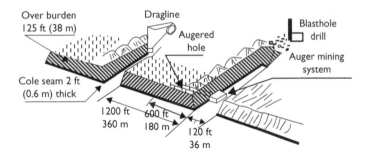

(d) Surface mining of soft ore (coal) using Augur.

Quarry – for dimension stones' mining

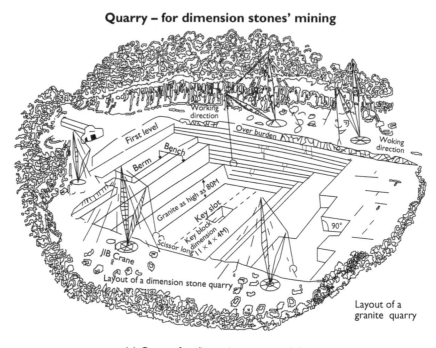

(e) Quarry for dimension stones mining.

Figure 5.5 Conceptual illustrations (a) to (e): Surface and underground mining methods/systems.[10]

the available land in terms of their techno-economical considerations. The figure also illustrates the direction of advance of these dumps. The term quarry used here represents any of the surface mines.

5.6.2 Dumping site

1   Separate dumping site for:

  a   Waste dump
  b   Sub-grade mineral
  c   Mineral of economic interest later on
  d   Ore (temporary).

2   Site selection: Care should be taken in the following aspects:

  a   Favorable surface topography
  b   Near to the working pit
  c   Devoid of any mineralization of economic interest
  d   Devoid of any vegetation, plantation, forest area, agriculture land etc.
  e   Not an area of public utility and infrastructures
  f   Not the source of water or obstruction to a source of water. Also not a water body.

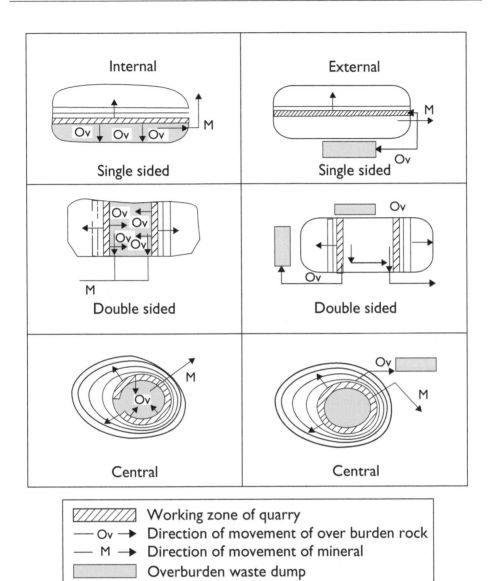

Figure 5.6 Various schemes of waste rock dumps.[10]

3   Procedure of dumping:

   a   Keeping the height of dump to 2.5 m to 3 m, and angle of repose to 35–40°
   b   Leveling the dumps so formed using a bulldozer
   c   Dumping over the leveled heap and again leveling till the area permits
   d   Keeping a record of dumping area-wise, date-wise and mineral-wise
   e   Putting fencing around the dump and also keeping big boulders near the fencing to prevent the passage of silt during the rainy season
   f   Stabilization of dumps by growing vegetation. This will help in improving the environment in the area.

Types of dumping sites

1   Constructed on a flat terrain
    a   Heaped dump constructed in successive layers
    b   Heaped dump constructed in a single layer.

The manner in which, the dump-yard could be developed is shown in Figure 5.6.

2   Using a Valley:

    a   Valley filled by terracing
    b   Head-of-shallow fill – right up to full depth
    c   Cross-valley fill – in an extensive valley area.

3   Hill side dumps:
    a   Side hill fill – from a side of the hill starting at the predetermined elevation
    b   Both sides of a ridge.

5.7   WASTE MANAGEMENT IN PETROLEUM INDUSTRY – A CASE STUDY[1, 2, 5, 9]

5.7.1   Waste management

This is the reduction of hazardous wastes generated or subsequently treated, stored or disposed of. It includes any source reduction or recycling activities undertaken by a generator that results in either:

1   The reduction of total volume or quantity of hazardous waste.
2   The reduction of toxicity of the hazardous waste or both, so long as such reduction is consistent with the goal of minimizing present and future threats to human health and the environment.

5.7.2   Waste in the petroleum industry

The following paragraphs describe practices that are followed at some of the leading petroleum companies, operating in the Gulf region. In Figures 5.7 and 5.9 the photographs show an oil field.

The greatest environmental impact of petroleum activities arises from the releases of some wastes into the environment in a concentration that is not naturally found. These wastes could be categorized as:[5]

• Gaseous Emissions (refer table 5.1)
• Liquid & semi – liquid and
• Solid wastes.

While some of these wastes can have a significant adverse effect on the environment, some have little impact. In virtually all cases, the adverse impact can be

Figure 5.7 Oil industry and the environment (Photograph: A working site in an oil field in the Gulf region.).

minimized or eliminated through the implementation of a proper waste management plan. The following are the steps:

5.7.2.1 Audit

Types of wastes generated at a particular site: One of the first steps in developing this plan is to conduct environmental audits to identify all of the waste streams at a particular site and to determine whether those waste streams are being managed in compliance with all applicable regulations.

Once an audit has been conducted, a written waste management plan for managing each waste stream should be developed. This plan identifies how each waste stream is to be handled, stored, transported, treated, reused, recycled and disposed of. The plan also gives the communities flexibility to design waste systems that meet local needs and realities and at the same time protect public health and the environment. The effective waste management plan is based on a hierarchy of preferred steps (as described in preceding sections and includes *Source reduction, Reuse, Recycle, Incineration and Landfill.* Refer to Figure 5.3).

5.7.2.2 Waste management plan

Waste management plans identify exactly how each waste stream should be managed. They ensure that appropriate engineering controls, proper waste management options, adequate record keeping and reporting systems, and ongoing employee training are in place.

The information obtained from environmental audits can be used in developing a waste management plan.

One of the first steps in developing a waste management plan is to identify the region and scope to be covered. All materials generated within the region must be identified, quantified, and characterized. These data must include chemical and toxicological characteristics, health impacts, proneness to fires and explosions, and information on reactivity.

They should also include first aid procedures to be used in the event of human exposure to the material. Material Safety Data Sheet (MSDS) provides much of this information and can be obtained from chemical suppliers.

A critical factor that must also be considered in developing waste management plans is the regulatory status of each material at a site.

A critical aspect of good waste management plans is to develop and maintain good bookkeeping practices. This bookkeeping must include a waste tracking program, which identifies where the waste was generated, the date the waste was generated, the type of waste and its volume, any transportation of the waste, the disposal method and location, and the contractor employed. Figure 5.8 is the pro-forma used at an oil field in the Gulf region describing details of a waste consignment ready for dispatch. A waste management plan must also identify which personnel are responsible for the proper management of all wastes produced at the targeted facilities.

5.7.2.3 Waste consignment note

The first and most important action in the waste management hierarchy is to reduce and/or recycle the volume of wastes generated. The next action is to reuse the wastes or materials in the wastes. Only after those actions have been completed, should the remaining wastes be treated and disposed of. By following this hierarchy, both the volume of waste to be disposed and the ultimate disposal cost will be minimized.

A number of examples of waste management activities for drilling and production operations are discussed according to the hierarchy of waste management principles. These activities include ways to minimize the volume and/or the toxicity of wastes generated and ways to reuse or recycle, or treat and disposal options.

5.7.3 Waste minimization

The most effective way to reduce the environmental impact associated with exploration and production of oil and gas is to minimize the total volume and/or the toxic fraction of wastes generated. The primary waste minimization activities are making changes in how chemical inventories are managed, how operations are conducted, which materials and chemicals are used, and how equipment is operated.

The merits of waste minimization include:

- Avoidance of waste transportation and disposal costs
- Elimination of expensive pollution control equipment
- Improvement of product quality and less administrative record keeping
- Lowering on-site handling costs and a smaller waste storage area
- Reducing wastes and tax obligations

Section A: WASTE DETAILS

1 Please tick (√) box below to indicate the type of waste you are transferring
NON-HAZARDOUS WASTE TYPE

☐ Domestic Waste (kitchen refuse)   ☐ Domestic waste (tree cuttings)   ☐ Office waste   ☐ Scrap Metal

☐ Non-Hazardous waste chemicals   ☐ Empty Drums (non hazardous)   ☐ Electrical Cable

☐ Waste wood materials   ☐ Plastic Drums (non-hazardous)   ☐ Construction material

If other please describe waste accurately here

_____

_____

HAZARDOUS WASTE TYPE

☐ Oily sand/soil   ☐ Oily sludge   ☐ Waste Lubricants   ☐ Pigging sludge

☐ Hazardous waste chemicals   ☐ Sewage sludge   ☐ Batteries   ☐ Clinical waste

☐ Hazardous lab. Chemicals   ☐ Empty Drums (hazardous)   ☐ NORM   ☐ OBM & or OBM Cuttings

☐ Tyres (used)   ☐ Other

If other please describe waste accurately here

_____

_____

2 Quantity:

3 Please give any other additional information including details of any problems your waste may present, that will affect containment, transport, treatment or disposal of the waste by any subsequent holder, e.g. type of premises waste comes from, full analysis, process that produced the waste:

Section B1: WASTE ORGINATOR (Asset)

1 pl. give name of the site:

2 Location of   Destination:

Return white copy to Originator by.   a) Fax/Fax No:_____   b) Mail: _____   c) Driver:_____

d) monthly waste returns:_____

3 Signature (waste originator):_____   Date: _____

4 Full Name (Please print):_____

Section C: WASTE TRANSPORTER

1 Company Name: _____

2 Signature of Driver: _____

3 Full name of Driver (Please print): _____

Section D: WASTE DISPOSAL FACILITY

1 Waste disposal Location/Site:_____

2 Name of Waste Disposal Site Operator:_____

3 Date & time waste received:_____

4 Signature of Waste Disposal Operator:_____

Distribution:
Yellow Copy: **Originator**   Blue Copy: **Disposal Site**   White Copy: **Waste Contractor Return to Originator**

Figure 5.8 Waste Consignment Note (Front Page) for Naturally Occurring Radioactive Materials (NORM). Water based drilling mud (WBM); Water based drilling mud cuttings (WBMC); Oil based mud (OBM).[5, 9]

- Improvement of public image
- Lowering potential environmental impacts, and reducing future liabilities.

Nevertheless, many opportunities are available for minimizing wastes, which have been described below.

5.7.4   Inventory management

One aspect of waste minimization is to carefully monitor inventories of all materials at a site. Accurate written records of all raw and processed materials and their volumes should be kept for every stage of handling and production. Better management of materials inventories provides significant environmental benefits. It allows a material balance to be conducted on all materials at all stages of usage. A detailed material balance can help identify where unwanted losses and waste may be occurring.

5.7.5   Improved operation

Another important method for minimizing the amount of potentially toxic wastes generated is to change the operating procedures at the various sites. All operations should be carefully planned in advance to minimize the use of materials. Materials storage, handling, and transportation procedures should be reviewed to minimize losses. A very important step in improving operations is to keep different types of wastes segregated. Waste streams should never be mixed, to allow the best disposal options to be selected for each waste. A number of operational changes during production can also be implemented to minimize the total of waste generated. Routine inspection and/or pressure testing of all tanks, vessels, gathering lines, and flow lines should be scheduled.

Any company should specify activities that are prohibited while on site. Such activities can include unnecessary rig washing, painting of the contractor's equipment or changing lube oil during downtime. This will minimize the probability that excess water, painting wastes, or used oil gets dumped into reserves pits. An environmental activity review should be conducted with all the concerned crews just prior to the start of activities. This review should include waste handling and minimization procedures.

5.7.6   Material substitution

Another important method for minimizing the amount of potentially toxic wastes generated is to use less toxic materials for the various operational processes. A variety of opportunities are available during production operations to substitute less toxic materials for more toxic, traditional materials. For example, less toxic detergents can be used to wash rigs. A better solution, however, is for contractors to install closed-loop wash water systems for washing rigs at their own sites rather than at the wellhead. Whenever possible; unleaded water-based paints and non-solvent paint removers, cleaners, and degreasers can be used. Substitutes can be used for Halon gases in fire suppressants.[1]

Figure 5.9 The waste materials on the rig floor. (Photograph: A working site in an oil field in the Gulf region.)

One of the best opportunities for materials substitution is in wells where oil-based mud is needed. Two alternatives to the use of diesel oil as a base fluid are being studied: using a less toxic oil-based mud and using water-based mud with an improved additives package. These alternative mud systems, however, are providing significantly improved environmental protection.[1]

5.7.7 Equipment modifications

Another important method for minimizing the volume of potentially toxic wastes generated is to ensure that all equipment is properly operated and maintained. Inefficient equipment should be replaced with newer, more efficient equipment.

One of the first steps to be taken is to eliminate all leaks and spills from equipment. Fugitive emissions from leaking valves, flanges, and such fittings can be minimized by replacing leaking equipment.

If the volume of waste generated cannot be sufficiently reduced with the existing equipment, newer equipment should be installed. Important environmental features of newer equipment should ensure that they are easy to monitor and clean up, as well as facilitating waste recovery and recycling. Now equipment with automated process controls can be installed to ensure optimal operations.

5.7.8 Waste reuse

Many materials in petroleum waste streams can be used more than once. If materials are intended for future use they are not wastes. The following materials have a potential for reuse: acids, amines, antifreeze, batteries, catalysts, caustics, coolants, gases, glycol, oils, plastics, solvents, water and wax. Water has a considerable potential for reuse. For example, water from mud can be cleaned and used as rig wastewater. Rig wastewater can be collected and reused. Produced water, after treatment, can be re-injected for pressure maintenance during water floods or for steam injection in heavy oil recovery. Installing equipment that allows reuse can facilitate material reuse. For example, closed loop systems can be installed so that solvents and other materials can be collected and reused in plant processes. Many wastes could be used at other sites or be returned to the vendor. For example, some used chemical containers can be returned to the vendor for refilling.[1]

In some cases, only part of a particular waste stream contains valuable materials that can be reused. It may be possible to recover or reclaim the valuable materials, reducing the net volume of waste. For example, crude oil tank bottoms, oil sludge, and emulsions can be treated to recover their hydrocarbons. Oily materials can also be burned for their energy content. Gravel and cuttings can be washed and used in construction of roads and other sites.

5.7.9 Waste recycling

True recycling means diverting a waste product from disposal, processing it into a raw material, manufacturing a new product from it and having that product purchased for use. This is known as closing the loop. Recycling has grown dramatically, especially in recent years. Recycling starts with waste generators, a separation step that typically occurs in the effort to produce relatively clean consistent bundles of material for manufacturing.

5.7.10 Waste treatment

Most wastes require some type of treatment before they can be disposed of. Waste treatment may include reducing the waste's total volume, eliminating its toxicity, and/or altering its ability to migrate away from its disposal site. A variety of treatment methods are available for different types of wastes. These treatment methods selected, however, must comply with all regulations, regardless of their cost.

One of the most important steps in waste treatment is to segregate or separate the wastes into their constituents, e.g., solid, aqueous, and hydrocarbon wastes. This isolates the most toxic component of the waste stream in a smaller volume and allows the less toxic components to be disposed of in less costly ways. Primary separation occurs with properly selected and operated equipment, e.g., shale shakers and separation tanks. Using hydro-cyclone filter presses, gas flotation systems, or decanting centrifuges can improve separation.

A number of methods are available for treating hydrocarbon-contaminated solids like drill cuttings, produced solids, or soil. These treatment methods include distillation, solvent extraction and incineration.

5.7.11 Incineration

Waste incineration offers (as described in preceding sections) a way to reduce the volume of waste 70–90%. The waste is burned at extremely high temperatures, producing ash that is used in roadbed, landfill cover or artificial reefs. Incineration is considered a very clean process, where as any incineration has a special system for cleaning the flue gases from dangerous dust, and absorbing the harmful gases by chemical treatment.

5.7.12 Waste disposal

A described in the preceding section; that numbers of disposal methods are available for petroleum industry wastes. The method used depends on the type, composition, and regulatory status of the waste. Disposal can occur either on or off-site.

- The primary disposal method for aqueous wastes is to inject them into abandoned wells. If the quality of wastewater meets the regulatory limits, permits to discharge it into surface waters may be obtained in some areas.
- The primary disposal methods for solid wastes are to bury (landfill) them or to spread them over the land surface. Underground injection of slurries has also been used for solids disposal in some area.
- **Landfill:** Landfill is the most popular method of disposal due to its relatively low cost. There are many cases where waste materials cannot be disposed of on the site where they are generated. It is then necessary to use the services of specialist contractors for the transport and disposal of wastes.
- When external transport or disposal contractors are employed, we must be satisfied that they can deal with waste materials safely, effectively and legally, and confirm that waste consignments reach the specified final disposal site and are disposed of in the agreed environmentally safety manner.

Environmental control is one of the most important aspects in any industry right from its exploration (pertaining to petroleum and mining), development/construction, production and liquidation phases to post closure/abandoned phase. Waste management techniques are developed and evaluated all over the world to save the environment in and around any industry area/setup. It can be concluded from the discussion above, that the different techniques of waste management should be studied and the best one should be selected for each particular waste solid, liquid or gas.

Environmental control is *regulated by rules and standards in all countries,* and for this reason at least, all companies should prepare a waste management plan. This would help in reducing the air and water pollution and land degradation to maintain the desired balance in ecology.

Policies and programs have been focused on waste management rather than preventing its generation at the first place, and therefore there is need of such a culture of prevention through education, training and awakening amongst those concerned.

Things to remember

What is the most environmentally friendly way to get rid of garbage? Source reduction is a basic solution to the garbage glut: using less material means less waste at the end. Because source reduction actually prevents the generation of waste in the first place, it comes before other options that deal with trash after it already exists. Recycling (or re-using) and composting are the next best options because they reduce the amount of waste going to landfills and also let materials be re-claimed and used again when possible. Landfilling is the last option, when waste and materials are simply discarded.

5.8 TIPS FOR REDUCING SOLID WASTE (AS ADVISED BY EPA)[11]

Reduce

- Reduce the amount of unnecessary packaging.
- Adopt practices that reduce waste toxicity.

Reuse

- Consider reusable products.
- Maintain and repair durable products.
- Reuse bags, containers, and other items.
- Borrow, rent, or share items used infrequently.
- Sell or donate goods instead of throwing them out.

Recycle

- Choose recyclable products and containers and recycle them.
- Select products made from recycled materials.
- Compost yard trimmings and some food scraps.

Respond

- Educate others on source reduction and recycling practices.
- Be creative – Find new ways to reduce waste quantity and toxicity.

5.9 A CLASSIC EXAMPLE FROM LORD BUDDHA'S DISCIPLE AS HOW TO REUSE!!!!![7]

Point to Ponder

Buddha, one day, was on deep thought about the worldly activities and the ways of instilling goodness in human. The following is the text of conversation between him and his disciple.

One of his disciples approached him and said humbly "Oh my teacher! While you are so much concerned about the world and others, why don't you look into the welfare and needs of your own disciples also?"

Buddha: "All right. Tell me how I can help you"
Disciple: Master! My attire is worn out and is beyond decency to wear. Can I get a new one, please?"
Buddha found the robe indeed was in a bad condition, which needed replacement. He asked the storekeeper to give the disciple a new robe to wear. The disciple thanked Buddha and retired to his room. Though he met his disciple's requirement, Buddha was not all that contented with his decision. He realized he missed out some point. After a while,

he realized what he should have asked the disciple. He went to his disciple's place and asked him "Is your new attire comfortable? Do you need anything more?"

Disciple: "Thank you my Master. The attire is indeed very comfortable. I need nothing more."

Buddha: "Having got the new one, what did you do with your old attire?"

Disciple: "I am using it as my bed spread."

Buddha: "Then, I hope you have disposed of your old bed spread."

Disciple: "No... no... Master. I am using my old bedspread as my window curtain."

Buddha: "What about your old curtain?"

Disciple: "Being used to handle hot utensils in the kitchen."

Buddha: "Oh...I see... Can you tell me, what did they do with the old cloths they used in the kitchen?"

Disciple: "They are being used to wash the floor."

Buddha: "Then, the old rags being used to wash the floor...???"

Disciple: "Master, since they were torn off so much, we could not find any better use, but to use them as a wick in the oil lamp, which is right now lit in your study room...

Buddha smiled in contentment and left for his room.

If not to this degree of utilization, can we at least attempt to find the best use of all our resources – at home, office, or any industrial setup?? It becomes imperative in a critical time of 'Recession' if it happens to any nation.

Treat the earth well; our parents did not give it to us, our children loaned it to us. We did not inherit the earth from our ancestors; we borrowed it from our children.

5.10   CONCLUDING REMARKS

Let us fight 'Wastage' like enemy at any organization you belong to including your own home. It includes resources, which could be machines, equipment, materials and energy. It is equally applicable for the natural resources that include flora, fauna, sun (effective use of light and heat), water (without wasting even a drop), air (least pollution) and minerals (maximum recovery). And last but not the least man himself (effective utilization of time, talent and capabilities).

The solution lies in a proactive outlook, bringing precision, going for renovation wherever feasible, prompt attention to abnormalities, up-to-date knowledge of the subject-matter, creative attitude and self-discipline.

QUESTIONS

1   'We did not inherit the earth from our ancestors; we borrowed it from our children.' What underlying message does this convey to us? And how it could be achieved?

2   Classify waste disposal techniques based on its management hierarchy.

3   Define: Waste, hazardous and non-hazardous wastes.

4   How far is incineration useful to deal with solid waste?

5   How could we succeed in maintaining the desired balance in ecology?
6   Is bookkeeping practice a critical aspect of a good waste management plan? Describe how it could be achieved.
7   List the causes of Land degradation and Soil degradation/pollution due to industrialization.
8   List the merits of waste minimization mechanisms.
9   List the reasons for increase of wastes the world over.
10 List resources and how you can minimize their wastage.
11 List tips as advised by EPA to reduce solid waste.
12 Prepare a note on: Waste Generation & its Management in Mining & Excavation (Civil) Industries. Define the term: Stripping Ratio.
13 Prepare a classification of wastes.
14 Recycling is the best way of treating waste? Why?
15 Suggest the most effective way to reduce the environmental impact associated with an Oil and Gas Field.
16 What are the environmental hazards due to solid, aqueous and gaseous wastes? And possible impacts due to them?
17 What is 'Landfill' technique? List the requirements of a 'Modern Landfill'.
18 What is compost? How it is made and what is its use?
19 What is the most environmentally friendly way to get rid of garbage?
20 What does it indicate when waste accumulation is beyond a certain limit?

REFERENCES

1   Hassan, M. & Kamal, K. (2000) *Waste Management in View of Environment*, SPE 61473.
2   John, C. (1996) *Environmental Control in Petroleum Engineering*.
3   Kupchella, C.E. & Hyland, M.C. (1993): *Environmental Science*. Prentice-Hall International Inc. pp.559.
4   Mahfoudha Al-Belushi. (2000) Term project – Literature survey on Industrial waste. Sultan Qaboos University, Oman.
5   Petroleum Development of Oman – HSE management, course material, reports and documents and interaction with those concerned during period: 1995–2004.
6   Shaaban, S. (2000) *Environmental Control*. SPE 61474.
7   Sharing Best Practices (2009) The Manager Mentor, Aditya Birla Group, India.
8   Stephen, R Smith. (2002) Municipal solid waste management techniques: the UK perspective. *Oman International Conference on Waste Management, Muscat, Oman; 16–18 Dec.*
9   Tatiya, R.R (2000–2004) *Course material for Petroleum and Natural Resources Engg; Chemical Engg. and Petroleum and Mineral Resources Engg.* Sultan Qaboos University, Oman.
10 Tatiya, R.R.(2005) *Surface and underground excavations – methods, techniques and equipment.* A.A. Balkema, & Taylor and Francis, Netherlands. pp. 1–17.
11 www.epa.gov/epawaste.wycd. Also EPA Q & A. [Accessed Aug. 2009]

Chapter 6

Industrial hazards

'We must be compatible with the systems of nature; otherwise, the systems of nature will be unable to support us. We must learn ourselves as living within the systems of nature'.[8]

Keywords: Hazards, Fires, Fire-fighting Equipment, Fire-fighting Department, Confined & Unconfined Explosions, Dow Index, Toxicity, Hazardous Materials, Hazard Analysis Methods, Inherent Safer Design and Strategies, Mine Hazards, Breathing Apparatus, Machinery hazards, Disaster, Health Risk, Fuels, Oxidizer, Ignition Source, Flash Point, Fire Point, Auto Ignition, Flammable Liquids, Flammable Gases, Self-Oxidation.

6.1 INDUSTRIAL HAZARDS[11]

Hazard means danger, risk; Hazardous means dangerous, risky. Hazard is a condition with the potential of causing harm or damage to resources of any kind: man, property, air, water, land, flora and/or fauna. Natural products are not hazardous under a given set of conditions but when any material is synthesized to produce a chemical substance having any one of these properties:

Flammable, explosive, corrosive, toxic, or if it readily decomposes to oxygen at elevated temperature; it is considered as hazardous. Some specific examples of hazardous materials include:[11]

- Chlorine is toxic when inhaled
- Sulfuric acid is extremely corrosive to (eating into or gradually wearing away) skin
- Ethylene is flammable
- Steam confined in a drum at 600 psig contains a significant amount of potential energy
- Acrylic (synthetic) acid can polymerize (process of joining two or more like molecules), releasing large amount of heat.

These hazards cannot be changed and they are the basic properties of the materials and the conditions of usage. The inherently safe approach (described in the following sections) is to reduce the hazard by reducing the quantity of the hazardous material or energy, or by completely eliminating the hazardous agent.

A hazard in the process industry is the escape of process material, which may be inherently dangerous (toxic or flammable) and/or present a high pressure and high or low temperature. Large and sudden escape may cause explosion, toxic clouds and pollution whose effects extend beyond the premises of a factory or an industrial establishment.

Examples:[7, 11]

- In 1984 the explosion of liquefied petroleum gas in Mexico City causing 650 deaths.[11]
- The release of toxic methyl isocyanate gas in Bhopal, India causing 2000 deaths and 200,000 injuries.[7]
- Methane – coal dust explosions in many underground coalmines all over world.

Hazards differ from industry to industry and even from process to process within the same industry. Their magnitude depends mainly on the materials involved, their quantity and their quality.

6.1.1 List of hazards[5–8, 11]

- Acute toxicity
- Chronic toxicity
- Flammability
- Reactivity
- Instability
- Extreme conditions (temperature or pressure)
- Environmental hazards including

  - Air pollution
  - Water pollution
  - Ground water contamination
  - Waste disposal.

- The hazards associated with normal plant operations such as normal stack emissions and fugitive emissions, as well as **specific incidents such as spills, leaks, fires and explosions, should be considered.** It is also necessary to consider business and economic factors in making the process selection, particularly 'the demolition and future clean-up and disposal costs' while considering the total economics of the project that include:

  - Capital investment
  - Product quality
  - Total manufacturing cost
  - Operability of the plant
  - Demolition and future clean-up and disposal costs.

The preceding paragraphs and Figure 1.6 classify hazards, which may lead to disasters (defined below).

6.1.2 Disaster

Disaster: This means a major accident or natural event or natural calamity involving loss of lives (human and other creatures), property and resources. It could be a natural or manmade disaster. The definition differs from country to country.

6.1.3 Health risk

A hazard ultimately causes risks to health as shown in Figure 10.9.

6.2 FIRES

Fire or Combustion is a chemical reaction in which a substance combines with an oxidant and releases energy. Part of the energy released is used to sustain the reaction.[5] The Fire Triangle concept as shown in Figure 6.1, is used to understand the mechanism of fires.

6.2.1 The fire triangle concept[5, 14]

All materials (solid, liquid or gas), which will burn are known as combustibles. There must be a thorough balance amongst three elements: fuels, oxidizer and ignition source, like the three sides of a triangle, for a fire or explosion to onset (Figure 6.1), while the absence or withdrawal of any of them would not allow the fire or explosion to set or continue.

- FUELS—Almost all organic chemicals and mixtures in the solid, liquid or gaseous state are flammable, and as such they are potential fuels.
  - Solids: Wood, cotton fabrics, papers, coal, fibers.
  - Liquids: Oil, fats, petrol, gasoline, acetone, ether, pentane.[5]
  - Gases: LPG, acetylene, propane, carbon monoxide, hydrogen.

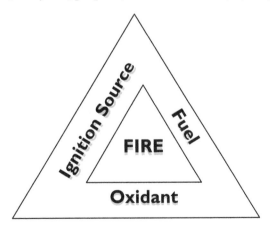

Figure 6.1 'Fire' Triangle and the three components that constitutes its sides.

- Most of the inorganic substances are not combustibles and hence they are not potential fuels except some of them such as: hydrogen, sulfur, phosphorus, magnesium, titanium, and aluminum.

- OXIDISER: Oxidant can be solid, liquid or gaseous.

  - Gases: air, which contains 21% oxygen (Oxygen-rich atmosphere). Other examples include[11] Chlorine, Nitrogen chlorine trifluoride, Nitrogen oxides.
  - Liquids: hydrogen peroxide, nitric acid, perchloric acid, chromic acid.
  - Solids: metal peroxides, ammonium nitrate, permanganates.

- AN IGNITION SOURCE: Spark, flame, jolt, friction, shock, incandescent sparks, or heat from a light bulb. Electrical discharge from making or breaking of an electric circuit or from static electricity also constitutes a possible means of ignition.[14] In addition, an open fire, hot surface, non-intrinsically safe electrical equipment, mechanically-generated sparks, static electricity and smoking materials could be added to this list.[11]

6.2.2 Concepts – mechanism of fire[5, 22]

The fire triangle, Figure 6.1, contains no information on the speed of reaction, or the amount of energy given off, or it's upper and lower flammable limits. In addition, to understand the mechanism of fire, the following concepts should be born in mind:

- The '*flash point*' of a liquid is the lowest temperature at which it gives off sufficient vapor to form an ignitable mixture with the air, which is capable of ignition under prescribed test conditions.
- The '*fire point*' is the lowest temperature at which vapor above a liquid will continue to burn once ignited, the fire point temperature is higher than flash point.
- Ignition temperature is the lowest temperature at which the substance will ignite spontaneously. It is applicable to both liquids as well as gases.
- *Auto ignition temperature* – a fixed temperature above which adequate energy is available in the environment to provide an ignition source.
- Flammable liquids must vaporize before they ignite.
- Flammable gases (or vapors) and air ignite readily only when their composition lies within a certain range – the flammability limits, known as '*lower-flammable (explosive) limit*' and '*upper-flammable (explosive) limit*'. 'Lower-flammable (explosive) limit' is defined as the smallest concentration of flammable gas or vapor in air, which is capable of ignition and subsequent flame propagation under prescribed test conditions. The 'upper-flammable (explosive) limit' is defined as the greatest concentration of flammable gas or vapor in air, which is capable of ignition and subsequent flame propagation under prescribed test conditions.
- The combustion process is described by P. Waterhouse (1999).[14] Consider first a pool of liquid. Immediately above the surface level of liquid there is a layer of vapor; its concentration depends upon the vapor pressure of the liquid at the temperature in question.

  - If this is ignited and it continues to burn, the heat from the flame causes more of the liquid to evaporate and the burning is sustained. Close examination would indicate that there is a thin gap between the liquid surface and the flame. In this gap the vapor concentration is within the flammable limits.

- In the solids there is no vapor above the solid to ignite, so initially the heat from a flame raises the surface temperature of the solid, which causes more decomposition and the burning is sustained. Again, as in case of liquid, there is a thin gap between the flame and the surface of the solid, and in this gap the vapor concentration is within flammable limits. The hot products of combustion rise and it would increase the temperature of every thing with which they come in contact.

- Well known parameters: *Conduction, Convection and Radiation* are responsible for the *spread of fire*. In addition, fire can spread by the collapse of hot or burning materials. They can spread by the transport of burning debris in the updraught of the fire, or by the flow of burning liquids down a slope.[14]
- Flammable gases and vapors of flammable liquids are in many ways more dangerous than the liquids themselves. This is because they are invisible, cannot be easily contained, and may travel a considerable distance before they reach an ignition source – either a leak or exposed surface of a flammable liquid. *Flammable liquids, therefore, should not be kept or handled in open containers.*
- Underground ore oxidation.

Ore containing more than 40% sulfur in the form of pyrite or pyrhotite, etc. are liable to fires, and thus are considered as most hazardous. Fire may become intensive if ore fines and dust get mixed with some timber. Even in the absence of timber the pyrite ores after their fragmentation and prolonged storage, may cause intensive heating due to their *self-oxidation,* as described in 6.1.

6.2.3   Ignition sources of major fires[1, 5, 14]

Table 6.1 Describes major ignition sources for fires based on experience gained in this aspect.

Table 6.1 Major ignition sources of fires.[5]

Sources	%*	Remark/examples/brief description
Electrical	23	This is arcing when a circuit is made or broken, or heat is generated by the passage of electricity through a conductor, that creates an ignition source[14]
Smoking	18	Based on statistics smoking is the major cause of fires. Self-discipline is the remedy to prevent fires due to this cause.
Friction (bearings or broken parts)	10	Heat generated by friction can be transmitted both by conduction and convection to combustible material nearby. Regular maintenance and correct lubrication could reduce the likelihood of overheating due to friction[14]
Overheated materials	8	Where heat is an integral part of the process; suitable insulation and keeping flammable materials away from high temperature areas should be an established practice.

(Continued)

Table 6.1 (Continued)

Sources	%*	Remark/examples/brief description
Hot surfaces	7	Hot pipes and ducts should be lagged where they pass close to combustible materials.
Burner flames	7	They should be fenced to prevent combustible materials coming too close to them.[14]
Combustion sparks	5	All IC Engines have hot surfaces and generate hot exhaust gases. Wherever there is a possibility of flammable gases or vapors the use of petrol driven engines should be banned.[14]
Spontaneous ignition/self oxidation/auto	4	*Self-Oxidation:* In mines apart from sulfide ores, any other ores or coal (particularly the brown coal called lignite) after its fragmentation, if stored for a prolonged period, may get oxidized resulting in heat generation and rise in temperature of the surroundings, this phenomenon is known as '***spontaneous heating or self-oxidation***'. It can cause problems during ore beneficiation/concentration. Another examples where this could happen are: oil-soaked rags or overalls, oil-soaked lagging, hay and straw,[14] insulation on steam pipe saturated with certain polymers and filter aid saturated with certain polymers.[5]
Cutting and welding	4	Where flame cutting, welding or grinding are carried out, the area should be well defined, and a non-inflammable curtain should suitably barricade it.
Exposures – fires jumping into new areas	3	Fires can be spread by the collapse of hot or burning materials, which can be spread by the updraft of the fire, or by the flow of burning liquids down a slope.[14]
Incendiaries (fires maliciously set)	3	This could be intentional.
Mechanical sparks	2	Other than what have been mentioned above.
Molten substances (hot spills)	2	Where these substances are produced.
Static sparks (electricity) and lighting	2	Best method to prevent a charge building up in the first instance is by way of an effective earthing or by the use of static eliminators.[14]
Chemical action	1	Particularly exothermic.
Miscellaneous	1	Not covered above.

Source: Accident Prevention Manual For Industrial Operations; Chicago National safety Council 1974.[1]
* % (percentages) shown are indicative representing relative reasons for the fires; they could vary from place to place, or from one situation to another.

6.2.4 Classification of fires

Table 6.2 describes different classes of fires and type of extinguishers to deal them.

6.2.5 Fire protection[2, 9, 17, 22, 23]

Adhering to the guidelines given below could minimize the incidence of fire.

*   Fire detection and protection equipment should be provided in accordance with laid out norms and prevalent regulations. Fire-blankets should be made available

Table 6.2 Fire classification and type of extinguishers.[2, 14, 23]

Fire type (Class)	Description (Fire involving)	Extinuisher type	
		Type	Color
A	Ordinary combustibles, organic solid such as wood, paper, plastic, natural fiber (cloth), rubber etc. (Water is the best means to extinguish)	Water Foam CO_2 Dry Powder Vaporizing Liquid	Red Cream Black Blue Green
B	Flammable or combustible liquid or liquefiable gases such as petrol, oils, fats etc. (Immiscible with water)	Foam CO_2 Dry Powder Vaporizing Liquid	Cream Black Blue Green
C	Gases and liquefied gases such as methane, propane, butane etc.	Shut off gas supply. Then use dry powders.	
D	Metals such as Mn, Na, K, Al, titanium etc.	* Cover them using 'Dry powders' such as talc, soda ash, dry sand etc. as they are the best means of control.	
Electrical Fires	Due to Short circuits, leakage etc. from energized electrical equipment, electrical motors, power tools, breaker boxes, computers etc.	First Switch off	Then use proper extinguisher.

* No need of fire extinguisher.

in kitchens. Such fire-blankets should be manufactured from woven glass fiber, or equivalent.

- A facility for general audible alarm should be provided in all areas of risk.
- Plans and procedures should be put in place for

  - Fire prevention
  - Building evacuation and muster points
  - Fire fighting
  - Maintenance (including periodic testing) of fire protection equipment.

- All fire escape routes and exit doors, alarm points and fire fighting equipment should be kept clear of obstructions at all times.
- Fire wardens (officials) should be appointed for all accommodation and office buildings, in sufficient numbers such that in the event of a fire, control and safe evacuation of personnel to allocated muster points could be efficiently accomplished.
- All personnel should be familiar with the fire emergency procedures, alarms and equipment available, personal responsibilities and evacuation procedures in the event of a fire alarm. Regular fire drills should be performed to ensure this.
- All flammable liquids, such as photocopier toners, cleaning solvents and draughts-man's sprays, should be stored away from sources of heat and ignition or naked flame, in metal cabinets. Only quantities in direct use should be brought into the workplace.

- Empty containers and aerosols, which have contained flammable liquids, should be disposed of forthwith in a secure lidded refuse container and in accordance with the laid out procedures and prevalent regulations.
- 'No Smoking' signs should be strictly obeyed. Where smoking is permitted, cigarette butts and spent matches should be disposed of in specifically designated ashtrays or sand containers.
- Appliances such as electric heating rings should not be permitted in normal office space, but should be confined to designated cooking areas. However, domestic electric kettles and coffee percolators may be placed in the general office space provided they are positioned securely, and are supplied directly with a power point. They should not be powered via an extension cord.
- On-line gas bottles for use in kitchens or laboratories should be located outside. If the bottles are placed closer than 5 m from combustible materials, a block-work separation wall should be constructed. Any enclosure for the gas bottle(s) should be freely ventilated.
- Storage cupboards for stationery and other flammable materials should be metallic and, when not in a dedicated storage room, fitted with doors.

6.2.6   Fire and emergency[2, 10, 22, 23]

In the event of a large uncontrollable fire:

- Close or stop the source of fire, if you can.
- Sound the alarm by breaking the glass.
- Telephone emergency number (1234–**As applicable to the place where fire occurred.)
- Leave the building using nearest exit.
- Go to nearest Assembly Point.

If you hear a FIRE ALARM:

- Leave the building using nearest exit.
- Do not run.
- Do not stop to collect personal belongings.
- Do not re-enter the area before fire brigade permits.

If the FIRE is SMALL:

- Use a Fire Extinguisher to extinguish it.
- Don't operate Fire Alarm or Fire Bell.

The following is the procedure for using a Fire Extinguisher. It is denoted by: '*PASS*' that signifies:

- *P*ull the pin
- *A*im the extinguisher at the base of the flames
- *S*queeze the trigger, holding the extinguisher upright
- *S*weep the extinguisher from side to side, covering the area with the extinguishing agent.

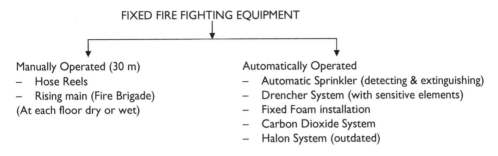

Figure 6.2 Fixed-type fire fighting equipment.

6.2.7   Fixed fire fighting equipment[2, 10, 17, 22, 23]

Figure 6.2 classifies 'Fixed-type fire fighting' equipment.

6.2.8   Fire fighting department[10, 11, 22, 23]

6.2.8.1   Introduction

Every year, fires kill thousands of people, injure thousands more, and destroy billions of dollars worth of property. In olden days, Whenever a fire broke out, all the people in the community used to rush to the fire area, and they used to form a row/queue from a source of water to the fire and passed buckets of water from one person to another to put the fire out. As cities and towns grew larger, volunteer and paid fire departments were organized. Today, fire departments in most industrialized nations have well-trained crews and a variety of modern firefighting equipment.

6.2.8.2   Functions

- Fire departments fight fires that break out in homes, factories, office buildings, stores, and other places.
- Firefighters risk their lives to save people and protect property from fires.
- The men and women who work for fire departments also help people who are involved in many kinds of emergencies besides fires.
- Fire departments work to prevent fires by enforcing fire safety laws.

  Procedure to be followed in the event of outbreak of fire:

- Leave the building immediately
- Children should not attempt to fight fires
- Adults may use a fire extinguisher on a small fire if it seems safe to do so
- Before opening any door, touch it briefly with the back of your hand
- Never open a door that feels hot
- Try another escape route or wait for help
- Crawl on the floor when going through a smoky area

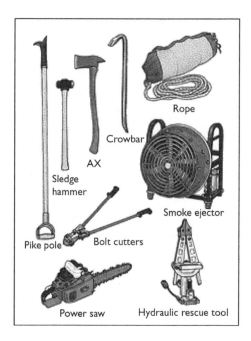

Figure 6.3 Fire fighting tools with Fire Department.[22]

- Do not return to the building
- Call the fire department at the earliest.

The Work of Fire Department

- The two basic firefighting units in most fire departments are *engine crews* and *ladder crews*
- Engine crews operate trucks called engines, which carry a pump and hoses for spraying water on a fire
- Ladder crews use ladder trucks, which carry ladders of various lengths and a hydraulically extendable ladder or elevating platform to rescue people through windows or to spray water from a raised position
- Both kinds of trucks also have other rescue equipment and firefighting tools
- At a fire, engine and ladder crews work together under the direction of an officer of the fire department
- The officers in command quickly assess the situation and direct the firefighters for action. Their first and most important task is to rescue people who may be trapped in the building
- The members of the ladder crew search for anyone who may be trapped.
- Meanwhile, members of the engine crew connect a hose from their pump to a nearby fire hydrant. Their first concern is to keep the flames from spreading
- Ladder crew members ventilate the building to let out the smoke, heat, and gases that build up during a fire

- Finally, the officer in charge makes out a report that gives all the important facts about the fire.

Emergency Rescue Operation

- Large fire departments have rescue crews to handle non-fire emergencies
- Rescue workers may be called to free people trapped under the wreck of a fallen building or in a car after an accident
- At a building fire, the rescue workers help the ladder crew to get people out of the building; they give first aid to people overcome by smoke or suffering from burns.

Emergency Medical Operation

- Fire departments provide medical care in non-fire emergencies before the patient is taken to the hospital
- They provide quick pre-hospital care to victims of injury or sudden illness, and transport the patient to the hospital
- Paramedic units operate ambulances and use communication equipment to stay in touch with a nearby hospital.

Public Building Inspection

- Most cities have a fire safety code that applies to such buildings as theaters, department stores, schools, and hospitals
- These codes specify that the buildings should not be made of materials that burn easily
- The codes also require portable fire extinguishers, a certain number of exits, and other fire safety features in public buildings
- Many fire safety codes require large buildings to have built-in sprinkler systems and special water lines to which fire hoses can be attached.

Public Education Programme

- Fire departments should work with other local agencies to train people as how to prevent fires and what to do during a fire
- They may use a portable model of a house to show peoples as how to safely leave a burning building
- Fire departments advise people to install smoke detectors in their homes
- Fire protection experts recommend at least one detector for each floor of a residence
- Fire departments also recommend that people have portable fire extinguishers in their homes.

To conclude it could be said that:

- The Fire department is one of the most important organizations in a community
- The public have to be trained how to deal with fire
- To make one's home safer, check the heating and air-conditioning systems and the cooking equipment, and avoid overloading them.

6.3   EXPLOSIONS (information from various sources)[2, 4, 5, 12, 17, 20, 22]

It is a phenomenon in which there is a sudden widespread expansion/increase of gases in rapidly moving pressure or shock wave. The expansion could be mechanical, or it could be result of a rapid chemical reaction. It is a noisy outburst. The damage is caused by the pressure or shock wave (as described in sec. 6.3.1). The basic difference between fires and explosions is the rate of release of energy. It is faster in the case of explosions compared with fires. Fires can result from explosions, and vice versa is also true. Figure 6.4 classifies explosions, which could be due to various reasons.

6.3.1 Classification

6.3.1.1   Mechanical[5]

For such an explosion a reaction does not occur, and the energy is obtained from the energy content of the contained substance. If energy is released rapidly, an explosion may result, for example sudden failure of a tire full of compressed air, and sudden catastrophic rupture of a compressed gas tank.

Another example, which could be considered as a mechanical explosion is when pipes and equipment containing liquids are liable to solidify. This usually happens in water pipes in a cold climate. When water freezes and becomes ice, which then melts,

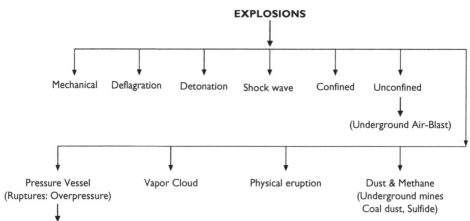

Figure 6.4 Classification – Explosions.[2, 5, 12, 17, 20, 22]

there is an increase in volume, sometimes causing bursting. In many cases where a small proportion of water pipes are exposed to a cold atmosphere, the water should be kept flowing. Sometimes, based on a weather forecast predicting frost, the pipeline should be completely drained off.

6.3.1.2   Detonation, Deflagration and Shockwaves[5, 18]

To understand these three terms, let us understand as how an explosive works. An explosive is a substance or mixture of substances which, with the application of a suitable stimulus such as shock, impact, heat, friction, ignition, spark etc., undergoes an instantaneous chemical transformation into an enormous volume of gases having high temperature, heat energy and pressure. This, mixture when initiated, it undergoes chemical decomposition. This decomposition is a self-propagating exothermic reaction, which is known as an '*Explosion*'. The gases of this explosion with an elevated temperature are compressed at a high pressure. This sudden rise in temperature and pressure from ambient conditions results in shock or detonation waves traveling through the unrelated explosive charge. Thus, *Detonation* (Figure 6.5(left)) is the process of propagation of the shock waves through an explosive charge. The pressure rise could be greater than 10 atm within 1 millisecond. The velocity of detonation is in the range of 1500 to 9000 m/sec. well above the speed of sound. *Deflagration* (Figure 6.2 (right)) is the process of the explosive's ingredients burning at an extremely rapid rate, but this rate or speed of burning is well below the speed of sound. The pressure rise could be 1–2 atm within many milliseconds.[5]

Mechanism of explosion:
All mixtures of gases or vapors with air, which are within the flammable range, are liable to explode on ignition, particularly if the mixture is partly or wholly confined. Such explosions are generally deflagrations, which can lead to pressure rise up to eight times more than the original pressure. Accidental explosions of this type are too common.

Some flammable gases or vapors (for example as given in Table 6.3), when mixed with air within a still narrower concentration range, can detonate when strongly ignited. Such detonations can cause peak pressure rise to even up to 20 times the original pressure to create shock waves which are far more damaging than the blast waves created by explosive deflagration.

Deflagration ⟶ Explosive deflagration ⟶ Detonation.

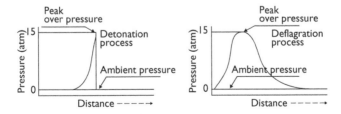

Figure 6.5 Detonation and deflagration.

Table 6.3 Flammability & detonability for some gases and vapors.[11]

| | % volume in air | | | |
| | Flammability | | Detonability | |
Gas or vapor	Lower	Upper	Lower	Upper
Acetylene	1.5	80	4.2	50
Hydrogen	4.1	74	18.3	59

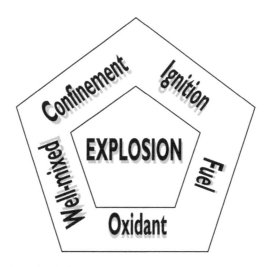

Figure 6.6 Ingredients Fuel, Oxidant and Ignition-source, when well mixed and having proper confinement result a successful explosion or blasting operation.

6.3.1.3   Confined & unconfined explosions[17]

Confined – an explosion occurring within a confined space, vessel, structure or building can rupture them, resulting in the projection of debris over a wide area. This debris or missiles causes extensive damage and results in injuries to those involved. 'Blasting Operations' undertaken at mines, tunnels or civil construction sites are confined explosions. The ingredients as shown in Figure 6.6, namely: fuel, oxidant and ignition-source, when well-mixed and properly confined result in a successful explosion or blasting operation.

Unconfined explosions also create missiles by blast wave impact and subsequent translation of structures. Air blast in underground mines is an example of unconfined explosion.

6.3.1.4   Air blast[18, 20]

In underground mines roof pressure over a worked out space depends upon the texture of rock constituting the roof/back, its coherence, dip of the deposit, span, the rate of mining and the duration of its exposure. The rock constituting the roof, and also the dip of the deposit cannot be changed but all other parameters can be controlled. The roof pressure can be counter-acted using supports filling the worked out space, or by controlled caving. In some cases the roof is so firm that it stands for a longer duration and the worked out room becomes too extensive. In such cases *induced caving* becomes essential to avoid large pressure building in and around such excavations. Excessive pressure-built around these large excavations, if allowed, may result their sudden failure (collapse) and that in turn displaces the air that was filled in the worked space so rapidly that escaping air is heard as a big blast damaging anything coming its way; this phenomenon is known as '**Air Blast**'. It is common in coal as well as metal (non-coal) underground mines.

6.3.1.5   Pressure Vessel Ruptures (over-pressure)[5]

Pressure Vessel Ruptures (over-pressure) could be due to any of these reasons:

- Exothermic runaway reactions
- Physical overpressure of pressure vessels
- Brittle fractures
- Polymerizations
- Decompositions
- Undesired reactions catalyzed by materials of construction or by ancillary materials such as pipe dope (lubricant) and lubricants.
- Boiling Liquid Expanding Vapour Explosion (BLEVE)

    Boiling Liquid Expanding Vapour Explosion (BLEVE) – this occurs when external fire heats the contents of a tank of volatile material. As the tank's contents heat, the vapor pressure of the liquid within the tank increases and tank's structural integrity is reduced due to heating. If the tank ruptures, the hot liquid vaporizes explosively.[5] It is special type of phenomenon that can release a large quantity of material, which could be flammable or toxic. When flammable, a VCE might result (described in the following sections).

6.3.1.6   Rock burst and bumps[18, 20]

Deepening of mine workings below 600–800 m is often accompanied by a considerable rise in rock pressure, which is transferred to the rock mass known as pillars surrounding the worked out spaces/areas. An abrupt inrush of rock burst from the stressed pillars (in mines other than coal) is a phenomenon known as 'Rock Burst'. It usually occurs at the very deep horizons. The similar phenomenon which occurs in underground coalmines is known as 'Bumps'.

6.3.1.7   Vapor Cloud Explosions (VCE)[5]

This is considered to be the most destructive and damaging explosion in the Chemical Process Industry. Such an explosion follows the following sequence as described by Crowl and Louvar:[5]

- Sudden release of large quantity of flammable vapor (typically this occurs when a vessel containing a super-heated and pressurized liquid, ruptures)
- Dispersion of vapor throughout the plant site while mixing with air, and
- Ignition of the resulting vapor cloud.

The accident of Flixborough, England in June 1974, a sudden failure of a 20″ cyclohexane line between reactors led to vaporization of an estimated 30 tonnes of cyclohexane C_6H_{12}.[7] The vapor cloud dispersed through out the plant site and was ignited by an unknown source 45 seconds after the release. It demolished the plant, killed 28 workers, injured 36 others, and damaged 1821 houses and 167 shops and factories. A massive fire after the explosion continued for 10 days.[5]

A summary of 29 VCEs which occurred during period 1974–1986 shows property loss for each event between $5 millions to $100 millions and 140 fatalities; say 13/year. This is mainly due to an increase in the inventories of flammable materials, and also operations were under more severe conditions. Any process containing quantities of liquefied gases, volatile superheated liquid or high-pressure gases is considered a likely scenario for a VCE.[5]

6.3.1.8   Physical explosion or eruption[4, 5, 11]

This is a sudden vaporization of liquid in a closed or partly closed vessel. Eruption is a more appropriate term that can be used for such events. The pressure rise from a physical expansion is usually too fast for any standard pressure relief device to handle.

Physical explosions have frequently occurred during the start-up of distillation columns (especially vacuum columns) of oil refineries, when hot oil has come in contact with a pocket of water lying in a low point of the column or its associated pipework.

It is also possible for a physical explosion to occur in an unstirred tank or vessel containing water and a hydrocarbon layer above it, when one or other of these layers is heated.

6.3.1.9   Methane and coal dust explosions[14, 18]

Almost all coal deposits release certain amount of methane although the quantity varies; from a few cubic centimeters to 100 m³ per tonne of coal mined. Methane when mixed with air is explosive within a range of about 5 to15% with air, and most explosive when it is 10%;[18] it may produce as much as 5200 kJ/kg of mixture.[14]

While mining coal, underground dust is generated. Coalmine explosions are invariably initiated by ignition of methane through a variety of causes, usually at or near the coal face.[18] Usually, the ignition of methane raises a coal dust cloud and the burning of methane ignites it, the flame then propagates away from the point of ignition and may soon become a propagating coal dust explosion.[20] The explosive mixture may contain a range of 50 to 5000 gms of coal dust/m³; and once coal dust has become involved in an explosion, the rate of advance can quickly accelerate to the order of six times the speed of sound in the ambient air.[16] In addition to this

Table 6.4 Categorizing flammable gases & vapours based on ease of ignition, as per British Standard.[11]

Flammable gases & vapors	Category
Methane	I (ignition energy 0.29Mj)
Flammable gases & vapors	IIA
Diethyl ether, ether	IIB
Hydrogen	IIC (Ignition energy 0.019Mj)

concentration, for the coal dust explosion to occur the coal dust must possess some of the following properties: it should have sufficient volatile matter, be of necessary fineness, be in a dispersible condition, raised into a cloud by some agency, and ignited by a flame of sufficient intensity.[20]

The prudent, judicious and liberal use of *stone dust* generally minimizes the occurrence of coal dust-based explosions. The stone dust acts as breakdown path for the combustion process. In table 6.4 Categorization of flammable gases & vapours based on ease of ignition, as per British Standard has been shown.

6.3.1.10 Sulfide dust explosions[20]

Blasting in pyrite ($CuFeS_2$) mines generates sulfide dusts, which is liable to result in hazardous conditions. Dust which is about 0.1 mm in size is considered to be the most explosive, while dusts finer than 0.1 mm are almost inert. Sulfide dust, like coal dust, also explodes only when it is found in a suspended state in the mine air. The sulfide dust-air mixture possesses an explosion danger if dust concentration is between[20] 0.25 and 1.5 g/m^3. Such an explosion takes place only during blasting, as a large amount of dust is generated with potential for such an event. Wet drilling, sprinkling water at the blasting site prior to blasting, and effective ventilation are the preventive measures which should be enforced to minimize/avoid such an occurrence.

6.3.1.11 Explodable dusts[5]

Some explodable dusts include: sugar, coal, polyethylene, starch, cellulose, maize, dextrin, organic pigment, aluminum and few others.[2, 11] A series of grain silo explosions[5] in Westwego near New Orleans in 1977 killed 37 people.

For most dusts the lower explosion limit[5] is between 20 g/m^3 and 60 g/m^3 and the upper explosion limit is between 2 kg/m^3 and 6 kg/m^3.

6.4 DOW INDEX TO ACCESS DEGREE OF HAZARDS[5-7, 10]

The Chemical, Petroleum and Natural Gas industries are prone to fire and explosion hazards. Dow Index by Mond Division of Imperial Chemical Industries (ICI) provides separate indices for fire (F), internal explosion (E) and aerial explosion potential (A) along with overall hazard rating (R) as shown in Table 6.5.

Table 6.5 Dow index range for various degree of hazards.[6,7] *

Potential hazard category	Fire F	Internal explosion E	Aerial explosion A	Overall hazard R
Mild to Light	0–2	0–1.5	0–10	0–20
Low	2–5	1.5–2.5	10–30	20–100
Moderate	5–10	2.5–4	30–100	100–500
High	10–30	4–6	100–400	500–2500
Very High	20–50	Exceeding 6	400–700	2500–12500
Extreme	100–250		Exceeding 700	12500–65000
Very Extreme	Exceeding 250			Exceeding 65000

* F. E. A. and R are defined as
- Fire index (F) relates to the amount of flammable material in the Industrial Unit. Its ratings are shown in column 2 to describe degree of hazard (Light to Extremely Hazardous).
- Internal Explosion Index (E) is the measure of the potential for explosion within the Industrial Unit. Its ratings are shown in column 3 to describe degree of hazard (Light to Extremely Hazardous).
- Aerial Explosion Index (A) relates both to the risk and magnitude of a vapor cloud explosion originating from a release of flammable material, usually present within the Industrial Unit as a liquid at a temperature above its atmospheric boiling point. Its ratings are shown in column 4 to describe degree of hazard (Light to Extremely Hazardous).
- Overall Hazard Rating (R) is used to compare different Industrial Units with different types of hazards. Its ratings are shown in column 5 to describe degree of hazard (Light to Extremely Hazardous).

In this process Dow index (D) is calculated using following formula (6.1):

$$D = B (1 + M/100) (1 + P/100)(1 +[S + Q + L + T]/100) \qquad (6.1)$$

D – The equivalent Dow Index
M – Special hazard
P – General process hazard
S – Special process hazard
Q – Quantity hazard
L – Layout hazard
H – Acute health hazard
B – Material factor**

** B Material Factor[6,7] – This number is generally from 1 to 60, which denotes the intensity of the energy release from the most hazardous material or mixture of materials present in significant quantity in a unit or plant.

Thus, to arrive at the numerical value (the quantitative assessment) of Dow Index D; weighting is given to special hazards, process hazards, quantity of hazardous materials, plant layout and the health hazards that are associated with the various operations involved in the manufacturing process. It also includes material factor B, which is given by the formula 6.2.

$$B = \Delta H_c/2326 \qquad (6.2)$$

ΔH_c – Net heat of combustion (kJ/kg)

Table 6.6. Dow fire and Explosion Index (D) and Degree of hazard.[5-7, 11]

Degree of hazard	Dow fire & explosion index
Mild to Light	1–60
Moderate	61–96
Intermediate	97–127
Heavy	128–158
Severe	159 & above

In the computation of Overall Hazard Rating (R), formula 6.3 is used

$$R = D \left[1 + [0.2 \times E \times \{(AF)^{1/2}\}] \right] \tag{6.3}$$

6.5   INCIDENTS RESPONSIBLE FOR ONSET OF HAZARDS AND ACCIDENTS[1, 2, 5, 7, 12, 19]

6.5.1   Spillage – which could be caused by:

• Overflow, backing up, blowback, air lock, and vapor-lock;
• Failure of control or major service;
• Surging, priming, foaming, puking, spitting;
• Condensed products in vapor, change in normal discharge;
• Malicious intent, vandalism.

It is the best practice as not to spill any oil on the ground. Put used motor oil in a clean plastic container with a tight lid. Never store used oil in a container that once held chemicals, food, or beverages. Do not mix the oil with anything else, such as antifreeze, solvent, or paint. Take used motor oil to a service station or other location that collects used motor oil for recycling.

6.5.2   Leakage (also refer sec. 9.6)

Which could be caused by:

• Broken, damaged or badly fitted pipe, vessel, instrument, glass, gasket, gland, seal, flange, joint or seam-weld;
• Internal leaks, overpressure of pipe or vessel;
• Deterioration of bursting disc (pin holing).

6.5.3   Unintended venting (also refer sec. 9.6)

Which could be caused by:

• Evaporation through open line, drain, cover;
• Relief valves leaking, bursting discs blown, lutes blown;

- Valve struck, scrubber overloaded, ejector failure;
- Equipment failed/out of service (e.g. scrubbers, flares), excessive pressure, wrong routing, loss of vacuum;
- Vessel damaged, tilted, collapsed, vibrated, over-stirred;
- Overloading of open channel/conveyor.

6.5.4   Failures at normal working pressure (also refer sec. 9.6)

Which could be caused by:

- Inadequate design, materials, construction, support, operation, inspection or maintenance;
- Deterioration due to corrosion, erosion or fatigue;
- Mechanical impact.

6.5.6   Equipment failure due to excessive pressure – (also refer sec. 9.6)

Which could be caused by:

- Overfilling, over-pressurizing or drawing vacuum;
- Overheating or under-cooling;
- Internal release of chemical energy;
- Exposure to fire or other sources of external heating (e.g. radiation).

6.6   LOSSES IN THE CHEMICAL INDUSTRY DUE TO FIRES AND EXPLOSIONS[5]

An analysis made for a period 1978–80 for the USA's Chemical Industries for losses incurred reveals that:

- Most frequent and severe losses are due to fire and explosions.
- An explosion causes more severe losses than fire.
- Main causes of explosions are accidental and uncontrolled chemical reactions.
- Most explosions occur in closed buildings and involve batch reactions.
- Rupture of vessels, pipes and equipment contribute greatly to the magnitude of fire and explosion losses.
- Release of flammable gases and liquids results from most of the fires.

6.7   HAZARDS WITH FLAMMABLE LIQUIDS, AND PRECAUTIONS[2, 4, 7, 11, 12, 19]

- The resistance of many flammable liquids including ethanol, acetone and crude oils is sufficiently low to prevent risk of their developing dangerous static charges when they are stored in conducting and earthed containers.

- The inflammable hydrocarbons including gasoline, naphtha and toluene, have high resistance which enables dangerous static charge to develop during their storage even when their containers are conducting and earthed (unless they contain an antistatic additive). Flow velocity in the pipe for such liquids should be restricted to 7 m/sec, and if no second phase is present, to about 1 m/sec. If a second phase is present. A complete guide should be referred.
- Even low resistance liquids can accumulate dangerous charge when they are in contact with conducting and earthed pipes, tanks and equipment. *This applies to liquid droplets produced when a jet of falling liquid breaks up in air.*
- Proper containers, which are earthed or stand on the earthed metal surface, should always be used for small scale handling of flammable liquids. When pouring any flammable liquid, containers as well as any hoses and nozzles used should be bonded and earthed. Such liquid containers should be of metal, except for small sizes smaller than 2 liters, where plastic may be used.
- Liquids flowing in pipes produce electrostatic charges at rates which increase rapidly with velocity. The presence of a second liquid phase (such as water) and constrictions such as valves and filters increase the rate of charge generation.
- Liquids falling freely into a tank also acquire considerable charges. The presence of a flammable atmosphere in the vessel or tank should be prevented if at all possible, e.g. by use of floating roof tank or by inert gas blanketing.
- Positioning liquid inlets: Flammable liquids should enter the tank or vessel through the bottom inlet to avoid free fall.
- Precautions against static electricity:[5, 22, 14]

  - Proper bonding, grounding, humidification, ionization, or their combinations should be the established practice when static electricity is a fire hazard. Drums, scoops, and bags should be physically bonded and grounded.
  - Clothing that generates static electricity should be prohibited.
  - Free-fall filling generates static charge and discharge.
  - Use of nonconductive hoses is the source of static build up.
  - Large voltages are generated when crumpling and shaking an empty polyethylene bag.
  - Shoes with conductive soles are required when handling flammable materials.
  - Conductive grease should be used in bearing seals that need to conduct charges.
  - Stainless steel centrifuges must be used when handling flammable material.
  - Flanges in piping and duct system must be bonded.
  - Re-circulation lines must be extended into liquid to prevent static build up.

6.8   STATIC HAZARDS ASSOCIATED WITH AMMONIUM NITRATE FUEL OIL MIXTURE (ANFO) LOADING[18]

6.8.1   Blasting agent ANFO

Ammonium nitrate (AN), which was earlier known as an oxidizer in the manufacturing of explosives, has become the principal ingredient of the commercial explosives in use in the mining industry. When AN is mixed with 5–6% fuel oil, the mixture is known as ANFO. ANFO has become an indispensable explosive for most of the surface mines and underground non-coal mines. When loaded into blast-holes pneumatically in underground mines, the electrostatic hazards pose problems as outlined below:

Static charge is built up on the loading hoses and equipment and the problem has been the greatest so far in the usage of ANFO underground. A certain degree of safety can definitely be achieved by using conductive or semi-conductive hoses while using ANFO during underground blasting operations. The salient points to safeguard against the generation of electrostatics charge during loading (charging) the explosive pneumatically in the blast-holes may be summarized as under:

a   The loading hose should be semi-conductive with a resistance high enough to insulate any stray current yet conductive enough to bleed off any static charge build-up. Such loading hoses are called 'LO STAT' and a yellow stripe runs throughout their length to identify them.

b   The electrical characteristic of the loading hose should be uniform throughout its length. Resistance ranges from 17,000 to 67,000 ohms/m; electrical capacity of typical PVC hose is 4 p.f. and available discharge energy is 24MJ. Basically it should have sufficient resistance to corrosion from oil and stiff enough to avoid too much kinking.

c   Except in case of non-electric detonators like Anodets etc., bottom priming should not be done and priming should be done at the collar at the end of loading, allowing sufficient time for the hole and operator to get discharged of any electric static charge.

d   The entire system and the operator should be effectively grounded to the earth. Only approved semi-conductive loading hoses should be used. In case of hoses lined with a non-conductive material, pneumatic loading should not be adopted.

e   Maximum resistance of the drill hole to the ground must be less than 10 Mega ohms and the maximum resistance of the loader operator to the ground must be less than 100 Mega ohms, while electric detonators are to be used.

f   After every loading operation some time should be allowed for the charge to leak away and the detonator should be placed from the collar side, and the operator should also ground himself before handling the detonators.

g   Detonator continuity tests must be made religiously. This will prevent discharge between the shell and the lead wires, which may be of corona type, or a simple spark.

h   Leg wires should be shunted and not connected to ground during pneumatic loading.

i   An effective earth line should be connected with the entire assembly of ANFO loader.

j   Synthetic fibers like nylon, teryline etc. should not be on the body of the persons doing ANFO loading since they have tendency to accumulate and retain charge. For similar reasons rubber-soled boots should also not be used. The quantity of electric charge has been seen to depend largely upon humidity conditions and general conductivity of the rocks. When the rock is fairly conductive, the charge is dissipated as soon as developed.

6.8.2 Case History: On static electricity hazards[21]

Two plant operators were filling a tank car with vinyl acetate. One operator was on the ground, and other was on top of car with the nozzle end of a loading hose. A few seconds after the loading operation started; the contents of the tank exploded. The operator on top of the tank was thrown to ground; he sustained a fractured skull and multiple body burns and died from these injuries.

The accident investigation indicated that a static spark that had jumped from the steel nozzle to the tank car caused the explosion. The nozzle was not bonded to the tank car to prevent static accumulation. The use of a nonmetallic hose probably also contributed.[21]

6.9 TOXIC GASES[5, 11, 13, 15, 17, 19]

Toxic gases are those which, on breathing in a sufficient quantity for a sufficient time, will seriously disable and possibly kill the person. They act as a poison. They can be grouped into three classes as detailed below, based on their action on the body:

6.9.1 Asphyxiate gases[13, 18]

These are either simple in nature which exclude oxygen from lungs, for example, CH_4 and CO_2. First symptom is fast breathing and hunger for air. With time there may be nausea, vomiting, lying flat on ground, loss of consciousness and finally convulsion, deep coma and death.

- CO is a chemical asphyxiate. Its affinity for hemoglobin is 210 times greater than O_2.[18]
- Thus, CO reduces the O_2 carrying capacity of blood. The seriousness can be judged by the presence of carboxyhaemoglobin in the blood. This can be calculated by:[18]

b = 4ate/100

b = carboxyhemoglobin content in the blood, %
a = concentration of CO in air, ppm
t = time of exposure in hours
e = factor 1 for resting, 2 for walking and 3 for working.

If value of b is:

- below 20% – no symptoms
- 20–30% – headache
- 30–50% – dizziness, nausea, muscular weakness and danger of collapse.
- 50% and above – Unconsciousness and death.

6.9.2 Irritant gases[13, 18, 20]

- These gases induce inflammation to tissues such as skin, conjunctiva of eyes, the membranes of the respiratory tract when they come into contact with them.

- If the gases are not soluble into the moist upper respiratory tract, they enter into the lungs and cause exudation of fluid from the lungs, which may lead to suffocation.
- **Nitric oxide, Nitrogen dioxide and Sulfur dioxide** are the most common irritant gases. Nitrogen dioxide can cause inflammation of lungs, which is great concern to health.

6.9.3 Poisonous gases[13, 15, 18]

These gases destroy tissues with which they come in contact. Nitric oxide, Hydrogen sulfide and Sulfur dioxide are not only irritant but also poisonous.

The Bhopal disaster in 1984 in India has demonstrated the hazards associated with liquefied gases. Prominent liquefied toxic gases include: Hydrogen chloride (HCl), Hydrogen sulfide (H_2S), Chlorine (Cl_2), Ammonia (NH_3), Sulfur dioxide (SO_2), Phosgene ($COCl_2$), Hydrogen fluoride (HF) and Hydrogen cyanide (HCN). In addition to these there are about dozen more, which are considered to be toxic.

6.9.4 Portal of entry[13]

- These occupational poisons gain entry to the body via the lungs, skin or sometimes the gut.
- Absorption of a poison depends upon its physical state, particle size and solubility. Of those substances entering into the lungs, some may be exhaled, coughed up and swallowed, be attacked by scavenger cells and remain in lung, or enter the lymphatic system.
- Soluble particles may be absorbed into the blood stream.

6.9.5 Remedial measures

1 Prevention of formation of gases.
2 Prevention of exposure of persons.
3 Dilution of gases.
4 Removal of gases.

6.9.6 Toxicology[13, 21]

Toxicity is the potential of any substance to cause harm on contact with body tissues. Toxicology is the scientific study of the chemical effects of poisons on living beings. To determine these effects, which may be acute or chronic (continuous, recurring or repetitive types), tests can be carried out on animals, man (in vivo tests) or in test tubes (in vitro tests). These are carried out in three stages:

- To establish acute effects – Lethal dose (LD) or Lethal concentration (LC) carried out on mice or rats; the degree of toxicity is indicated by the perecntage killed in one dose. LD50 means 50% of animals are killed by one dose given to them. These are expressed as mg or gram of poison/kg of body weight as shown in Table 6.7.

Table 6.7 Details of degree of toxicity based on amount of dose.[5]

Degree of LD50 toxicity	Dose
Extreme toxic	1 mg or less
Serious toxic	1–50 mg
Highly toxic	50–500 mg
Moderately toxic	0.5–5 g
Slightly toxic	5–15 gm
Relatively harmless	15 g or more

- Degree of LD50 toxicity can be assessed based on amount of dose as shown in Table 6.7. From this it is possible to calculate the toxic dose of persons of known weight.
- Similarly LC20, LC50 etc. are expressed when exposed to a chamber of controlled atmosphere of a particular gas, fumes, vapor, dust etc. Inhalation tests require the concentration and exposure times to be recorded to ascertain the uptake by the animals.

Such tests are carried out on animals for 90 days to show the chronic effect. Observations are made on growth, food intake, urine, faeces, biochemistry of blood, urea, sugar, electrolytes, fat etc.

Effects: effects may be acute i.e. rapid onset and short duration, or chronic i.e. gradual onset and prolonged. They may be local, at the site of contact only or general following absorption. Toxic substances may disturb normal cell functions, damage cell membranes, and interfere with enzymes and the immune system. Pathological response may be irritant, corrosive, toxic, fibrotic, allergic, asphyxiate, narcotic and anesthetic (loss of sense).

Metabolism: most substances carried by the blood stream will be carried to the liver where they may be rendered less harmful by a change in their chemical composition. However, some may be made more toxic e.g. naphthylamine which is responsible for bladder cancer and tetra-ethyl lead which may be converted into the tri-ethyl form and is toxic to the central nervous system.

Excretion:[13] the body eliminates the harmful substances in the urine, lungs and less commonly the skin. Some are also excreted in the faeces and milk. The time taken to reduce the concentration of a substance in the blood by 50% is known as biological half-life.

Factors influencing toxicity:[13]

- The inherent potential of a substance to cause harm.
- Its ease of body contact and entry: work method, particle size and solubility.
- Doses received (concentration and time of exposure)
- Metabolism in the body and its half-life

- Susceptibility of the individual, which depends upon a number of factors:

  - Body weight
  - Extremes of age in the working population are more prone to skin damage
  - Failure to reach set health standards
  - Inadequate level of training, information, supervision and protection.

Toxic effects can be divided into two types:[13] carcinogenic & non-carcinogenic.

Carcinogenic types induce or promote tumors (cancer), that is, abnormal or uncontrolled growth and division of cells. Many carcinogens are site-specific that is, particular chemicals attack a particular organ. The few known human carcinogens are benzene, vinyl chloride, asbestos, chromium. There are many probable carcinogens such as carbon tetrachloride, cadmium etc. and hundreds of possible carcinogens.

Non-carcinogenic effects include all other toxicological responses, of which there are countless examples: organ damage (kidney, liver), neurological damage, suppressed immunity (resistance power), and adverse effects on birth and development . Chemicals identified as endocrine disruptors may cause breast cancer in women, for example pesticides (such as DDT), industrial chemicals such as some surfactants and some prescribed drugs such as dioxins. Dose and duration are very important to realize its effects.

Safety precautions:

- Workers should be thoroughly trained to work in a hazardous environment and to follow safety practices. For example one should not work alone and unattended.
- In many cases standard practices are not enough. Check lists should be developed.
- Any one who works in a hazardous industry should *learn to stop breathing instantly when an emergency happens.* A single breath of smoke, fire or any of the toxic gases may be fatal. Don't take that breath. In one minute you can probably reach safety. If you can't reach safety, breath very shallowly through your nose so that sinuses may absorb most of the toxic agents before the air gets into lungs.[15]

6.9.7   Summary: Classification – toxicity-related hazards[4, 5, 11, 15]

- Environmentally toxic to plants, animals or fish:

  - Chronic or acute
  - Toxic to individual species or broadly hazardous
  - Pesticides, fungicides, insecticides, fumigants.

- Toxic to Humans

  - Chronic or acute
  - Reversible injury or irreversible injury or death
  - Carcinogens (any substance that produces cancer)
  - Endocrine modifiers (e.g. estrogen mimics)
  - Persistent bio-accumulative toxins (PBTs).

- Long-term Environmental Hazards:

  - Greenhouse Gases
  - Ozone Depletory.

- Product Hazards

  - Customer injury
  - Waste Disposal Environmental Hazard.

6.10 HAZARDS WHILE USING MACHINERY[11, 14, 18]

'A cracked bell can never sound well, nor will equipment run smoothly without preventive or predictive maintenance'. Every industry invariably uses machines, equipment, appliances, tools and tackle for carrying out various unit operations, and is liable to have hazards due to the following:

- Hazards while operating machines and equipment; and working in a construction site
- Crushing
- Shearing
- Cutting or severing
- Entanglement
- Drawing-in or trapping
- Impact
- Stabbing or puncture
- Friction or abrasion
- High pressure fluid ejection
- Electrical shock
- Noise and vibrations
- Contact with extremes of temperature
- Falling from height.

Ways to eliminate or reduce hazards[14, 18] are listed below:

1 Operating as per the laid out rules, regulations, norms, procedures, and best practices.
2 Making those involved aware of all these aspects through effective training programs including supervisors and managers (as per the guidelines given in Sections 9.13 and 10.4.1 and table 8.2).
3 Selection of equipment that is safe, simple to operate and maintain. User-friendly machinery is always advantageous.
4 Proper layout to house, maintain and up keep equipment is equally important.
5 Periodic check-ups including preventive and predictive maintenance of plant, machines, equipment, structures and buildings.
6 Effective illumination, ventilation, hygiene, drainage, welfare amenities, first aid, fire extinguishers at strategic locations, provision of safety wear and appliances including PPEs as appropriate in a given situation.
7 Display of procedures, caution – boards wherever required in languages prevalent at the workplace.

Table 6.8 Dos and Don'ts while operating plant, machinery and equipment.

Remember–	Never–
• Prior to operating any machine, make sure that you know how to stop it. • Make sure that all guards are fitted properly and they are in working order. • Materials if any to be used are clear of working/moving parts of the equipment • The space around the equipment is free of any obstruction • Inform the competent person including the concerned supervisor in case of its failure or breakdown. • Make sure required PPEs are used.	• Use/operate the equipment unless you are authorized to do so, and are fully conversant with its operating procedure. • Attempt to clean a machine when it is in motion • Use a machine if it has been tagged with a danger sign. The sign should be removed by the authorized person who is satisfied that its operation is safe. • Wear loose clothes, dangling chains, loose rings, or keep long hairs, which could be caught up in the moving parts. • Distract people who are operating the equipment.

8 It has been found that most accidents occur due to moving machinery, falling and rolling of materials, people falling from heights, and not following do's and don'ts as listed in table 6.8. Provision for effective guards, fencing, barriers and dykes wherever required is mandatory to guard against such hazards. Equally important is regular checkups and maintenance.

9 Effective security to prevent unauthorized entry. Preventing unauthorized operation of any equipment is equally important.

10 Automation of the operations that are repetitive in nature and bring excessive fatigue and strain to workers. They are potential source of accidents.

11 Checking for their effectiveness and proper working conditions of the tools and tackle such as lifts, cranes, chains, rope slings etc. before their use; must be ensured.

12 Use proper ladders, lifts, bridges, and crossovers wherever required.

13 For any operation identify hazards, try to reduce or eliminate them and at the same time carry out a risk analysis to understand the potential hazards, and formulate an action plan to deal with them.

14 Guidelines given in the Table 6.8 could be a useful guide.

6.11 SURFACE OR SUBSURFACE (UNDERGROUND) MINE HAZARDS[17, 18, 20]

• In an underground situation the working space is inherently tight, distorted, congested, isolated and inaccessible, of poor quality, and deteriorating. These adverse conditions endanger personnel, damage mobile equipment, and affect all activities.

• It is not only the confined space underground but also adverse working conditions such as darkness, heat, humidity; gassy and watery conditions that make the miner's job difficult and risky.

- This is also the case while working at surface mines under adverse climatic conditions.
- The miners are also liable to occupational diseases such as asbestosis, silicosis and a few others. In addition, the risks of fire, explosion, inundation and ground failure are part and parcel of this industry.
- *Environmental impacts*: Mining and processing to a get the final product is a complex operation. To produce any end product from a mineral deposit is a long process. First ore is mined from surface or/and underground mines. There are several stages after mining, as described in sec.1.4 chapter 1; and they are: concentration (crushing, grinding, separation, classification, leaching, thickening, drying, etc.), smelting, refining and casting etc. Mining as well as extractive metallurgical operations are detrimental to the environment. Land degradation, water and air pollution, and changes in land use, disturbance to flora and fauna in and around the area occupied by the mining lease, are some of the inherent features of mining which cannot be avoided, but their adverse impact can be minimized. It may be noted that any mining venture, which is not able to meet the costs of mitigating adverse environmental impacts and land reclamation of the mined out areas, is not feasible.

6.12 CLASSIFICATION OF HAZARDOUS MATERIALS[11, 16, 18, 19]

The hazardous substances/materials could be classified as follows:

- Explosive materials
- Compressed gases
- Flammable liquids and solids
- Chemically reactive materials
- Biologically active materials
- Radioactive materials (described in Chapter 7)
- Toxic materials.

The labels describing the presence of these materials are shown in Figures 6.7(a) to 6.7(f) based on UN Classification of hazardous substances.

In our day-to-day life, consumable hazardous substances can be in various forms; such as liquids (paints, cleaners, solvents); dusts and fibers (from vacuum machines) fumes or smoke, bacteria (such as those causing legionnaires' diseases), vapors (such as petrol) or gases.

How these substances could cause harm would depend upon the port of entry into the body, which could be any of these or their combination: breathing, swallowing or absorbed through the skin. In addition, contact with some substances can cause irritation or corrosive burns.

Many substances affect health, not just safety. The risk to harm from a substance would not only depend upon the properties of the substance but its concentration and the way it is used. For example, cleaning with solvents can be a risk, especially in a confined space with poor ventilation – people become ill or could even die from the harmful vapors. People engaged in working with hazardous substances should answer these questions:

- Do I have clear instructions as how to use this chemical?
- Have I read the SHOC (Safe Handling Of Chemicals) Card's details? and do I understand what it says?
- Have I clear instructions on what PPE I need?
- Do I know what to do in the event of something going wrong?

In the event of a substance affecting skin after contact, or splashes into eyes, wash with plenty of water, and seek first aid.

6.12.1 Explosive materials[16, 19]

Explosive material is defined as a chemical compound or mixture of compounds, which suddenly undergoes a very rapid chemical transformation, with the simultaneous production of large quantities of heat and gases. It has been classified as follows and labeled as shown in Figure 6.7(a).

- Class A explosives are materials that are capable of detonation by means of a spark or flame or even with a small to moderate shock wave.
- Class B explosives are materials that pose a hazard principally because they are rapidly combustible, as opposed to their potential detonation.
- Class C explosives are materials that do not ordinarily detonate in restricted quantities and, thus, a minimum explosion hazard.

6.12.2 Compressed gases[16, 19]

A compressed gas is defined as any material or mixture that is in a container with either an absolute pressure exceeding 40 psi at 70°F or an absolute pressure 104 psi at 130°F.

As required by the DOT (Department of Transportation, USA); the cylinders and other containers should be labeled as under (Figure 6.7(b)):

- Flammable gases – Red.
- Nonflammable gases – Green.

Cylinders of compressed or liquefied gases present a hazard in a fire since an increase in temperature causes an increase in pressure. Leakage of a compressed gas

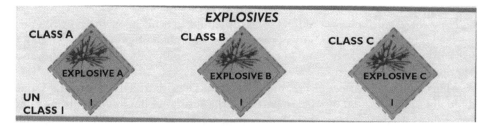

Figure 6.7(a) The labels describing explosive materials.

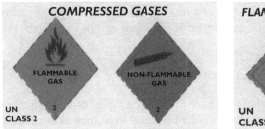

Figure 6.7(b) The labels describing compressed gases (left). (c) Flammable liquid (right).

Figure 6.7(d) Label describing an oxidizing substance (left). (e): Corrosive substance (right).

Figure 6.7(f) Label describing flammable solid.

can produce an explosive atmosphere. Danger of 'rocketing' during a fire is also there. Proper storage compartments are, therefore, essential.

6.12.3 Flammable liquids and solids[16, 19]

A flammable liquid is one that evolves flammable vapors in air at a temperature of 80°F or below as determined by a specific method. The vapour from flammable liquids ignite when they are mixed with air in an enclosed space. They bear the label as shown in 6.7(c).

6.12.4   Chemically reactive materials[16, 19]

DOT defines an oxidizing substance as a material that yields oxygen readily to stimulate the combustion of organic matter. Such reactions can be explosive. Figure 6.7(d) displays its label.

6.12.5   Corrosive material[16, 19]

A corrosive material (as labeled in Figure 6.7(e)) is a liquid or solid that is capable of causing visible destruction or irreversible alterations in human skin tissue at the area/site of contact.

6.12.6   Flammable solids[16, 19]

Flammable solids include materials (as labeled in Figure 6.7(f)) other than explosives, which are likely to cause fire by self-ignition through friction.

6.12.7   Controlled materials[16, 19]

The substances as described above in sections 6.12.1 to 6.12.6; are also known as *'Controlled products'*. Position table 6.9 below this from its current position.

6.12.8   Workplace Hazardous Materials Information System (WHMIS)

This is a uniform system designed to ensure that a standard method of identifying controlled products is used throughout Canada. This could be used in your own plant wherever it is located. Three key elements of WHMIS are:

1   Labeling of all controlled products – Each supply of the controlled product must be labeled using one or more of the eight hazards symbols, which have been described in Table 6.9.
2   Material Safety Data Sheet (MSDS) – It must be made readily available to workers. Employers must identify products created or blended in their facilities, which are controllable products.

Table 6.9 Classification of controlled products.[19]

Class	Description	Remark
A	Compressed Gas	Hazardous materials are
B	Flammable and combustible materials	usually labeled by their
C	Oxidizing Material	manufacturer with such terms
D	Poisonous and infectious material	and phrases as: **'Caution'**
	1 Materials causing immediate and serious toxic effects	**'Danger' 'Warning' and 'Handle with care'.** Classes
	2 Materials causing other effects	'D' have three symbols and
	3 Bio-hazardous infectious material	rest each one symbol. This makes a total of 8 symbols.
E	Corrosive material	
F	Dangerously reactive material	

3   Training to all concerned – Those who are handling or in contact with the hazard-ous materials must be fully trained. Training involves proper labeling, emergency procedures, health hazards, personal protection and MSDS inspection.

6.13   HAZARDS ANALYSIS METHODS[2–5, 7, 11, 19]

Every industry has hazards of various kinds, and there are a number of methods/ techniques that are applied to identify them. Figure 6.8 outlines guidelines as which technique could be applied during a particular phase of an industrial setup. Figure 6.9

Different phases of an industrial setup	Hazards Analysis Methods											
	Safety Review	Check List	Relative Ranking	PHA	What-If Analysis	What-if -Checklist	HAZOP	FMEA	FT	ET	OCA	HRA
R&D	O	O	•	•	•	O	O	O	O	O	O	O
Conceptual Design	O	•	•	•	•	•	O	O	O	O	O	O
Pilot plant operation	O	•	O	•	•	•	•	•	•	•	•	•
Detailed Engineering	O	•	O	•	•	•	•	•	•	•	•	•
Construction/Start up	•	•	O	O	•	•	O	O	O	O	O	•
Routine Operation	•	•	O	O	•	•	•	•	•	•	•	•
Expansion or Modification	•	•	•	•	•	•	•	•	•	•	•	•
Incident investigation	O	O	O	O	•	O	•	•	•	•	•	•
Decommissioning	•	•	O	O	•	•	O	O	O	O	O	O

O – Rarely used or inappropriate.   • – Commonly used

Figure 6.8 Application of hazards analysis methods during various phases of an industrial setup.

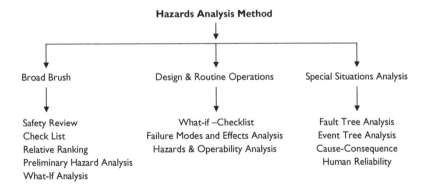

Figure 6.9. Classification: hazards analysis techniques/methods.[2–5, 7, 11, 19]

Table 6.10 Summary and overview of selected HA methods.[2,3,7,11,12,19]

Name	Purpose	When to use	Procedure	Types of results	Assessment outcome	Data requirement	Limitations; Comments
Safety check list	Identification of safety issues and concerns that need to be addressed.	Early in conceptual or preliminary design phase.	Check off applicable safety items or pre-designed list	Checklist of items of concern	Qualitative only	Gross knowledge of system and applicable safety standards	Success limited to the experience of users & breadth of list.
Job safety analysis	Provide safety requirements for simple job tasks	For existing job procedures with annual update	Step by step review of job tasks	List of specific requirements to do tasks safely	Qualitative only	Written job instructions are helpful	Only good for well defined, noncomplex job tasks
What if analysis	Identification of likely things that could go wrong and possible controls	Popular approach that can be used in most situations, as system changes	Asking "What if" questions at each step of the process	List of potential problems and recommended controls	Qualitative only	Operating instructions, flow diagrams	Depends on team members' experience with similar situations
Hazard and operability studies HAZOP	Identification of problems that could compromise a system's ability to achieve intended productivity	Late design phase when design is nearly firm; also for an existing system when a major redesign is planned.	Examine installation diagrams, flowchart at each critical node; identifying operational deviations, causes and consequences.	List of hazards and operating problems, deviations from intended functions, consequences, causes and suggest changes.	Qualitative with quantitative potential.	Detailed system descriptions, flow chart, procedures, knowledge of instruments and operations.	Depends heavily for its success on data completeness and accuracy of drawings
Failure mode and effect analysis	Identification of all the ways a set of equipment can fail, and each failure mode's effect(s) on the system.	At design, construction, or operation to review every 3 to 5 years.	Collect up-to-date design data on equipment and relationship to the rest of the system; list all conceivable malfunctions; describe effects.	List of identified failure modes, potential effects, and required controls.	Qualitative although can be quantified if failure probabilities for components are known.	System equipment list: knowledge of equipment function, knowledge of system function.	Poor at interactive sets of equipment failures that leads to events; not useful for errors or common –cause failures
Fault tree analysis	Deduction of causes of unwanted events via knowing combinations of malfunctions.	At design, operation to update for significant changes.	Construct a diagram with logic symbols to show the logical relationships between situations.	List of sets of equipment and human errors that can result in a specific unwanted event.	Qualitative with quantitative potential with probabilistic data on components and subsystems	Complete understanding of system's functions	Enable ID and qualitative examinations of critical factors and interrupt modes for chain failures

classifies these techniques into 3 groups: Broad Brush (for preliminary assessments of hazards), Design & Routine Operations (to assess what could happen in the event of failures) and Special Situations (to analyze reasons for failures that have already occurred). Table 6.10[2, 3] describes each method in terms of Purpose, When to use, Procedure, Types of results, Assessment outcome, Data requirement and Limitations.

- Safety Review
- Check List
- Relative Ranking
- Preliminary Hazard Analysis (PHA)
- What-If Analysis
- What-if – Checklist
- Hazards and Operability Analysis (HAZOP)
- Failure Modes and Effects Analysis (FMEA)
- Fault Tree (FT) Analysis
- Event Tree (ET) Analysis
- Cause-Consequence Analysis (CCA)
- Human Reliability Analysis (HRA)

6.14   INHERENT SAFER DESIGN STRATEGIES [2–5, 7, 11, 21]

Safety technology must work right the first time. Usually, there is no opportunity to adjust or improve operations.[5] A well-planned plant would follow this concept. Inherent safety means built-in safety features that could ensure smooth running of a plant safely. This strategy advocates: why not remove the danger in the first place? Its principles incorporate the following features:

Minimize use of hazardous substances; if feasible substitute them by less hazardous materials. It also advocates moderating the process/design and making the process, equipment and procedure as simple as possible by using less aggressive conditions. In addition, due care should be given when locating the plant and its layout. Sections 6.14.1 to 6.14.5 describe these features.

6.14.1   Minimize

Use smaller quantities of hazardous substances (also called Intensification).

For example in a Chemical Plant, which involves use of hazardous material, the following checklist could be useful:[4]

- Have all in-process inventories of hazardous materials in storage tanks been minimized?
- Are all the proposed in-process storage tanks really needed?

- Has all processing equipment handling hazardous materials been designed to minimize inventory?
- Is process equipment located to minimize length of hazardous material piping?
- Can piping size be reduced to minimize inventory?
- Is it possible to feed hazardous material (for example, chlorine) as gas instead of liquid to reduce pipeline inventories?
- Is it possible to generate hazardous reactants on site from less hazardous materials, minimizing the need to store or transport large quantities of hazardous material?

When designing a plant, every piece of process equipment should be specified as large enough to do its job, and no larger. We should minimize the size of all raw materials and in-process intermediate storage tanks, and question the need for all in-process inventories, particularly hazardous materials. Minimizing the size of equipment not only enhances process safety but it can often save money.

Reactors:

These can represent a large portion of risk in a chemical plant. A complete understanding of reaction mechanism and its kinetics is essential to the optimal design of a reactor system. This includes chemical reactions and mechanisms, as well as physical factors such as mass transfer, heat transfer, and mixing.

The reactor configuration that maximizes yield and minimizes size, results in a more economical process reducing generation of by-products and waste, and increases inherent safety by reducing the reactor size and inventories of all materials. Some examples are:

Continuous Stirred Tank Reactor (CSTR): this is usually much smaller than a batch reactor for a specific production rate.

Tubular Reactors: often offer the greatest potential for inventory reduction. They are simple in design containing no moving parts and a minimum number of joints and connections.

Loop Reactor: this is a continuous steel tube or pipe, which connects the outlets of a circulation pump to its inlet. It is much smaller than a batch reactor producing the same amount of product, as shown in Table 6.11.

Table 6.11 Effect of reactor design on size and productivity for gas-liquid reaction.[4]

Reactor feature	Batch stirred tank reactor	Loop reactor
Reactor size (l)	8000	2500
Chlorination time (hrs)	16	4
Productivity (kg/hr)	370	530
Chlorine usage (kg/100 kg product)	33	22
Caustic usage in vent scrubber (kg/100 kg of product)	31	5

6.14.2   Substitute/Eliminate

This means replacing a material with a less hazardous substance. This aspect could also be understood using the checklist mentioned below, which is applicable for a Chemical Industrial Unit:

- Is it possible to completely eliminate hazardous raw materials, process intermediates, or by-products by using an alternative process or chemistry?
- Is it possible to completely eliminate in-process solvents by changing chemistry or process conditions?
- Is it possible to substitute less hazardous raw materials?

  - Non-combustibles rather than flammable solvents
  - Less volatile raw materials
  - Less toxic raw materials
  - Less reactive raw materials
  - More stable raw materials

- Is it possible to substitute less hazardous final product solvents?
- For equipment containing materials which become unstable at elevated temperature or freeze at low temperature, is it possible to use heating and cooling media, which limits the maximum and minimum temperature attainable?

Reactive Distillation[4, 11]
The combination of several unit operations into a single piece of equipment can eliminate equipment and simplify a process, but there may be inherent safety conflicts. A case-by-case basis decision should be taken. Reactive distillation is a technique combining a number of process in a single device. One company developed a reactive distillation process to manufacture methyl acetate that reduced the number of distillation columns from eight to three, and also eliminated an extraction column and a separate reactor. This resulted in a reduction in capital and operating costs.

6.14.3   Moderate

These aspects cover the operating conditions which are required for a process, and are advocated so as to use less hazardous conditions, a less hazardous type of materials, or facilities, which minimizes the impact of a release of hazardous material or energy (also called Attenuation and Limitation of Effects). The following checklist could be helpful in understanding this aspect.

- Can supply pressure of the raw materials be limited to less than the working pressure of the vessels they are delivered to?
- Can reaction conditions (temperature, pressure) be made less severe by using catalysts, or by using a better catalyst?

- Can the process be operated at less severe conditions? If this results in a lower yield or conversion, can raw material recycling compensate for this loss?
- Is it possible to dilute hazardous raw materials to reduce hazard potential? For example: dilute nitric acid instead of concentrated fuming nitric acid.

6.14.4 Simplify

The principle of 'KISS': Keep It Safe and Simple – Design, Layout, Equipment, Process and Procedure; is the essence of this concept. It should be applied in every plant. The design facilities that eliminate unnecessary complexity and have better tolerance for errors (also called *Error Tolerance)* should be preferred. Answering the following questions could be a useful guide[4, 7, 11]

- Can equipment be designed to be sufficiently strong to totally contain the maximum pressure generated, even if the 'worst credible event' occurs?
- Is all equipment designed to totally contain the materials which might be present inside at ambient temperature or at the maximum attainable temperature?
- Can several process steps be carried out in separate processing vessels rather than a single multipurpose vessel? This reduces the complexity and number of raw materials, utilities, and auxiliary equipment connected to a specific vessel, thereby reducing the potential for hazardous interactions.
- Can equipment be designed such that it is difficult or impossible to create a potentially hazardous situation due to an operating error (for example, by operating an improper combination of valves).

    Other examples[5]

- Pick equipment that requires less maintenance
- Select equipment with low failure rates
- Keeping the piping system neat and visually easy to follow.

6.14.5 Location/Sitting/Transportation[4, 5, 7, 11]

This aspect is with regard to location and layout of the plant and ancillary facilities. Guidelines given below could be useful in taking decisions on these aspects:

- Preferably the location should be away from other hazardous installations.
- Preferably the process unit should be isolated from housing or residential colonies and also from other plants, which could have their own adverse impacts.
- Least transportation should be preferred, by choosing the shortest route, particularly for hazardous materials.
- Work out the possibility that a multi-step process, where the steps are done at separate sites, be divided up differently to eliminate the need to transport hazardous materials.

6.14.6   Change to inherent safety strategy

It is wise to adopt this strategy right from conceptual, design and construction phases for any forthcoming industrial venture. Equally important is introducing this concept to existing plants whenever practical.

6.15   BREATHING APPARATUS[12, 17, 20, 23]

Table 6.12 classifies various types of breathing apparatus, which could be used during rescue and recovery operations, and also while dealing with fires or during any emergency.

Table 6.12 Details of breathing apparatus and their applications.[12, 17, 20, 22]

SN.	Type	Breathing apparatus brief description	Use
1	Self contained closed circuit breathing apparatus	Exhaled air is passed through the chemical, which replaces CO_2 with Oxygen.	Professional firemen and rescue workers mainly use it.
2	Self contained open circuit breathing apparatus	Supplies air on demand (through lung governed valve) or at a constant flow (through a pressure reduction and flow controller) to a full-face mask with an exhalation valve.	Used by professional firemen. A limited supply of compressed air is contained in a cylinder attached to the wearer's body.
3	Compressed air-line breathing apparatus	This is same as no. 2, except with a small diameter hose from a safe compressed air source replacing the compressed air cylinder	Used in industry for tasks such as tank cleaning and paint spraying where it is impossible to provide safe respirable atmosphere.
4	Fresh air hose breathing apparatus	Air suitable for respiration is drawn from an adjacent uncontaminated area through a large diameter hose by the breathing of the wearer, with or without blower. The latter is used when the length exceeds 9 m. Or the atmosphere is immediately hazardous to life. In this case the wearer should also wear a rescue harness attached to the rescue line.	
5	Escape breathing apparatus	A simple form of self-contained type of breathing apparatus for short duration use, which is usually open-circuit type, with its own small compressed air cylinder.	This is needed in works where there is a possibility of a large escape of toxic gas such as chlorine or ammonia, which could trap workers in an emergency. It may also be used where a risk of workers getting trapped may arise due to smoke and fumes from fire.

6.16 THE WAY FORWARD[9]

Working in the hazardous industries is a big challenge, which could be met by following these steps, and as illustrated in Figure 6.10.

- Working in hazardous conditions is a challenge – **accept it!!!**
- Understand it thoroughly. Bring perfection amongst those concerned through rigorous training based on past experience
- Consider whatever improvements could be incorporated, and
- Implement them
- Enjoy results
- Make the 'key learning points' standard practices and way of life. Inculcate them.

Figure 6.10 Logical steps to face any hazard. (After Latus, M. 2006).[9]

6.17 VOCABULARY[7, 8, 11, 18, 20]

- Adhesive – sticking.
- Asphyxiate gases – these are either simple in nature which exclude oxygen from lungs, for example, CH_4 and CO_2. First symptom is fast breathing and hunger for air. With time there may be nausea, vomiting, lying flat on ground, loss of consciousness and finally convulsion, deep coma and death.
- Cohesive – molecules of substance are held together. Cohesive means tendency to stick together.
- Combustible – flammable – that catches fire and burns easily. A flammable substance.
- Condensation – to change to a denser form, as from gas to a liquid. A condenser is an apparatus to convert gases into liquid.
- Cracking – the process of breaking down hydrocarbons by heat and pressure.
- Defuse – to remove fuse from bomb or to remove cause of tension from a crisis.
- Distill – to let fall in drops.
- Distillation is the process of heating a mixture and condensing the resulting vapor to a more nearly pure substance.
- Emulsion – a liquid formed by suspension of one liquid in another.
- Explosion – bursting with loud noise. Sudden, rapid and widespread increase.
- Flash point – the lowest temperature at which vapor will ignite.
- Herbicides – Commonly used as a weed killer. Ingestion may result in damage to the liver, kidneys and lung. There is no antidote and death occurs in about half the cases.
- Ignition – start burning
- Incineration – to burn to ashes.
- Incinerator – a furnace or other device for incinerating rubbish.
- Insecticide – any substance used to kill insects
- Melting – when a solid changes into liquid, it is in process of melting.
- Noxious – harmful to health.
- Osmosis – Tendency of a solvent to pass through a semi-permeable membrane so as to equalize concentration on both sides.
- Oxidize – to unite with oxygen as in burning or rusting. The process of oxidizing is called oxidation.
- Pesticide – any chemical used for killing insects (group of a small animals), weeds (undesired plant, grass)
- Reduction – to remove oxygen or to combine with hydrogen or to bring into metallic state by removing nonmetallic elements.
- Scrubber – a device for cleaning gas.
- Smelting – melting ores so as to separate impurities from pure metal.
- Suspension – particles dispersed in a liquid but not dissolved.
- Toxic – acting as poison. Toxic gases are those which on breathing in sufficient quantity for sufficient time, will seriously disable and possible kill a person. They can be grouped into three classes based on their action on the body: Asphyxiate gases, Irritant gases and Poisonous gases.
- Vapor – anything as smoke, fumes. The gaseous form of any solid or liquid.
- Viscous – cohesive and sticky.

QUESTIONS

1 'Working in hazardous industries is a big challenge' – how can we meet this challenge?
2 Can combustible dusts be responsible for fires and explosions? Can dusts be ignited by static electricity?
3 Classify hazards.
4 Define deflagration and detonation. In what connection are these terms used? What is auto-ignition temperature?
5 Define hazards. List common hazards associated with chemical and petrochemical industries. What are the sources of electrostatic hazards in these industries? How they can be minimized?
6 Define the terms flash point, ignition temperature, fire point.
7 Draw a 'Fire Triangle' and 'Explosion Pentagon', OR write main components of fire and explosions.
8 Give classification of controlled products.
9 Give color codes of various types of portable types of fire extinguishers.
10 How do explosions differ from fires? List the components that are responsible for causing explosions. Why are explosions more severe than fires? What are the main causes of frequent severe losses in the chemical and petrochemical industries?
11 How should flammable liquids be handled?
12 In a Chemical Industry some cylinders of liquefied gas are lying on the floor; in the event of fire what danger you anticipate?
13 In a process to identify hazards, what items/installation in your factory would you like to include in your checklist? (Hint: Name only the items and not their details)
14 In which event are rescue and recovery operations necessary? Name the apparatus that should be used during such operations.
15 List hazardous analysis methods and write their applications during various phases of an industrial setup.
16 List hazards associated with underground mining and tunneling operations.
17 List hazards associated with using machines, equipment, appliances, tools and tackles for carrying out various Unit-Operations and suggest measures/precautions to minimize them.
18 List precautions that should be taken against the generation of electro-static charge during the handling and storage of flammable liquids. Why are the gases and vapors of flammable liquids more dangerous than the liquids themselves?
19 List the type of layers that are usually provided between hazards that may arise by any process and the people, property and environment surrounding it. What does 'HAZOPS' stand for?
20 Name the most common irritant gases and asphyxiate gases.
21 Safety technology must work right the first time – Do you agree with this? If yes, how it could be achieved?
22 Suggest types of extinguishers for use on 'A' class & 'B' class fires separately. Can you suggest fire extinguishers for class C & D fires? If electrical equipment is on fire, how will you extinguish it?

23 What are the Do's and Don'ts while operating plant, machinery and equipment; tabulate them.
24 What does WHMIS stand for? What is its application? Classify controlled products.
25 What is 'Rocketing'; in what situation/scenario can it occur?
26 What is the concept of upper and lower flammability limits? Do you think fire or explosion would not occur beyond this limit?
27 What procedure should be followed in the event of outbreak of fire? Prepare 'Standing Orders' for the factory/organization you belong to.
28 What should process risk management strategies be? Or, how to classify safety strategies?
29 What types of dusts are responsible for fires and explosions? What is the effect of particle size of dust on fires and explosions?
30 What types of events are responsible for the major losses in Chemical Industries? Are explosions more severe in terns of financial (cost) losses than fires? Do fires occur more than explosions in the chemical industries? List the causes of explosions in the chemical industries.

REFERENCES

1 Accident Prevention Manual For Industrial Operations, Chicago, 1974.
2 Amyotte, P. (2000) Course on: Fundamentals of Loss Management, Dalhousie University, 2000. (Including review of HA methods by DiBeraradinis, 1999)
3 Amyotte, P. (1997) Orientation course, Dalhousie University. Overview of Hazards methods, Supplementary notes.
4 CCPS. (1992) Guidelines for Hazardous Evaluation Procedures, 2nd Edn. Center for Chemical Process Safety of the American Institute of Chemical Engineers, New York, NY. A Sample Inherently Safer Process Checklist: CCPS 1993a.
5 Crowl, D.A. & Louvar, J.F. (2002) Chemical Process Safety. New Jersey, Prentice Hall PTR. pp. 36, 54, 226, 227, 249–281, 538, 550.
6 Dow's Fire and Explosion Index Hazard Classification Guide, 7th ed. New York, American Institute of Chemical Engineers, 1994.
7 Gupta, J.P. (2000) Safety course: Safety in Chemical Industry including Dow Index by Mond Division of ICI (fire, explosion and toxic indices). Sultan Qaboos University, Oman.
8 Kupchella, C.E. & Hyland, M.C. (1993) Environmental Science. Prentice-Hall International Inc. pp. 558.
9 Latus, M. (2006) Leadership in Safety – One Company's Approach. Pres. In: International conference focusing on safety and health on 14–16 November, Johannesburg, South Africa. London ICMM. (Permission: Newmont Mining Corporation)
10 Loomis, I. & Mercer, D.B. (1997) Fires and fire extinguishers, Dalhousie University, Orientation course.
11 Pegg, M.J. (2003) Safety class-notes and course at Sultan Qaboos University, Oman.

12 Petroleum Development of Oman – HSE management (2000–4).

13 Ridley, J.R. (1994) Toxicology. In: *Workplace safety*. Ridley, J. and Channing J. Edts. Butterworth Heinemann. pp. 3.30–3.33.

14 Ridley, J.R. (1999) Safe use of machinery. In: *Workplace safety*. Ridley, J. and Channing J. Edts. Butterworth Heinemann. pp. 4.76, 4.77.

15 Robert, L. & Somerville, P.E. (1993) Control of liquefied toxic gas releases. In: *Prevention and control of accidental release of hazardous gases*. Fthenakis, Vasils M (Edtr.); Int. Thomsan Pub.

16 Sharifa Al-Rashdi. (2000) Term project – Literature survey on: Hazardous material including UN Classification of hazardous material. Sultan Qaboos University, Oman.

17 Stevenson, J.W. (2001) Mine fires and explosions. *Mine health and safety management*, Michael Karmis (edt.,) Littleton, Colorado, SME, AIME. pp. 371; www.smenet.org

18 Tatiya, R.R. (1996–2004) Class-notes, HSE for Petroleum, Chemical and Mining Engineering, Sultan Qaboos University, Oman.

19 Traves, T. & Amyyotte, P. (1997) Dalhousie University, Orientation course and Employee handbook on W.H.M.I.S.

20 Vutukuri, V.S. & Lama, R.D. (1986) *Environmental Engineering in Mines*. Cambridge University Press, 1986. pp. 87–100, 244, 264.

21 Washington DC; Manufacturing Chemists Association report, July 1962. pp.106.

22 Waterhouse, P. (1999) Fire precautions. *Workplace safety*. Ridley, J. and Channing J. Edts. Butterworth Heinemann. pp. 4.76, 4.77.

23 Yahya, Al Hajri. (2000) Term project – Literature survey on: Fire Fighting Department. Sultan Qaboos University, Oman.

Chapter 7

Occupational Health & Safety (OHS)

'HSE is not a commodity that can be bought but is like a shrub which must be planted, cultivated, fertilized and pruned regularly through vocational training, refresher courses, consistent checks and supervision.'

Keywords: Occupational Health, Industrial Hygiene, Exposure to: Dusts, Fibers, Noise, Vibrations, Welding, Salts, Diesel Emissions, Toxic Gases, Extreme Temperatures, Heat & Humidity, Radiation, Corrosive Liquids; Aqueous Effluents; House Keeping, Working Conditions, Organizational Culture and Commitment, Workplace Stress, Ergonomics, Medical Surveillance.

7.1 OCCUPATIONAL HEALTH AND SAFETY (OHS)

As described in chapter 1, every industry has hazards, but their magnitude (quality and quantity) and direction (as to whom they influence) differs. Figure 1.1 briefly outlines all these aspects. Hazards endanger the safety of men, machines, equipment and property. They degrade surroundings (the environment). All these impacts are detrimental to the health of people involved. The health of industrial workers is thus the first target, and then that of the people living in areas surrounding the industries. The Bhopal disaster in India in 1984 could be cited as an example in which not only the employees of the company Union-Carbide but also about 2000 people living in the surrounding areas were affected badly.[20] This example illustrates that *Occupational Health and Safety (OHS)* is not confined to industrial workers but also covers those living nearby, including flora and fauna. Chapter 8 details Industrial Safety, and this chapter is confined to the parameters/elements/factors other than the safety, that are responsible for **Occupational Health (OH)**. What could cause illness, deterioration in health, and inefficiency due to mental and physical stresses to industrial workers in first place, and then the people living nearby? The factors responsible for this are as shown in Figure 7.1.

7.2 ELEMENTS: OCCUPATIONAL HEALTH (OH)[1]

- Industrial Hygiene including efficient effluent discharge
- Housekeeping and working conditions

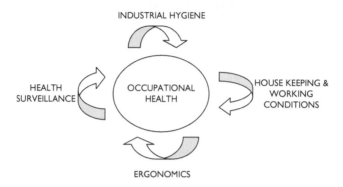

Figure 7.1 Elements: Occupational Health (OH).

- Ergonomics
- Occupational Health Surveillance.

7.3   INDUSTRIAL HYGIENE

The word hygiene means healthful/healthy, and industrial hygiene refers to a healthy-workplace. Industrial hygiene refers to those environmental factors or stresses in the workplace, which may cause sickness, impaired health or well-being, or significant discomfort among workers, and effects on local communities, flora and fauna.

7.3.1   Steps for managing industrial hygiene[1]

- Anticipated
- Identified/Recognized
- Evaluated: analyzed for quantity as well as quality (either by direct measurement or using appropriate methods of estimation)
- Recorded and reported, as appropriate
- Control measures taken to minimize them.

Exceeding the threshold limits has the potential to cause annoyance/nuisance and may, in some circumstances, present a risk to human health.

7.4   FUNDAMENTAL PRINCIPLES OF INDUSTRIAL HYGIENE[1, 23]

7.4.1   Anticipation

Referring to Figure 7.2, anticipation refers to the pre-production phase during which the concept is first developed; engineering studies and detailed design then follow. During these phases for any industry the likely hazards, work-stresses and unhygienic conditions that may arise, or encountered should be anticipated. A perfect design, if not eliminating them, should certainly minimize their magnitude. Figure 7.3 outlines the logical steps in assessing the workplace risk and remedial measures to combat/minimize them.

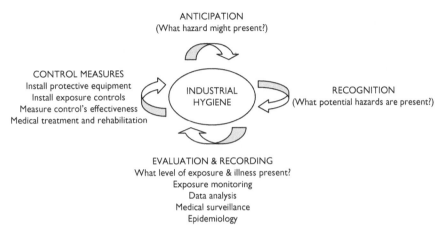

Figure 7.2 Steps for managing industrial hygiene.

Source: Mine health and safety management, Michael Karmis (edt.), SME, AIME, Littleton, Colorado, 2001. Permission: www.smenet.org. (Partly information used)[1]

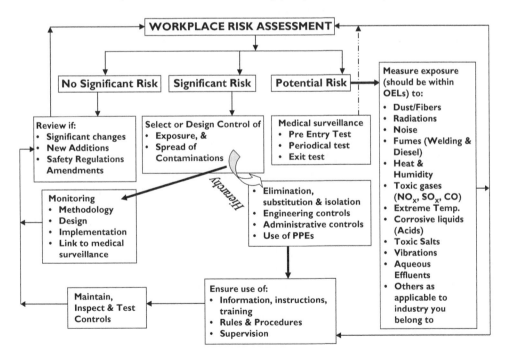

Figure 7.3 Logical steps in assessing workplace risk, and remedial measures to combat/minimize them. (Partially information used from Ref. [2, 14, 32]).

7.4.2   Identified/Recognized

During the regular production phase these conditions are felt, experienced and identified. Bulleted below are few examples:

- Dust generation
- Exposure to fibers – asbestos, non-asbestos, industrial fibers – Non-Asbestos Respirable Fibers (NARF), Man-Made Mineral Fibers (MMMF)
- Noise generation
- Vibrations
- Welding Hazard (Tables 7.1 and 7.2)
- Exposure to salts such as sodium cyanide
- Diesel emissions
- Emission of toxic gases such as CO, NO_x, SO_x
- Extreme Temperatures – Heat & Humidity
- Radiation Hazards
- Corrosive Liquids such as acids
- Aqueous Effluents.

7.4.2.1   Dust generation[3, 6, 16, 30]

Dust is generated by industrial operations such as mineral fragmentation (drilling & blasting), digging, excavation, loading, hauling, crushing, screening, dry grinding, storage, handling solids, transportation of finished goods and their unloading. This phenomenon is common during civil excavations, tunneling and mining operations, or even in those industries where bulk handling of solids takes place. This has been dealt with in detail in sec. 3.11.

7.4.2.2   Asbestos fibers[12, 13, 17]

This refers to a group of silicate minerals such as asbestos, which can be readily separated into thin, strong fibers that are flexible, heat resistant and chemically inert. Asbestos minerals are used in fireproof fabrics, yarn, cloth, paper, paint filler, gaskets, roofing composition, reinforcing agent in rubber and plastics, brake linings, tiles, electrical and heat insulation, cement and chemical filters. Fibers are dangerous when breathed. As such, measures should be taken against fibers becoming airborne. Asbestos is very well known for fibers, which may be trapped in the lungs of those who work with it. There are thousands of materials used in the construction industry that contain asbestos. These materials include, but are not limited to, pipe and boiler insulation, acoustical coatings, sprayed-on fireproofing, plaster, ceiling tiles, and floor tiles. There is a strong correlation between asbestos and cancer, particularly in people who smoke. A survey indicates that smokers have 100 times[17] more risk of death from lung cancer than an otherwise similar individual who neither smokes nor works with asbestos. Asbestos could enter through inhalation or orally affecting lungs, the pleural cavity, or the gastrointestinal tract.[17] There are other fibers that pose health risks, which are known as Non-Asbestos Respirable Fibers (NARF) such as gypsum, mineral wool, fiberglass, cellulose, Man-Made Mineral Fibers (MMMF), and Refractory Ceramic Fibers (RCF) etc. Generally, these are defined by their size (certain length and width ratios), chemical composition, and physical properties. EPA under the Clean Air Act controls man-Made Mineral Fibers. Often fibrous materials that fall into these categories are the by-products of manufacturing processes.

7.4.2.3   Noise generation[3, 6, 16]

This refers to generation of sound, while operating machines, equipment, plant and factories, that impairs hearing. Other unit operations such as: transportation, conveying, loading, and unloading within the plants' premises are equally responsible for noises which ultimately could lead to what is known as Noise Induced Hearing Loss (NIHL). Sec. 3.9 deals with this in detail.

7.4.2.4   Vibrations[3, 6, 16]

Vibration can be both a safety and health problem, as described in sec. 3.10.

7.4.2.5   Welding[23]

There are more than 80 types of welding; it is a common operation in any industry. Usually it is undertaken in a confined space, and the types of flame and fumes it produces are unhealthy. In addition to safety hazards posed by fire, explosion, projectiles, heat and burns, electrical and energy sources, welding produces a number of health hazards. Among these are physical and chemical hazards.

- Physical hazards: Physical hazards that may arise from welding include: Ultraviolet radiation – the bright light generated by the electric arc contains ultraviolet radiations. Without a shield or helmet fitted with the correct filter grade, exposure can produce painful conjunctivitis or photo-ophthalmic, known as "eye flash" or arc eye. Immediate medical attention must be provided because overexposure may also cause over heating and skin burn. Noise levels could be up to 90 dBA in several welding processes such as plasma, resistance and gas welding, and from the use of compressed air.[23]
- Chemical hazards due to fumes and foul gases from welding various metals and minerals, result in damage to various organs of the body of the affected person, as shown in Table 7.2. Table 7.1 describes welding hazards.

Table 7.1 Description and hazards of welding process. (*After Pon & Jones: Mine health and safety mangement*, Michael Karmis (edt.), SME, AIME, Littleton, Colorado, 2001. Permission: www.smenet.org).[23]

Welding process	Description	Hazards
Gas welding and cutting		
Welding	Torch melts metal surface and filler rod, causing a joint to be formed.	Metal fumes, nitrogen dioxide, carbon monoxide, noise, burns, infrared radiation, fire, explosions
Brazing	Two metal surfaces are bonded without melting the metal. Melting the filler metal above 450°C. Heating is done by flame heating, resistance heating and induction heating.	Metal fumes (especially cadmium), fluorides, fire, explosions, burns.
Soldering	Similar to brazing, except melting temperatures of the filler metal are lower. Heating is done using a soldering iron.	Fluxes, lead fumes, burns.
Metal cutting and flame gauging	The metal is heated by a flame and a jet of pure oxygen is directed into the point of cutting and moved along the line to be cut. In flame gauging, a strip of surface metal is removed but the metal is not cut through.	Metal fumes, nitrogen dioxide carbon monoxide, noise, burns, infrared radiation, fire, explosion
Gas pressure welding	Parts are heated by gas jets while under pressure and become forged together.	Metal fumes, nitrogen dioxide, carbon monoxide, noise, burns, infrared radiation, fire and explosion
Flux-Shield Arc Welding		
Shielded metal arc welding (SMAC); stick arc welding, manual metal arc welding (MMA), open arc welding:	Uses a consumable electrode consisting of a metal core surrounded by a coating.	Metal fumes, fluorides especially with low-hydrogen electrode, infrared and ultraviolet radiation, burns, electrical fire, noise, ozone, nitrogen dioxide.
Submerged arc welding (SAW)	A blanket of granulated flux is deposited on the work place, followed by a consumable bare metal wire electrode. The arc melts the flux to produce a protective molten shield in the welding zone.	Fluorides, fires, burns, infrared radiation, electrical, metal fumes, noise, ultraviolet radiation, ozone, nitrogen dioxide.
Gas-Shield Arc Welding		
Metal inert gas (MIG), gas metal arc welding (GMAC)	The electrode is normally a bare consumable wire of similar composition to the weld metal and is fed continuously to the arc.	Ultraviolet radiation, metal fumes, ozone, carbon monoxide (with carbon dioxide) gas, nitrogen dioxide, fire, burns, infrared radiation, electrical, fluorides, noise.

Process	Description	Hazards
Tungsten inert gas (TIG), gas tungsten arc welding (GTAW), heliarc.	The tungsten electrode is non-consumable and filler metal is introduced as a consumable into the arc manually.	Ultraviolet radiation, metal fumes, ozone, nitrogen dioxide, fires, burns, infrared radiation, electrical fire, noise, fluorides, carbon monoxide.
Plasma arc welding (PAW) and plasma arc spraying, tungsten arc cutting.	Similar to welding, except that the arc and stream of inert gases pass through a small orifice before reaching the workplace, creating a "plasma" of highly ionized gas which can achieve temperatures of more than 33000°C. This is also used for metalizing.	Metal fumes, ozone, nitrogen dioxide, ultraviolet and infrared radiation, noise, fire, burns, electrical fire, fluorides, carbon monoxide, possible X-ray.
Gas-shield Arc welding		
Flux core arc welding (FCAW), metal active gas welding (MAG)	Uses a flux-cored consumable electrode; may have carbon dioxide shield (MAG).	Ultraviolet radiation, metal fumes, ozone, carbon monoxide (carbon dioxide gas), nitrogen dioxide, fire, burns, infrared radiation, electrical fire, fluorides, noise.
Electric resistance welding		
Resistance welding (Spot, seam, projection or butt welding)	A high current at low voltage flows through the two components from electrodes. The heat generated at the interface between the components brings them to welding temperatures. During the passage of current, pressure by the electrodes produces a forged weld. No flux or filler metal is used.	Ozone, noise (sometimes), machinery hazards, fire, burns, electrical fire, metal fumes.
Electro-slag welding	Used for vertical butt-welding. The work pieces are placed vertically, with a gap between them and a copper plates or shoes are placed on one or both sides of the joints to form a bath. An arc is established under a flux layer between two or more continuously fed electrode wires and metal plate. A pool of molten metal is formed, protected by a molten flux or slag, which is kept molten by resistance to the current passing between the electrode and work pieces. This resistance-generated heat melts the sides of the joint and making a weld. As welding progresses, the molten metal and slag are retained in the position by shifting the copper plates.	Burns, fire, infrared radiation, electrical fire, metal fumes.

(Continued)

Table 7.1 (Continued)

Welding process	Description	Hazards
Flash welding	The two metal parts to be welded are connected to a low voltage, high current source. When the ends of the components are brought into contact, a large current flows, causing "flashing" and bringing the ends of the components to welding temperatures. A good forged weld is obtained by pressure.	Electrical fire, burns, fire, metal fumes.
Other welding processes		
Electron beam welding	A work piece in a vacuum chamber is bombarded by a beam of electrons from an electron gun at high voltages. The energy of the electrons is transformed into heat upon striking the work piece, thus melting the metal and fusing the work piece.	X-rays at high voltages, electrical fire, burns, metal dust, confined spaces.
Arc air cutting	An arc is struck between the end of a carbon electrode (in a manual electrode holder with its own supply of compressed air) and the work piece. Jets of compressed air blow away the molten metal produced.	Metal fumes, carbon monoxide, nitrogen dioxide, ozone, fire, burns, and infrared radiation, electrical fire.
Friction welding	A purely mechanical welding technique in which one component remains stationary while other is rotated against it under pressure. Heat is generated by friction and at forging temperature, the rotation ceases. A forging pressure effects the weld.	Heat, burns, machinery hazards.
Laser welding and drilling	Laser beams can be used in industrial applications requiring exceptionally high precision, such as miniature assemblies and micro techniques in the electronic industry. The laser beam melts and joins the work piece.	Electrical, laser radiation, ultraviolet radiation, fire, burns, metal fumes, decomposition product of work piece coatings.
Stud welding	An arc is struck between a metal stud (acting as the electrode) held in a stud welding gun and the metal plate to be joined, and raises the temperature of ends of the components to melting point. This forces the stud against the plate and welds it. A ceramic ferrule surrounding the stud provides shielding.	Metal fumes, infrared and ultra violet radiation, burns, electrical fire, noise, ozone, nitrogen dioxide.
Thermite welding	A mixture of aluminum powder and a metal powder (iron, copper etc.) is ignited in a crucible, producing molten metal with the evolution of intense heat. The crucible is tapped and molten metal flows into the cavity to be welded (which is surrounded by a sand mold). This is often used to repair castings or forgings.	Fire, explosions, infrared radiation, burns.

Table 7.2 Possible constituents of welding fumes and their impacts to health. *(After Pon & Jones: Mine health and safety management,* Michael Karmis (edt.), SME, AIME, Littleton, Colorado, 2001. Permission: www.smenet.org).[23]

Particulates				Gases	
Pneumoconiosis			Pulmonary Irritants or toxic inhalants	Primary	
Fibrotic	Non fibrotic	Relatively harmless		Pulmonary	Non pulmonary
Silica Asbestos Copper	Beryllium	Carbon Carbon Iron Aluminum	Cadmium Chromium Fluorides Lead	Ozone Phosgene Phosphine Oxides of nitrogen	Carbon monoxide Carbon dioxide
			Manganese Mercury Molybdenum Nickel Titanium Vanadium Zinc		

7.4.2.6   *Hazardous salts*

Cyanide salts such as sodium cyanide (NaCN) are used to extract gold and silver from their ores and have adverse impacts to the surroundings when exceeds allowable limits.

7.4.2.7   *Diesel emissions*[23]

Diesel Particulate Matter (DPM)

The emissions from diesel engines produce minute solid particles (DPM) due to incomplete combustion and impurities in the fuel. This matter consists of impregnated carbon and a variety of organic compounds, such as paraffin (wax), aldehydes, and polynuclear aromatic hydrocarbons. Some of these compounds are recognized carcinogens. Unfortunately, the standard catalytic scrubber (oxidation catalytic converter) is not efficient at removal of these particulates and moreover the particulates do not remain uniformly diffused in the exhaust air of the mine (they are subject to stratification). DPMs are measured in micrograms per cubic meter. (One milligram = 1,000 micrograms). Table 7.3 reviews diesel emissions control options.

7.4.2.8   *Foul gases*

Foul gases, such as diesel fumes, are a potential source for adverse effects on human health. Table 3.6 details all the relevant aspects.

Table 7.3 Diesel emissions control options. *(After Pon & Jones: Mine health and safety management,* Michael Karmis (edt.), SME, AIME, Littleton, Colorado, 2001. Permission: www.smenet.org)[23]

Risk control measure	Effect/Advantages	Disadvantages/Penalty
Engine selection		
More efficient engines and electronic engine control systems	Produces less emissions	Initial cost and/or replacement costs. Retrofit and re-engineering costs.
Totally enclosed operator compartment with filtered air	Significantly reduces operator exposure.	Initial cost. Increased ongoing maintenance.
Engine maintenance		
Improved tuning. Oil cleanliness.	Optimizes efficiency and lowers total emissions	None. Cost effective.
Chemical decoking of engines.	Reduces particulates, may increase power. The effects of decoking may be less for newer designed engines or engines operating on low emission fuel.	When decoking engines underground, exhaust emissions need to be effectively managed immediately following decoking
Diesel test station (Controlled air flow with vehicle while testing engine exhaust gases)	Reduces confounding influences. Eliminates detector tubes. Better engine monitoring. Improves fleet availability.	Initial cost
Fuel quality		
Use low emission fuels (less than 0.05% sulfur)	Reduces particulates, carbon monoxide and hydrocarbons. Reduces odors	Requires engine returning to maintain power. Mixing of different fuels may have adverse effects. This may be a problem with hired equipment that has been used across a range of sites. Some low emission fuels may freeze in winter. Cost premium.
Engineering controls		
Filters	Reduces particulate emissions by up to 85%	Ongoing cost. Backpressure may become excessive if maintenance/ replacement intervals are not strictly followed. Excessive backpressure may cause emissions, reduces engine life, lower water-based conditioner (scrubber) water levels.
Exhaust post treatment (e.g. catalytic converters, catalyzed filters, precipitation)	Lowers emissions in some situations	Cost penalty.
Advanced engine control systems	Reduces emissions if properly used.	Reliability may be an issue. May require retraining of maintenance personnel.
Ventilation		
	Currently the major method used to manage the exposure of personnel to emissions.	Not well understood. May not be adequate when multiple vehicles in use.

(Continued)

Table 7.3 (Continued)

Risk control measure	Effect/Advantages	Disadvantages/Penalty
Work practices		
Vehicle control systems	Has been shown to lower exposure to emissions in consumption with adequate ventilation.	May limit the most cost effective and flexible use of equipment.
Road maintenance		
Driving procedures		

Source: Based on: NSW Minerals Council 1999.

Emission of toxic gases such as CO, NO_x, SO_x could be natural from mines and oil and gas fields. But in other industries (including Chemical Industries) it could be the result of the processes involved. They could also be the exhaust emissions from some sets of equipment. Chapter 3 deals with these in detail. Table 7.4 details these natural gases from strata and their impact when mixed with air.

7.4.2.9   Metals

Metals/elements have very significant role in our body, as shown in the Table 7.5; we must have these quantities of elements in our body. In fact we take these in through our diet which includes taking supplements and drinks; but if deficient we need to take medicines to fulfill such deficiencies.

These metals which are found as dust or fumes in many industries have adverse health effects when exceeding allowable TLVs, as shown in Table 7.6. Heavy metals can take the place of metals that are a normal part of enzymes, or otherwise bind to enzymes rendering them inactive. There could be other effects depending upon the particular type of metal.[17]

Many of our pollution problems are the results of habits/practices/systems that consume resources but add nothing to our standard of living. Many of our packaging materials, disposable cans, containers, razors, utensils, toys, cameras, etc do not add to our living standard but add pollution.

Table 7.7(a) lists heavy metals and their concentration and permissible limits. Lead as dust and fumes is encountered in many industries. Table 7.7(b) briefly describes its adverse impacts.

7.4.2.10   Extreme temperatures – Heat & Humidity[29, 30]

Heat is a problem in many mines and subsurface openings including tunnels. Heat is added due to:

• Operation of machines
• Geothermal gradient
• Adiabatic compression
• Blasting operations.

Table 7.4 Natural gases from strata and their impact when mixed with air; sources, ill effects and safe limits shown.

Mine gases	Composition	Source	Detection	Ill effects	Safe limits*
Oxygen (O)	SG = 1.1056	Normal air	Breathing, FSL, DT, Electro-chemical detector, paramagnetic method.	Non-toxic, Oxygen deficiency can cause fatal at 6%.	Minimum 19%.
Nitrogen (N) Nitrous fumes	SG = 0.9673 Such as: NO, N₂O, NO₂ SG of NO = 1.04; NO₂ = 1.5895	Normal air, strata Explosives, diesel exhaust, incomplete combustion	Extinguishes FSL	Asphyxiating due to oxygen deficiency. 0.005% detected by smell; 0.01% – serious irritation and illness in half an hour; 0.15% – great discomfort, bronchi-pneumonia as after effect and death; 0.25% – fatal after short exposure.	Maximum 80% Oxides of nitrogen not exceeding 0.0002%
White damp (CO + Air)	Poisonous gas having no smell – SG of CO = 09672	Spontaneous heating, explosives, diesel engines, fire, explosion, etc.	Canaries, DT, Catalytic combustion	CO – ill effects: 0.02% – headache after 7 hrs. or after 2 hrs. if exerted; 0.04% – severe headache in 5 hrs. or in 0.5 hr. if exerted; 0.1% death in 3 hrs. or in 1 hr. if exerted; 0.2% – unconsciousness in 30 min.; 1% – unconsciousness and death in 3 min. and in 1 min. if exerted. Death is painless and face becomes flushed.	CO not exceeding 0.0006% TLV-TWA = .005 TLW-STEL = .04
Stink damp (H₂S + Air)	H₂S giving of evil smell of rotten eggs. SG = 1.1912	Reaction of acid on sulfur found in various forms.	Odor; DT, Electro-chemical detector	0.005% – no ill effect; 0.01% – irritation of eyes and throat, headache; 0.02% – intense irritation of eyes and throat; 0.06% – pain in chest in few minutes; 0.1% – death immediately.	H₂S not exceeding 0.00066% TLV-TWA = 0.001; TLW-STEL = 0.0015
Sulfur dioxide SO₂	Burning and choking effect. SG = 2.2636	Spontaneous heating, explosives, fire oxidation etc.	Odor; DT	0.003% detectable smell; 0.01% – uncomfortable; 1.0% bronchitis after exposure for half an hour.	Sulfurous gases not exceeding 0.0007% TLV-C = 0.0005
Methane CH₄	Odorless, Colorless Tasteless SG = 0.5545	Strata, Blasting, Diesel engine, Organic decay	Flame safety lamp, DT, Detectors: Optical, Thermal, Infra-red	Asphyxiating due to oxygen deficiency, explosive.	Below 1.25% 5.3–14 – Explosive

FSL – Flame safety lamp; DT – Detector tube. SG – Specific gravity. * Differs according to the safety rules and regulations of various countries.[25,26,30]

Table 7.5 Elements in the human body; illustrating the fact that each element fulfills a critical purpose.[19]

Sr. No.	Elements	Percentage
01	Oxygen	65%
02	Carbon	18%
03	Hydrogen	10%
04	Nitrogen	3%
05	Calcium	1.5%
06	Phosphorus	1%
07	Sulphur	0.25%
08	Potassium	0.20%
09	Chlorine	0.15%
10	Sodium	0.15%
11	Magnesium	0.05%
12	Fluorine	0.02%
13	Iron	0.006%
14	Zinc	0.0033%
15	Silicon	0.0020%
16	Rubidium	0.00170%
17	Zirconium	0.00035%
18	Strontium	0.00020%
19	Aluminum	0.00014%
20	Niobium	0.00014%
21	Copper	0.00014%
22	Antimony	<0.00013%
23	Lead	0.00011%
24	Cadmium	0.000043%
25	Tin	0.000043%
26	Iodine	0.000040%
27	Manganese	0.000030%
28	Vanadium	0.000030%
29	Barium	0.000023%
30	Arsenic	0.000020%
31	Titanium	<0.000020%
32	Boron	0.000014%
33	Nickel	<0.000014%
34	Chromium	<0.000009%
35	Cobalt	<0.000004%
36	Molybdenum	<0.000007%
37	Silver	<0.000001%
38	Gold	<0.000001%
39	Uranium	3×10^{-3}%
40	Cesium	$<1.4 \times 10^{-3}$%
41	Radium	1.4×10^{-13}%

Table 7.6 Adverse impacts of intakes of metals, dusts, gases and fumes on human body organs.[12, 13, 17]

Body parts affected	Exceeds allowable TLVs: Metals, Dusts, Gases, Fumes
Bones	Fluoride, Pb, Strontium 90
Brain	Organic-lead, CO
Bronchioles	SO_2, NH_3, Cd
Bronchus	Mn, Mg, Beryllium, Zn
Fat below skins	Chlorinated, Hydrocarbons
Gums	Fluoride, Selenium
Heart	CO
Kidney	Hg, Cd
Liver	Selenium, Chlorinated, Hydrocarbons
Lungs	Ozone, NO, Peroxyacetyl Nitrate (PAN); Dust, CO, Asbestos; Nickel Carbonate, H_2S
Mid brain	Hg, Mn, Pb
Nasal passage	Cr, Ni, As, Cd,
Peripheral arteries	Cd, Fluoride
Skin	As, Ni, Cr, Beryllium
Small intestine	Zn, Pb, As, Fluoride, Vanadium
Thyroid	CO, Iodine
Metals/Minerals	Adverse impacts to human body

Metal	Max. Concentration (mg/kg of dry solids)	Max. application rate, (kg/ha/yr)*	Max. permitted concentration in soil, (mg/kg of dry solids)
Cadmium	20	0.150	3
Chromium	1,000	10	400
Copper	1,000	10	150
Lead	1,000	15	300
Mercury	10	0.1	1
Molybdenum	20	0.1	3
Nickel	300	3	75
Selenium	50	0.150	5
Zinc	3,000	15	300

After the spreading of sludge there must be a minimum period of three weeks before grazing or harvesting of forage crops.
Sludge use is prohibited:
• On soils whilst fruit or vegetable crops, other than fruit trees, are growing or being harvested.
• For six months preceding the harvesting of fruit or vegetables, which grow in contact with the soil and which are normally eaten raw.
• On soils with pH < 7.0.
* Based on a 10-year average and a soil Ph > 7.0.

The presence of water and increasing depth adds to this problem significantly. It makes visibility poor and gives undue stress to the working crews.

The benefits of heat control by proper ventilation have been demonstrated in many instances. In deep mines, which have attempted a depth exceeding 3 kilometers

Table 7.7(a) Metals and their concentration and permissible limits.

Cd	Peripheral arteries, Kidney, Bronchioles
CO	Thyroid, Lungs
Hydrocarbons	Fat below skins, Liver
SO_2	Bronchioles
CO	Heart, Brain, Lungs
H_2S	Lungs
NO	Lungs
Ozone	Lungs
Metals	Bronchus, Bronchioles, Small intestine, Skin, Peripheral arteries, Mid brain, Lungs, Kidney, Bones
Fluoride, I	Small intestine, Thyroid, Bones

Table 7.7(b) Adverse impacts of Lead.[12, 13]

Metal: Lead – Sources	Adverse effects
Lead as a dust and fumes is encountered in many industries. It is heavy metal used in manufacturing of pipes.	Inorganic lead can enter the body by inhalation or ingestion. Up to 50% of that inhaled is absorbed and only about 10% is ingested. It is then transported to the bloodstream and deposited in all tissues. But about 90% of it is stored in bones. It is a cumulative poison and excretion is slow and occurs mainly in urine and faeces. Early symptoms are fatigue, loss of appetite and metallic taste in mouth. Constipation is a common complaint. Sometimes abdominal pain. Lead interferes with normal formation of hemoglobin causing anemia. Some times paralysis. Kidney damage is a long-term effect.
Organic lead Tetraethyl and tetra-methyl lead [TEL, $(C_2H_5)_4Pb$] are mostly used in industry and especially in petroleum industry to improve octane rating. It acts as an anti-knock agent so as to assist the even burning of fuel inside the engine. This is ultimately released to atmosphere, for example: each year, around 50,000 tones of lead is released to the atmosphere by UK. Much of this is deposited close to roads.	These substances can be absorbed via the lungs and the skin. In the liver they are changed respectively to tri-ethyl and try-methyl lead, which are much more toxic. They can cause headache, vomiting, dizziness, mania and coma. Excretion is mainly by the urine. The blood is less affected than with inorganic lead.

from the surface, air conditioning is required. For example in South Africa during gold mining:

- Virgin **rock temperatures** higher than 52°C (126°F) have been recorded in South African gold mines.[29]
- For every tonne of rock mined, nearly 15 tonnes of **ventilation air** is pumped underground.

- **Cooling plants** in South African gold mines have a capacity equal to nearly 3.5 million domestic refrigerators.

- The **deepest mine in the world** is Western Deep Levels gold mine on the Far West Rand, now approaching a depth of 4 kilometers. (Approx. 13000 feet). Mining does not as yet take place at that depth.

7.4.2.10.1   Air conditioning and refrigeration[29]

- In the Republic of South Africa, cooling is required when the natural rock temperature reaches the temperature of the human body (98.6 degrees F).
- A rough approximation of the cooling capacity required for a hot mine in North America is that the tonnes of refrigeration (TR) required per ton mined per day is 0.025 times the difference between the natural rock temperature (VRT) and 95 degrees F. For example, a 2,000 ton per day mine with a VRT of 140 degrees F. at the mean mining depth will require approximately $0.025 \times 45 \times 2,000 = 2,250$ TR.[29]

As mentioned above, underground air differs from the atmospheric air with respect to its humidity, temperature, pressure and density. The function of ventilation, apart from providing air to breathe, is to maintain normal temperature and humidity. In general, the *quality of air* that is warranted should have a composition as outlined below:

Quality of air: O_2 not less than 20%; CO_2 not more than 0.5%, and temperature, not more than 20°C. Safe limits of the mine gases: CO – 0.0006%; Nitrogen oxides – 0.0002%; Sulfurous gases – 0.0007%; H_2S – 0.00066%.

The impact of heat ultimately results in illness, as illustrated in the Tables 7.8 and 7.9.

7.4.2.11   Radiation Hazards[12, 13]

What is radiation? [12, 13]

Radiation is energy that travels in the form of waves or high speed particles. The sources of radiation include nuclear power plants, nuclear weapons, or radiation treatments for cancer. It also includes uranium and thorium mines, hydrometallurgical plants for processing uranium and thorium raw materials, plants for production of nuclear minerals by refining, isotope separation, manufacturing reactor fuel, transportation of reactor fuel and intermediate products, stores for spent fuel, plants for reprocessing spent fuel, and stores for radioactive wastes. These types of radiation have enough energy to break chemical bonds in molecules or remove tightly bound electrons from atoms, thus creating charged molecules or atoms (ions). These types of radiation are referred to as 'ionizing radiation.'

What is radioactivity?[12]

Radioactivity is the property of some atoms that causes them to spontaneously give off energy as particles or rays. Radioactive atoms emit ionizing radiation when they decay.[12]

To be able to understand radiation and radioactivity, we need to understand the language of atomic structure. An atom is the basic unit of any element, which cannot

Table 7.8 Heat impacts on human body. (After Pon & Jones: Mine health and safety management, Michael Karmis (edt.), SME, AIME, Littleton, Colorado, 2001. Permission: www.smenet.org).[23] Table provides the TLV WBGT values in °C and °F.

Classification, medical aspects, and prevention of heat illness

Category and clinical features	Predisposing factors	Treatment	Prevention
1 Temperature regulation			
Heatstroke: (1) hot, dry skin usually red, mottled or cyanotic; (2) rectal temperature 40.5°C(104°F) and higher; (3) confusion, loss of consciousness, convulsions, rectal temperature continues to rise; fatal if treatment delayed.	(1) Sustained exertion in heat by unacclimatized workers; (2) lack of physical fitness and obesity; (3) recent alcohol intake; (4) dehydration; (5) individual susceptibility; (6) chronic cardiovascular disease.	Immediate and rapid cooling by immersion in chilled water in message or by wrapping in wet sheet with vigorous fanning with cool, dry air, avoid over-cooling, treat for shock if present.	Medical screening of workers, selection based on health and physical fitness, acclimatization for 5–7 days by graded work and heat exposure, monitoring workers during sustained work in severe heat.
2 Circulatory Hypostasis			
Heat syncope: fainting when standing erect and immobile in heat.	Lack of acclimatization	Remove to cooler area, rest in recumbent position until recovery prompt and complete.	Acclimatization, intermittent activity to assist venous return to heart.
3 Water and/or salt Depletion			
(a) Heat exhaustion (1) Fatigue, nausea, headache, giddiness (2) skin clammy and moist, complexion pale, muddy or hectic flush; (3) May faint on standing with rapid thready pulse and low blood pressure; (4) oral temperature normal or low but rectal temperature usually elevated (37.5°–38.5°C) (99.5°–101.3°F); water restriction type: urine volume small, highly concentrated; salt restriction type: urine less concentrated. Chlorides less than 3 gm/L.	(1) Sustained exertion in heat; (2) lack of acclimatization; and (3) failure to replace water lost in sweat	Remove to cooler environment, rest in recumbent position, administer fluids by mouth, and keep at rest until urine volume indicates that water balances have been restored.	Acclimatize workers using a breaking-in schedule for 5 days, supplement dietary salt only during acclimatization, ample drinking water to be available at all times and to be taken frequently during work day.
(b) Heat cramps: painful spasms of muscles used during work (arms, legs or abdominal); onset during or after work hours.	(1) Heavy sweating during hot work; (2) drinking large volumes of water without replacing salt loss	Salted liquids by mouth, or more prompt relief by IV infusion.	Adequate salt intake with meal, in unacclimatized workers supplement salt intake at meals.

(Continued)

Table 7.8 (Continued)

Classification, medical aspects, and prevention of heat illness

Category and clinical features	Predisposing factors	Treatment	Prevention
4 Skin Eruptions			
(a) Heat rash (malaria rubra; "prickly heat"): profuse tiny raised red vesicles (blister-like) on affected areas prickling sensations during heat exposure.	Unrelieved exposure to humid heat with skin continuously wet with unevaporated sweat.	Mild drying lotion, skin cleanliness to prevent infection.	Cool sleeping quarters to allow skin to dry between heat exposures.
(b) Adiabatic heat exhaustion (miliaria profunda): extensive area of skin that do not sweat on heat exposure, but present gooseflesh appearance which subsides with cool environments: associated with incapacitation in heat.	Weeks or months of constant exposure to climatic heat with previous history of extensive heat rash and sunburn.	No effective treatment available for unanhydriotic area of skin, recovery of sweating occurs gradually on return to cooler climate.	Treat heat rash and avoid further skin trauma by sunburn, periodic relief from sustained heat.
5 Behavioural disorders			
(a) Heat fatigue-transient: impaired performance of skilled sensorimotor; mental or vigilance tasks, in heat.	Performance decrement greater in acclimatized and unskilled worker.	Not indicated unless accompanied by other heat illness.	Acclimatization training for work in heat.
(b) Heat fatigue- chronic: reduced performance capacity, lowering of self-imposed standards of social behavior (e.g. alcohol overindulgence), inability to concentrate.	Workers at risk come from temperate climates for a long residence in tropical latitudes.	Medical treatment for serious cases, speedy relief of symptoms on returning home.	Orientation for life in hot regions (customs, climate, living conditions)

Source: Adaptation of NIOSH 1986.

Table 7.9 Thermal stresses – interrelationship of policy, practices and ultimate impacts. [2, 14, 32]

Policy	To practice						Results
							Effects
Risk Assessment Heat/ Cold	System Controls, Planning, HTS, Medical Surveillance	Operational Control	Unsafe Acts – Failure to reduce temperature	Unsafe Conditions – Hot Environment	Incident: Exposure to Heat +32.5°C	Threshold Limit 32.5°C	Heat Related Illness
Snr. Management		Middle Management					Individual

be further divided by chemical means. The atom itself is an arrangement of three types of particles: Proton, neutron and electron. The proton is positively charged, having unit mass. Neutrons also have unit mass but do not carry any charge. An electron has a weight about 2000 times less than that of a proton and has negative charge. Neutrons and protons lie in the nucleus of an atom, and the electron in an orbit around the nucleus. In an electrically neutral atom, the number of electrons is equal to the number of protons. A particular isotope of an element is referred as a nuclide.

If the number of protons is not equal to the number of electrons, the atom has a net positive or negative charge, and it is said to be ionized. The process of losing and gaining electrons is called ionization and it occurs in many chemical and physical processes.

Some nuclides are unstable and spontaneously charge into other nuclides, emitting energy in the form of radiation, either particulate (e.g. α and β particles) or electromagnetic rays (e.g. γ rays). This property is called radioactivity and nuclides showing this property are said to be radioactive. Most of the nuclides occurring in nature are stable, but some are radioactive, for example, all isotopes (any of two or more forms of an element having same atomic number but different atomic weights) of uranium and thorium.

The radiation emitted during radioactive decay can cause the material through which it passes to become ionized and it is therefore called ionization radiation. X rays are a different type of ionizing radiation. Gamma rays, X rays are all electromagnetic radiations, similar in nature to ordinary light except that they are of much higher frequencies and energies.

Alpha radiation has a very short range and is stopped by a few centimeters of air or a sheet of paper, or the outer dead layer of the skin. Outside the body it does not present a hazard but inside the body α-particles lose their energy to tissues in a very short distance, causing local ionization. The impact of β-particles will depend upon their energy but they are less intense than α-particles.

The basic unit of tissue is the cell. Very high radiation can kill a number of cells. If the whole body is exposed, death may occur within a matter of weeks.

Protection against external radiation through three principles: shielding, distance and time. Shielding means some material is placed in between the source and the person to absorb the radiation, partially or completely. Plastic is a good material for β-radiations, while for X- and gamma rays a large mass of material is required, for example lead and concrete are commonly used.

Protection against exposure from internal radiation is achieved by preventing the intake of radioactive material through ingestion, inhalation and absorption through skin and skin-breaks.

7.4.2.11.1   Radiation hazards in Mining[25, 30]

Radon and Radon-daughters: Radioactive contaminants are not confined to uranium mines alone, but significant concentration of radon daughters have been measured in copper, gold, lead/zinc, phosphate, coal, limestone and over twenty other minerals (Anon 1987; Rock et al. 1975). Radon (in uranium mines), thoron (in uranium and thorium mineral mines) and their respective daughter products create airborne radiation problems. The uranium ores contain considerably higher amounts of radium than other types of ores. Radium is a link in the uranium-family decay chain, giving off radon gas while it disintegrates. Radon is a chemically inert gas, which diffuses in infinitesimally small amounts from the rock as, and after, it is formed by the decay of radium. The radioactive decay from radon gives rise to daughters such as: RaA, RaB, RaC, RaC'; of which RaA and RaC are α- and β-emitters; RaB is a β-emitter and RaC' is an α-emitter. These daughters are usually positively charged, atomic-sized particles. Because of their size and nature, the radon daughters tend to attach to respirable dust and to other free surfaces in the mine atmosphere.

When a miner inhales the air, radon diffuses from the lungs into the blood and part of the dust is deposited in the lungs and breathing passages, where the attached radon daughters continue to decay, emitting alpha radiation, which damages the lung tissue. Radon is eliminated from the body mainly in exhaled air. Most radon daughters decay in the body before they can be excreted. Thus, the radon daughters rather than the radon gas constitute the major health threat to the miners in mines with radon problems.

The radiation dose delivered to the lungs by inhaled radon daughters depends upon many factors, such as dustiness of the mine air, the length of time the air has been in circulation, the breathing rate, mucus in the bronchial passages, and the physiology of the bronchial passages as affected by smoking or infection.[30]

7.4.2.11.2   Ways and means to minimize the airborne radiation

- Confinement: These methods include the use of bulkheads (ventilation stopping), backfill, overpressure and sealants.
- Bulkheads: The most effective way to confine radon and radon-daughters is to isolate the mined-out stoping (a block of ore developed for the purpose of mining from underground) areas, and other similar openings, such as heading and raises by the ventilation stopping. It is even more effective if such areas are kept under vacuum by exhaust ventilation and bulkheads are kept leak proof.

- As radon is partially soluble in water and emanates from seepage upon exposure to the mine atmosphere, the sealing of such seepage by grouting, etc., and/or by diverting the water through pipes is another effective method of confining radiation.
- Back-filling the mined out areas:
    - Backfills can be used as stope support and to control the radiation. The back fill material prevents the radon and thoron flow from stope surfaces/walls.

- Over pressure: Forcing v/s exhaust ventilation system was examined in a uranium mine, and it was found that the total radon production was reduced by 20% by changing the ventilation pressure from −2.1" to +1.6" water (Bates 1992).
- Use of Sealant: A way to prevent radon escaping from the rock surface is to seal off the surface immediately after its exposure by the use of shotcreting (mortar or concrete conveyed though a hose and pneumatically projected with high velocity is known as shotcreting) or water based epoxy or acrylic latex sealants. But these steps affect the costs adversely.
- Ventilation: In uranium mines the fresh air is sent to working faces and the miners directly, and air containing radon is removed as quickly as possible to minimize the growth and accumulation of radon daughter products, which are many times more harmful than the radon gas. For this purpose air velocity should be kept high to allow the fresh air to travel to the active working faces. In fact, radioactive mines require high amounts of air. The exhaust in these mines is achieved by installing the exhaust fans at the surface usually at both the terminals of an orebody (diagonal system of ventilation). The fresh air is brought down through the shaft/opening located at the geometrical center of the deposit/mine.

The permissible radon daughter concentration and exposures are expressed in terms of working levels, WL. One WL is defined as 1.3×10^5 mev (million electron volts) of potential alpha energy per liter of air resulting from the decay of short-lived radon daughters. WL is an exposure level and not a dose rate. In USA the working level month (WLM) based on a working month of 170 hours has been adopted as the unit for dosage or cumulative exposure. The maximum cumulative dose of radiation permitted in USA is now 4 WLM per year. For example a miner may work a full 12 months at a WL of 0.33, or for 4 months at a WL of 1 and for the next 8 months in a radon free area.

Electromagnetic Radiation

In addition to the above, microwaves, radar, electrical power lines, cellular phones, and sunshine also emit radiations. There are many different types of radiation that have a range of energy forming an electromagnetic spectrum. These are known as nonionizing radiations having advese impacts as shown in Table 7.10.

As described in the preceding paragraphs, ionizing radiation is radiation which interacts with matter to form ions. Some high-energy electromagnetic radiations and particle radiations are capable of producing ions in their passage through matter. These types of ionizing radiation include alpha and beta particles, x-rays, gamma rays, etc. Table 7.11 shows the interrelationship of radiation and policy, practices and ultimate impacts.

Table 7.10 Electromagnetic radiation.[11] I – Ionizing radiations; NI – Non-ionizing radiations.

Form of radiation	Use	Health risk
Microwaves (NI)	Heat Treatment	Deep burns
Infrared waves (NI)	Drying and heat treatment	Burn to skin and eye tissues
Ultra-violet light (NI)	Welding	Skin & eyes
Lasers (NI)	Measurement, cutting; beauty industry for hair removal and skin exfoliation	Eye damage including blindness; skin damage
Ultrasound (NI)	Beauty industry for heat treatment and skin exfoliation	Over heating and burning of tissues
X-rays (I)	Metal inspection; treatment, geological, metallurgical and material science analyses	Whole body
Gamma rays (I)	Level measurement	Whole body

Table 7.11 Interrelationship of radiation and policy, practices and ultimate impacts. [2, 14, 32]

Radiations

Policy	To practice						Results
							Effects Cancer
Risk Assessment Radiations	System Controls, Planning, Design, Medical Surveillance	Operational Control	Unsafe Acts – Failure to Reduce Exposure	Unsafe Conditions – Radiations	Incident: Exposure to Radiations+50 mSv/a	Threshold Limit +50 mSv/a	
Snr. Management		Middle Management					Individual

Table 7.12 Definition for Threshold Limit Values (TLVs)**.

TLV type	Definitions
TLV-TWA	Time-weighted average for normal 8-hour workweek, to which nearly all workers can be exposed, day after day, without adverse effects. Excursions above the limit are allowed if compensated by excursions below the limit.
TLV-STEL	Short term exposure limit. The maximum concentration to which workers can be exposed for a period of up to 15 minutes continuously without suffering (1) intolerable irritation (2) Chronic or irreversible tissue change, (3) narcosis of sufficient degree to increase accident proneness, impair self rescue, or materially reduce worker efficiency, provided that no more than 4 excursions per day are permitted, with at least 60 minutes between exposure periods, and provided that the daily TLV-TWA is not exceeded.
TLV-C	Ceiling limit. The concentration that should not be exceeded, even instantaneously.

** TLV should not be used for (1) a relative index of toxicity; (2) air pollution work, or assessment of toxic hazard from continuous, uninterrupted exposure.[5]

7.4.2.12   Vapors[5]

Exposure: TLVs refers to airborne concentrations, which correspond to conditions under which no adverse effects are expected during a worker's lifetime. The exposure occurs only during normal working hours, eight hours per day and five days/week. The American Conference of Governmental Industrial Hygienists (ACGIH) has established threshold doses, called threshold limit values (TLVs) for alarge number of chemical agents. Table 7.12 gives definitions for Threshold Limit Values (TLVs) and related terms used in practice.

7.4.2.13   Liquids[23]

Sulfuric acid is a widely used acid, particularly in metallurgical (refinery) and chemical industries. It can cause corrosive effects to tissues. Its mist is a skin, eye and respiratory irritant. The OSHA & MSHA permissible exposure limits for sulfuric acid is 1 mg/m^3 as an 8-hour TWA.[23]

7.5   AQUEOUS EFFLUENTS – PERMISSIBLE QUALITY & EFFICIENT DISCHARGE[21, 25]

Aqueous effluent discharges have the potential to be hazardous to human health andcan harm the environment. They result from industrial activities, products or services, and include the following:

*   Production water; in the case of the oil industry – discharges of formation water that is co-produced with oil and gas; in mines – as mine-water which could be acidic or with some contamination.
*   Sewage effluent.
*   Process water (pertaining to any industry), and drainage water (discharges of any used water that has to be disposed of, and any other drainage water, including water resulting from maintenance and cleaning activities).
*   Reverse osmosis plant discharges.
*   Ballast water (discharges to the marine environment of water that is pumped out of ballast tanks before loading marine tankers).
*   Hydro-test water (discharges of water used to pressure test flow-lines and pipelines).
*   Any other source not covered above.

7.5.1   Parameters concerning effluent discharge

The effluent discharges above mentioned should be

*   Identified
*   Sampled and analyzed, as appropriate
*   Quantified (either by direct measurement or using appropriate methods of estimation)
*   Recorded, as appropriate (refer to Section 4.11)
*   Reported, as appropriate (refer to Section 4.11)

7.5.2   Performance standards[20, 21, 25]

The *permissible quality* of effluent discharges is governed by the *prevalent regulations,* and in their absence best practice should be applied as per the aqueous effluent discharge standards, to ensure that effluent discharges do not adversely affect the receiving environment (Section 4.3.1).

In an Oil Company in the Gulf, guidelines are given with regard to production of water and its subsequent disposal are as under:[20, 21, 25]

1   Minimize the volumes of water produced during oil extraction
2   Maximize reuse of such produced waters
3   Phase out the use of shallow disposal wells and prevent disposal into useable or exploitable aquifers.
4   Return production water to the producing reservoir.
5   Dispose surplus waters to formations which have a salinity >35,000 mg/l, in conjunction with case-specific monitoring programmes.

7.5.3   Effluent discharges receiving environment

Effluents could be discharged into, or the receiving environment/end use, could be:

- Surface water-bodies, which could be used for:

    - Freshwater
    - Drinking water
    - Irrigation of crops
    - Watering of livestock

- Marine/Sea
- Sewage Treatment Systems.

7.5.4   Effluent discharge/disposal – Surface water-bodies[21, 25]

7.5.4.1   Water Quality Standards (WQS)

In the event that there are no local regulations, the principle of best practice should be applied to ensure that the effluent discharges do not adversely affect the receiving environment. For this, water quality standards (WQS) should be formulated and adopted by any industrial setup . The WQS are a measure of the concentration of specific substances in the receiving water following an initial period of dispersion and dilution. Table 4.3 depicts a guideline that has been followed by an Oil Company in the Gulf.

In addition to the discharge water quality limits set in the prevalent regulations, the following requirements should also be met:

- Discharge points should be designed and located to maximize the rate of mixing.
- Discharges should have no adverse effects on the general visible amenity of an area
- Efforts should be made to eliminate acute toxic effects on organisms in the mixing zone
- Discharges should not impinge on any shoreline where there is unrestricted access by the public
- Discharges should not impinge upon ecologically sensitive habitats
- Table 7.13(a) depicts guidelines for the discharge of metals, minerals and other elements into water-bodies based on the standards followed at an Oil Company

Table 7.13(a) Guidelines – Discharge of metals, minerals and other elements into water-bodies; Standards followed at an Oil Company, in Gulf.[21,25]

Parameter	Standard A*	Standard B*
BOD (5 days @ 20°C)	15.0 mg/l	20.0 mg/l
COD	150.0 mg/l	200.0 mg/l
Suspended solids	15.0 mg/l	30.0 mg/l
Total dissolved solids	1500.0 mg/l	2000.0 mg/l
Electrical conductivity	2000.0 micro S./cm	2700.0 micro S./cm
Sodium absorption ratio	10.0	10.0
pH	6–9	6–9
Aluminium	5.0 mg/l	5.0 mg/l
Arsenic	0.1 mg/l	0.1 mg/l
Barium	1.0 mg/l	2.0 mg/l
Beryllium	0.1 mg/l	0.3 mg/l
Boron	0.5 mg/l	1.0 mg/l
Cadmium	0.01 mg/l	0.01 mg/l
Chloride	650.0 mg/l	650.0 mg/l
Chromium	0.05 mg/l	0.05 mg/l
Cobalt	0.05 mg/l	0.05 mg/l
Copper	0.5 mg/l	1.0 mg/l
Cyanide	0.05 mg/l	0.10 mg/l
Fluoride	1.0 mg/l	2.0 mg/l
Iron	1.0 mg/l	5.0 mg/l
Lead	0.1 mg/l	0.2 mg/l
Lithium	0.07 mg/l	0.07 mg/l
Magnesium	150.0 mg/l	150.0 mg/l
Manganese	0.1 mg/l	0.5 mg/l
Mercury	0.001 mg/l	0.001 mg/l
Molybdenum	0.01 mg/l	0.05 mg/l
Nickel	0.10 mg/l	0.10 mg/l
Nitrogen – ammoniacal	5.0 mg/l	10.0 mg/l
Nitrogen – nitrate	50.0 mg/l	50.0 mg/l
Nitrogen – organic	5.0 mg/l	10.0 mg/l
Oil and grease	0.5 mg/l	0.5 mg/l
Phenols	0.001 mg/l	0.002 mg/l
Phosphorous	30.0 mg/l	30.0 mg/l
Selenium	0.02 mg/l	0.002 mg/l
Silver	0.01 mg/l	0.01 mg/l
Sodium	200.0 mg/l	300.0 mg/l
Sulfate	400.0 mg/l	400.0 mg/l
Sulfide	0.1 mg/l	0.1 mg/l
Vanadium	0.1 mg/l	0.1 mg/l
Zinc	5.0 mg/l	5.0 mg/l
Faecal coliform bacteria	*200.0 per 100 ml*	*1000.0 per 100 ml*
Viable nematode ova	*<1.0 per litre*	*<1.0 per litre*

* The definitions relating to Standards A and B are detailed below.

Table 7.13(b) Definitions relating to Standards A and B are detailed below[21, 25]

	Standard A	*Standard B*
Crops	Vegetables likely to be eaten raw. Fruit likely to be eaten raw and within 2 weeks of any irrigation.	Vegetables to be cooked or processed. Fruit if no irrigation within weeks of cropping. Fodder, cereal and seed crops
Grasses and Ornamental Areas	Public parks, hotel lawns, recreational areas. Areas with public access. Lakes with public contact (except places which may be used for praying and hand washing).	Pastures. Areas with no public access
Aquifer Recharge	All controlled aquifer recharge.	
Method of Irrigation	Spray or any other method of aerial irrigation not permitted in areas with public access unless with timing control.	
Any other Reuse Application	Subject to the approval by the regulatory authorities	

in the Gulf. Drainage exceeding the limits (as shown in Tables 7.13(a) and (b) or as per the prevalent regulation in the country of operation) should not be discharged directly to water-bodies.

Except in the event that alternative limits are specified within individual licenses or permits issued by the Regulatory Authorities, these discharge limits are:

7.5.5   Effluent discharges/disposal – Marine[21, 25]

In addition to the regulatory requirements, effluent discharges should not result in:

- Visible oil or grease on the surface of receiving waters
- A change in color of receiving waters
- Emission of foul smells
- Any harmful effect or change which may lead to a harmful effect, on marine life or the marine environment.

Table 4.8 depicts Water Quality Discharge limits for Discharges to the Marine Environment in a Gulf country.

7.5.6   Effluent discharges – Sewage treatment systems[21, 25]

A Sewage Treatment Plant (STP) usually caters for domestic raw sewage where treatment by septic tanks systems is inadequate. STPs may, in addition to accepting raw sewage from domestic activities, accept industrial waste streams within certain limits. The guidelines below, as used at one oil field, could be followed. They restrict discharges to STPs of:

- Petroleum hydrocarbon products (gasoline, naphtha, fuel oil, or mineral oil or other flammable or explosive liquid, solid or gas).
- Garbage.
- Waters or waste which contain grease, oil, or other substances that will solidify or become viscous (prevent or block sewage flow) at temperatures between 32 and 150° F (65°C).
- Any water or waste containing emulsified oils and grease, exceeding 100 ppm at any one time, only wastewater from hand washing containing residual hydrocarbon and detergent shall be allowed.
- Aqueous wastes containing heavy metals (copper, chromium, cadmium, cyanide, lead, mercury, molybdenum, nickel, selenium and zinc) in excess of 2 ppm by weight, as these will suppress biological action in the STP.
- Any ashes, cinders, sand, mud, glass, rags, feathers, plastics, wood, entrails, chemical residues, paint residues, or any other solid or viscous substance capable of causing obstruction to the flow in sewers, or other interference with the proper operation of the STP.
- Any liquid or solid hazardous waste described in the Waste Management Guideline overall total influent to the STP which will exceed the STP design influent, and organic load in kg BOD/day as specified in the EIS approved by the MRME.
- Acidic and alkaline waters outside the range pH 6.5–8.5.
- Any aqueous waste having a temperature higher than 150° F (65°C).
- For combined drainage (sewage and industrial waste) an oil water separator must be installed upstream from the STP, which must be capable of maintaining an effluent with not more than 100 ppm total oil and grease at any one time.

Where toxic substances and/or biological inhibitors are present in the industrial waste stream, prior treatment or isolation and separate treatment shall be carried out.

Septic tanks and holding tanks
Septic tanks would be allowed only for discharge of domestic wastewater from an equivalent population of not more than 150 people.

7.6   HOUSE KEEPING[21, 25, 31]

House keeping: A term encompassing all those activities that are necessary for cleanliness, orderliness, and neatness in all areas of work. House keeping is seen as an inherent part of every employee's job nowadays.

Good house keeping is the key to preventing accidents and in maintaining good health standards in any industrial setup, which usually consists of many buildings encompassing: offices, workshops, laboratories, factories, plants, warehouses, vocational training centers, residential accommodation, community/recreation centers, health centers, clubs and all other buildings within its premises. It is considered to be the first-line of defense against any illness or injury that could be caused while undertaking any operation.

7.6.1   Aspects to be adhered to[21, 25, 31]

The following aspects should be adhered to:

- All buildings should be kept cleaned. They should be maintained clear of debris, waste and other rubbish, which should be disposed of in accordance with the manufacturer's instructions (if any) and the Company's Waste Management Manual/Guidelines.
- Wherever appropriate waste should be stored in wastebaskets, which should be regularly emptied.
- All floor coverings should be regularly inspected by a competent supervisor giving due attention to any loose or damaged items, which should be replaced or re-secured immediately.
- All furniture should be regularly inspected for broken or loose legs, castors, wheels; damaged or cracked perspex or glass work surfaces; non-functioning or ill-fitting drawers or doors. This also includes cupboards, filing cabinets and shelves. Any damaged items should be repaired or replaced immediately.
- All emergency walkways, passages and exits, fire doors, break-glass alarm points, fire fighting equipment, first aid and other emergency stations should be kept clean, unobstructed and in good working order.
- Stockpiles of raw materials should be made at the areas earmarked for this purpose and not allowed to accumulate in the workplace.
- Particular care should be taken to ensure that:
  - Quantities of chemicals likely to react adversely should not be disposed of together. Chemicals requiring disposal should be placed in clearly labeled containers with a description of the contents and International Hazard Sign.
  - Waste should be suitably segregated to minimize hazards to disposal personnel. Particular attention should be given to separating all broken glassware and other sharp objects from regular waste.
  - The proper disposal of waste should be ensured by assigning the responsibility to a competent person.
  - No odorous substances should be emitted to the environment of residences or compounds for more than 150 hours/year, or the limit prescribed by the Competent Authorities/Regulatory Agencies.
  - Disposal areas should be earmarked and they should be of the specification/norms as laid out by the regulatory authorities.
  - Smoking while working should not be permitted in any workshop or industrial area, except in designated smoking zones.
  - Consumption of foodstuffs should be restricted to designated areas during work-breaks.

7.6.2 Dealing with spillage[21]

Any spillage of liquid, especially of toxic, corrosive or otherwise hazardous liquids; should be:

- Immediately cleaned up in accordance with manufacturer's instructions or safety data sheet.
- All concerned should be familiar with the procedures for dealing with it.
- Any material used to mop up spills should be immediately removed to a safe place and stored in closed containers for its safe disposal.

7.6.3 Administrative controls

Administrative controls include:

- Any industrial setup should have a 'Waste Management Manual/Guidelines', which should be strictly adhered to.
- Allocation of adequate time for house keeping: time could be during off working hours, or if carried during normal working hours it should not interrupt regular operations/activities.
- Allocation of adequate resources: men, appliances, cleaning tools and consumables. Specialized cleaning personnel should be engaged and a competent supervisor should supervise the operation.
- It should be ensured that cleaning personnel are neither at risk from chemicals, nor cause damage to the machines, equipment, appliances, tools and tackles.

House keeping in a chemical plant would aimed at keeping dust and toxicants contained by:[5]

- Dykes around tanks and pumps
- Providing water and steam facilities for cleaning
- Providing lines/mechanisms for efficient cleaning and flushing
- Incorporating a well designed sewer system with emergency containment.

7.6.4 The 5S Concept[31]

'5S' is a Japanese concept involving the words that begin with 'S': **Seiri** – Sorting; **Seiton** Straighten or Set in Order; **Seisō** Sweeping or Shining or Cleanliness; **Seiketsu** Standardizing; **Shitsuke** Sustaining the discipline/Selfdiscipline.

5S is a method for organizing a workplace, especially a **shared** workplace (like a shop floor or an office space), and keeping it organized. It's sometimes referred to as a housekeeping methodology; however this characterization can be misleading, as workplace organization goes beyond housekeeping.

5S is a philosophy and a way of organizing and managing the workspace and work flow with the intent to improve efficiency by eliminating waste, improving flow and reducing process unevenness. Its phases are as follows:

- Phase 1 – **Seiri** Sorting: Going through all the tools, materials, etc., in the plant and work area and keeping only essential items. Everything else is stored or discarded.
- Phase 2 – **Seiton** Straighten or Set in Order: Focuses on efficiency. When we translate this to "Straighten or Set in Order", it sounds like more sorting or sweeping, but the intent is to arrange the tools, equipment and parts in a manner that promotes work flow. For example, tools and equipment should be kept where they will be used (i.e. straighten the flow path), and the process should be set in an order that maximizes efficiency. For every thing there should be place and every thing should be in its place. (Demarcation and labeling of place.)
- Phase 3 – **Seisō** Sweeping or Shining or Cleanliness: Systematic Cleaning or the need to keep the workplace clean as well as neat. At the end of each shift, the work area is cleaned up and everything is restored to its place. This makes it easy to know what goes where and have confidence that everything is where it should be. The key point is that maintaining cleanliness should be part of the daily work – not an occasional activity initiated when things get too messy.
- Phase 4 – **Seiketsu** Standardizing: Standardized work practices or operating in a consistent and standardized fashion. Everyone knows exactly what his or her responsibilities are to keep to the above 3S's.
- Phase 5 – **Shitsuke** Sustaining the discipline: Refers to maintaining and reviewing standards. Once the previous 4S's have been established, they become the new way to operate. Maintain the focus on this new way of operating, and do not allow a gradual decline back to the old ways of operating. However, when an issue arises such as a suggested improvement, a new way of working, a new tool or a new output requirement, then a review of the first 4S's is appropriate.

A sixth phase, "Safety," is sometimes added. In fact following the 5S's correctly will result in a safe work environment.

There will have to be continuous education about maintaining standards. When there are changes that will affect the 5S program – such as new equipment, new products or new work rules – it is essential to make changes in the standards and provide training. Companies embracing 5S often use posters and signs as a way of educating employees and maintaining standards.

The key targets of 5S are improved workplace morale, safety and efficiency. The assertion of 5S is, by assigning everything (that is needed) a location, time is not wasted by looking for things. Additionally, it is quickly obvious when something is missing from its designated location. Advocates of 5S believe the benefits of this methodology come from deciding *what* should be kept, *where* it should be kept, *how* it should be stored and most importantly *how* the new order will be maintained. This decision-making process usually comes from a dialog about standardization which builds a clear understanding between employees of how work should be done. It also instills ownership of the process in each employee.

7.6.5   Sanitation

Ill health caused by poor sanitary conditions, consumption of unsafe drinking water, indoor and outdoor air pollution, inadequate vector control and unhealthy life style – 13 million deaths/year; out of which 2.6 millions/yr in India and 5 million deaths in China (38%).[28]

In India 68 years (on an average basis) of ill health per 1000 inhabitants are caused by environmental health issues, for Russia it is 54, Brazil 37; China 34; the worst rates are: Angola, Burkina, Faso, Mali and Afghanistan. The best rated are Israel and Iceland with 14 years, Italy 16 years, Germany, Spain and France 17 years and USA 19 years. The common causes of diseases are: diarrhea, lung cancer, cardiovascular etc.[28]

A WHO study indicates that low-income countries are losing 20 times more healthy years of life than high-income countries.

7.7   WORKING CONDITIONS[5]

This includes perfection in the following aspects

- General layout: Adequate working space and perfect layout includes provision for fencing, guards, dykes, barriers, emergency access, aisle ways, enough headroom and sufficient clearance. Buildings should have adequate ladders, stairways, escape ways, fire doors, safety glasses, fire-proof structure, overhead power lines, etc. wherever required. Also refer to section 8.3.1.
- Effective ventilation and air conditioning, for example in a Chemical Plant this includes[5]
  - Contain and exhaust hazardous substances. Keep exhaust system under negative pressure
  - Use properly designed hoods; use hoods for charging and discharging
- Wholesome water – Potable as well as non-potable
- Effective/Perfect drainage and sewerage
- Proper lighting and illumination
- Warning signals and alarms
- Safety and security
- Proper sanitation and housekeeping; For example[5]
  - Use wet method to minimize contamination with dust.
  - Use water sprays for cleaning.
  - Clean areas frequently.
- Effective communication and information systems
- Adequate recreation facilities and amenities
- Use of IT and software/computing wherever applicable
- Automation wherever applicable
- Use and availability of PPEs; For example for effective hygiene control[5]
  - Use safety glasses and face shields
  - Use aprons, arm shields and space suits

- Wear appropriate respirators, airline respirators are required when oxygen concentration is less than 19.5% (ref. Sec. 6.15 for breathing apparatus)
- Effective supervision and trained staff (ref. Sec. 8.9 for training)

- Adequate measures to minimize pollution
- Noise abetment measures (ref. Sec. 3.9.4)
- Well maintained equipment, appliances, tools and tackle
- Effective services – transportation, communication, connectivity, power, water and sanitation
- Health and hygiene care, for example in a Chemical Plant[5] it could be achieved by:

  - Enclosing room or equipment and place under negative pressure
  - Enclosing hazardous operations such as sampling points
  - Shielding high temperature surfaces
  - Pneumatically conveying dusty material

- Clarity of job profile, role and responsibility
- Administrative controls – disciplinary actions, encouragement and rewards for exceptionally good performance; delegation of responsibilities. This can be described as Organizational Culture. A proactive culture brings Organizational Commitment (OC) that reduces workplace stresses, as described in sec. 7.7.2
- Exposure to changes in technology and renovations
- Adequate medical facilities including effective first aid, rescue and recovery
- Routine checkups and medical surveillance
- Emergency preparedness and measures
- Security – personal as well as property.

All these aspects have been dealt in detail in various chapters.

7.8 ERGONOMICS[3, 6, 16, 27]

7.8.1 Introduction

The term ergonomics has been derived from the Greek words *ergon* (meaning "work") and *nomos* (meaning "rules"). Thus, its literal meaning is 'the rules of work' and in practice

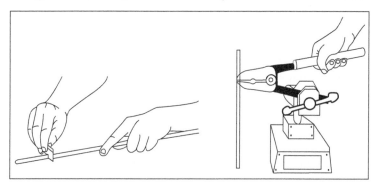

Figure 7.4 Use of one hand instead of both hands to carry out work faster and less painfully.[6]

we need to follow the 'rules' that can optimize any work, task or operation. Example: One of the rules of work is that you shouldn't use your hands as fixtures. The illustration (left) shows using a razor blade to remove a coating from a tube used in medical instruments, thus using both hands as fixtures. The work was slow and painful. In this instance, Dan, (a specialist of ergonomics)[6] developed a contraption for less than $30 that cut the motions by 80% and the time needed to do the job by 60%. As shown in Figure 7.4 use of one hand instead of both hands to carry out the work faster and less painfully.

In fact, the formal definition of this field of ergonomics is "optimizing the interface between humans and systems."

"Ergonomics" is a broad field, but the basic goal of an ergonomics program is injury prevention. This injury prevention is accomplished by fitting the job to the worker instead of fitting the worker to the job.

Another way to understand this field is that it can help us to work smarter, and not harder. But, we can ask ourselves: If we were to look at some task – whether at work or at home – and try to figure out a smarter way of doing the work, how would we go about it? How do we actually figure out a smarter way of working?

The answer is ergonomics. Ergonomics provides a method for finding smarter ways of working – the principles and techniques by which we can improve ways to work.

7.8.2   Making things user-friendly[3,6,16,27]

A good way to understand what ergonomics means is to think about the term "user-friendly." The two terms are actually synonymous; anything that is user-friendly is ergonomic, and anything that is user-unfriendly is un-ergonomic.

Ergonomics aims at making things more human compatible, which can bring improvement in productivity – a key to save money, and making the operation *'user friendly'*. In the workplace, the focus should be on making tools, appliances, equipment, work-methods/procedures/techniques, layouts and working environment *user-friendly*.

Humans have been doing "ergonomics" for a long time (that is, reducing the physical demands of jobs). Good ergonomic improvements include switching over from stone-age tools to modern tools and appliances (Figure 7.5).[26]

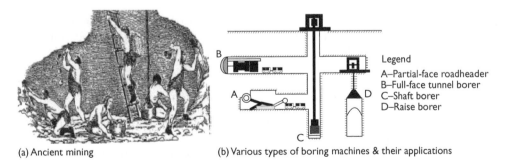

(a) Ancient mining   (b) Various types of boring machines & their applications

Legend
A–Partial-face roadheader
B–Full-face tunnel borer
C–Shaft borer
D–Raise borer

Figure 7.5 Human's efforts since long to overcome physical demands of job by inventing modern equipment for digging rocks and ground.[26]

7.8.3   Impacts of poor ergonomics[3,6,16,27]

Poor ergonomic conditions typically involve the bones, muscles, joints, tendons, and nerves. Symptoms include:

- Painful joints
- Pain, tingling or numbness in hands or feet
- Pain in wrists, shoulders, forearms, knees, etc.
- Back or neck pain
- Fingers or toes turning white
- Shooting or stabbing pains in arms or legs
- Swelling or inflammation
- Stiffness
- Weakness or clumsiness in hands
- Burning sensations
- Heaviness
- Sustained muscle exertion, which reduces blood flow to the muscles, causes muscle strains and sprains
- Contact stresses, which are injuries that occur due to repeated contact with a hard surface

These symptoms could also be the result of other medical conditions, so checking with the doctor is important if there is any concern about any of these.

7.8.4   Impacts of good ergonomics

7.8.4.1   Improved labor relations

Ergonomics issues are often good ones for joint problem solving between management and labor. Redesigning the workplace using the principles of ergonomics is a "win-win" situation for management and labor. Companies such as GM, Ford etc have benefited by this approach.

One reason why workers are absent is that they are experiencing the early stages of a musculoskeletal disorder.[6] Work that hurts doesn't exactly encourage people to come everyday.

Dissatisfaction caused by fatigue, working in uncomfortable postures, and the pain and discomfort created by overexertion may easily lead to increased employee turnover (increased man-hours).

7.8.4.2   Safeguarding skilled and experienced human resources

In any organization forced absenteeism on account of pains to any organ of the body is common. It is very much hurting to smaller organizations, as it causes disruption due to the loss of a key person – even if this loss is just a few days because of back pain. Furthermore, we may not be able to replace that person, even temporarily. Even if a replacement is found, skill levels may not be the same or the learning curve may be extended. Please refer to section 7.9.4 also.

7.8.4.3   Offsetting limitations on age of employees

Older employees have more experience, tend to be more reliable, and are already trained and educated. When ergonomic adaptations are made, older workers can be as productive as younger workers, if not more so.

7.8.4.4   Reduced maintenance downtime

All the tools and techniques of workplace ergonomics can be applied to maintenance tasks. You can eliminate barriers and thus speed the time in which operations can be brought back on line. It is about providing clearance, reducing exertion, and reducing motions. Chap. 9 details these aspects.

7.8.5   Work in neutral postures[3, 6, 16, 27]

- The best positions in which to work are those that keep the body "in neutral". For example: It is important to maintain the natural S-curve of the back, whether sitting or standing (Figure 7.6(a)). Holding a car's steering in 10 and 2 O'clock position (Figure 7.6(b)).[6]
- How long one can shift goods in a bending position? Use of a suitable tilter or lifter is the viable solution (Figure 7.6(c)).[6]
- Use a suitable mounting for an instrument rather than bending your neck (Figure 7.6(d)). When a product is not at a suitable height for working, pain in the waist is obvious (Figure 7.6(e)); but when the same product is placed in a lower position the shoulders and elbows are at their relaxed posture. This aspect has been illustrated in Figure 7.6.

i   Improved efficiency with better working posture. Working in awkward postures can directly reduce efficiency in three ways that ergonomics can help to remedy:

- *Reduced strength* – Think of bending at the waist and reaching out across a large object and then trying to exert. You have little or no strength in an out-stretched position like this. Consequently it takes you longer to complete a task than it would be if you were working in a proper position.
- *Less accuracy in your motions* – Again, think of reaching out across a large object and trying to do something intricate. You make a lot of mistakes and it takes a lot longer time, if indeed you can do it at all.
- *Quicker fatigue* – When you work in an awkward posture, you tire much more easily, which slows you down.

ii   Reduce Excessive Force
Excessive force invites fatigue and could result in injury. Illustration (Figure 7.6(e))[6] depicts rough floor and improper wheels for heavy load requiring excessive force on body (back and shoulders). Likewise there could be thousands of examples in our day-to-day life. The solution lies in reducing such excessive forces.

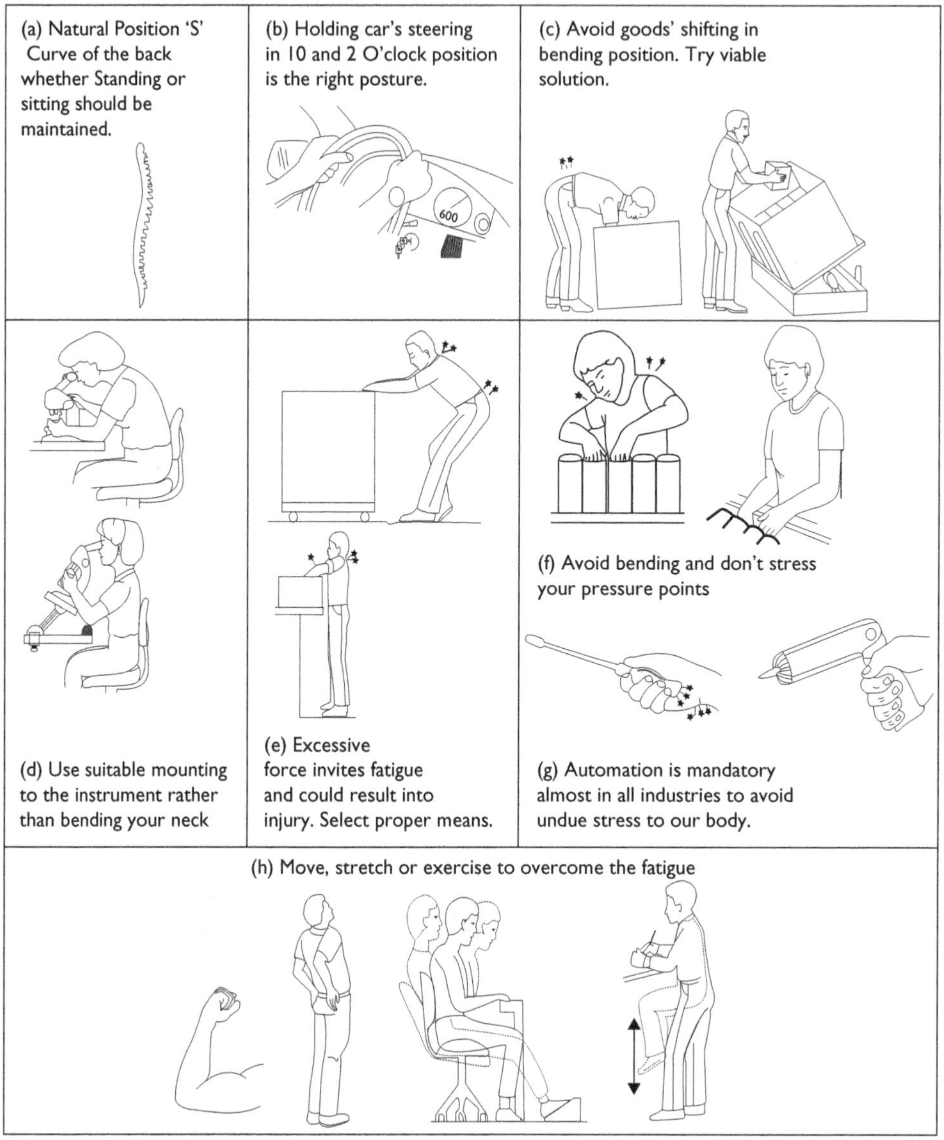

(a) Natural Position 'S' Curve of the back whether Standing or sitting should be maintained.

(b) Holding car's steering in 10 and 2 O'clock position is the right posture.

(c) Avoid goods' shifting in bending position. Try viable solution.

(d) Use suitable mounting to the instrument rather than bending your neck

(e) Excessive force invites fatigue and could result into injury. Select proper means.

(f) Avoid bending and don't stress your pressure points

(g) Automation is mandatory almost in all industries to avoid undue stress to our body.

(h) Move, stretch or exercise to overcome the fatigue

Figure 7.6 How to avoid bending neck, waist, shoulder and other parts of our body to avoid pains.[3, 6, 16, 27]

iii   Keep Everything Within Easy Reach

As described, one of the important features of '5S' (section 7.6.4); things that we use frequently should ideally be within the reach envelope of our full arm. Things that we use extremely frequently should be within the reach envelope of our forearms. Similarly, referring to the illustration, which is a better reach? Obviously, the solution lies in reducing the height of the reach.

iv   Work at Proper Heights

As a rule of thumb, most work should be done at about elbow height, whether sitting or standing. But there are exceptions to this rule, i.e. heavier work is often best done lower than elbow height. Precision work or visually intense work is often best done at heights above the elbow, as could be seen in all survey instruments such as theodolite, levels etc.

Poor heights and reaches can affect productivity. If we can't reach an object at all, we may have to stop productive work, and fetch a step stool or taking time to remove an obstruction. An inappropriate height or a long reach causes an awkward posture; we end up losing productivity for that reason.

v   Reduce Excessive Motions

Excessive motions throughout a day, whether that of fingers, wrists, arms, or back. They are a potential source of fatigue and stress. The solution lies in automation or use of powered tools. Figure 7.6(g) illustrates that reducing repetitive movements could be achieved through proper automation.[6] Automation is mandatory in almost in all industries to avoid undue stress to our body.

vi   Minimize Fatigue and Static Load

Holding the same position for a period of time is known as *static load* (Figure 7.7 (left)). It creates fatigue and discomfort and can interfere with work. Note the difference of using a fixture, as shown in Figure 7.7(right) that can avoid undue stress to our organs.

One of the core areas of ergonomics is to understand the causes of excessive fatigue and ways to effectively reduce or eliminate them. Fatigue caused by working in static positions, is a problem that has increased in recent decades.[6]

vii   Minimize Pressure Points

Analyzing the pressure points (Figure 7.6(d)) while we sit on a cushioned chair; A particularly vulnerable spot is behind knees, which happens if the chair is too high or when we dangle our legs. Another pressure point that can happen when we sit is between our thigh and the bottom of a table. This is known as "contact stress." The idea is that the involvement of pressure points should be minimum.

viii Provide Clearance

Work areas including cabins need to be set up in such a way that there is sufficient room for our head, knees, and feet. Adequate clearance from all corners is a must for working smoothly.[6]

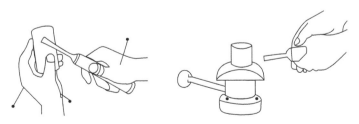

Figure 7.7 Minimize Fatigue and Static Load.[6]

ix   Move, Exercise, and Stretch

Based on the nature of the job, different exercises on the job can be helpful (refer Figure 7.6(h)):

- If we have a physically demanding job, we may find it helpful to stretch and warm up before any strenuous activity.
- If we have a sedentary job, we may want to take a quick "energy break" every so often to do a few stretches.
- If we sit for long periods, we need to shift postures:
    - Adjust the seat up and down throughout the day.
    - Move, stretch, and change positions often.

x   Maintain a Comfortable Environment

Poor environment could result in:

- Extreme temperature, which can reduce sensitivity to pain and reduce blood flow
- Vibration, which can reduce blood flow and sensory response
- Lighting and glare.

One good way to solve lighting problems is by using task lighting; that is, having a small light right at your work that you can orient and adjust to fit your needs, as could be seen in an airplane, trains etc.

xi   Results[3, 6, 16, 27]

The application of ergonomic principles can result in:

- Increased productivity. It is common for ergonomic improvements to increase productivity by 10–15%.
- Improved health and safety. The smaller your organization, the greater risk you have of disruption due to the loss of a key person – even if this loss is just a few days because of back pain.
- Increased employee turnover. Dissatisfaction caused by fatigue, working in uncomfortable postures, and the pain and discomfort created by overexertion injuries may easily lead to increased employee turnover.
- Increased work quality.
- Today, ergonomics is all about efficiency, giving special focus on studying the interface between humans and systems.
- Less lost time at work.
- Lower worker's compensation claims.

7.8.6   Identifying wasted activities

By evaluating items such as fatigue, motions, and exertion through a task step by step, it is possible to identify wasted activities.

7.8.7   Fresh insights on your operations

The tools of ergonomics are especially useful because they focus on ways to eliminate problems like fatigue, awkward working positions, and excessive motions.

7.9   OCCUPATIONAL HEALTH SURVEILLANCE[4, 15, 18]

The preceding sections have dealt with the important elements that are essential for the good health of industrial workers. Managers paying full attention to the following practices is also equally important:

- Organizational culture and workplace stresses[4]
- Lost performance at work (presenteeism)[18]
- Occupational hygienic risk – exposure assessment and control measures[15]

7.9.1   Organizational culture and workplace stresses

7.9.1.1   Organizational culture and commitment[4]

"Organizational culture is what employees perceive a company's values to be. Everyone knows what is important and not important. What's important is what management focuses on." In a healthy Organizational Culture the first step is to locate the sources of stresses and find solutions to reduce them. In fact we can understand stresses. Some of them could be removed whereas for some we may have little control over them. In such cases change in behavior and attitude, as described in sec. 8.13; should be exercised. Mick Cairney (2007)[4] in the following paragraphs has proposed an integrated model, as shown in Figure 7.8, which states:

- The organizational culture can be positive or negative; compliant or convergent
- Workplace stress (Figure 7.10) can be caused by negative or compliant work cultures
- Reduced stress improves Organizational Commitment (OC)
- Improved OC develops positive or convergent organizational culture

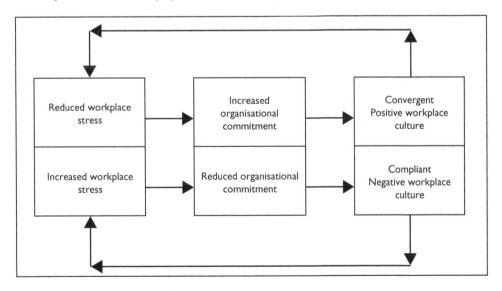

Figure 7.8 An integrated approach to bringing about changes in Organizational Culture leading to an increase in Organizational Commitment (OC).[4] (After Cairney, M. 2007).

- This culture is desirable for working together for common high values and standards.

Through research it has been established that the following six reasons are the main causes of stresses; these are known as 'Six Stressors':

- Uncertainty
- Novelty (continuing change and ambiguity)
- Lack of input
- Lack of feedback
- Lack of resources
- Lack of reward/recognition.

A survey was conducted (a voluntary and anonymous exercise) with 61 respondents to test variables as shown in table 7.14, using questions on a 1 to 5 scale for:

- The six stressors
- Commitment and culture
- Stress and culture and
- Any other stressors.

Table 7.14 Survey score of integrated model variables.[4] (After Cairney, M. 2007).

Testing the integrated model survey variables	Score
1 Uncertainty	3.0
2 Novelty (continuing changes or ambiguity)	2.6
3 Lack of input	3.1
4 Lack of feedback	3.7
5 Lack of resources	3.9
6 Lack of reward/recognition	3.2
7 OC Lead to positive culture	4.1
8 Stress causes negative culture	4.2
9 Other stressors (due to lack of resources)	8 out of 10 were resource-based

Table 7.15 Comparison of positive and negative work-cultures and their impacts.[4] (After Cairney, M. 2007).

Convergent – A positive work culture and its characteristics	Compliant – A negative work culture and its characteristics
• Democratic	• Bureaucratic
• Good communication	• Insufficient feedback/input
• Role clarity	• Poor communication
• Adaptable	• Uncertainty
• Committed	• Under performing
• A convergence of ideas	• Do the minimum
	• Creates stress

Thus, as shown in table 7.15, simplifying workplace culture leads to positive results. Workplace stress could also be due to the following circumstances:

- Organizational change
- Company mergers
- Uncertainty
- Think how your workplace has been when change is made, or employees 'kept in dark' – the mushrooms?

7.9.1.2   Workplace stress, its adverse impacts and ways to avoid[4, 10, 31]

Stress is part and parcel of our behavior and varies from person to person (figure 7.9), time and circumstances. By and large nobody is stress-free. It pervades all layers of life.

Stress is an arousal response the body makes, when a situation is perceived as being threatening. The impact of this arousal affects emotional as well as cognitive thinking. The impact is seen in behavioral and physiological change.

Figure 7.9 Workplace stresses – Should any one live with them (first 4 from left)? Or Like the first two from right?[4] (After Cairney, M. 2007).

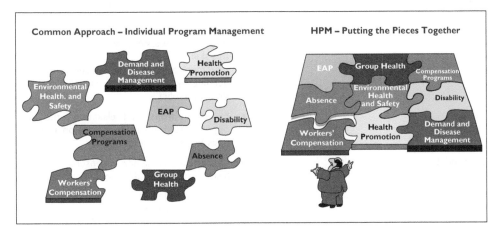

Figure 7.10 Health Promotion Management (HPM) is an integrated management aim of health related issues. (After McDonald, R. 2006).[18]

When stress is long term and chronic, it leads to chronic psychosomatic ailments. While there is an undeniable link between stress and illness, optimal stress is a prerequisite for success in every task. Stress is the wind beneath our wings pushing us towards achievement. If stress can trigger off psychosomatic ailments in those living in the fast track, it can be a trigger even to those who suffer from monotony, boredom or frustration.[4, 10] It is therefore very essential to bring a balance between too much and too little stress (which has been described as an 'optimal stress'). Apart from the solutions proposed by Mick Cairney (2007), the following are very useful guidelines:[4]

- Any exercise programme done regularly utilizes the body's stress hormones. Exercise not only makes the body fit, but also acclimatizes the heart and lungs to increased activity, as in stressful situations.
- Relaxation and meditation techniques result in calming brain waves, and reducing the effects of stress. This is especially effective with stress-related ailments such as hypertension, headaches, digestive ailments, cardiac ailments, sleep problems to name just a few.
- One of the best gifts of nature is sleep. Daily sleep of 6 to 8 hours helps in relaxation and repair of the body and mind. It provides sufficient time for restoring physical and mental health. Sleep well and we will have renewed energy to face the next day's stresses. Check if the time schedule as tabulated in Table 7.16 could suit you?[10]
- Table 7.16 gives a 24 hours time management schedule.[10]

Table 7.16 A 24 hours time management schedule.[4, 10, 31]

Activity	Hours	Remarks
Sleep	6–8	• By sleeping well we will have renewed energy to face the next day's Stresses.
Work, occupation, study	8–10	• Planning our career well gives us challenge, satisfaction & security. • An assertive personality is responsible to self & others. We should learn this skill. • Communication is the key to relationship building. By listening better our conflicts will slowly dissolve.
Exercise	1	• Exercise is arousal and relaxation is the opposite. Yet both are necessary on a daily basis. We should practice both.
3 big and 2 small meals & quality family time	1	• Quality time for family will ensure that we have a retreat called home.
Relaxation, meditation, personal hygiene, social, quiet personal time, and others	6	• Planning a little idleness & quietness each day enable us to recharge. • Develop a sense of humor. We should learn to laugh at life, its paradoxes and at ourselves.

(Continued)

Activity	Hours	Remarks
Total time	24	• Planning our time well, as we have 24 hours/day and many important areas to fit in/cover. One could personalize it to suit his/her own needs. • Remember!!! We can't delete any activity or reduce time for health factors. • George Bernard Shaw famously remarked, "Better never than late." So punctuality is very much essential in our life.

Time related best practices:[4, 10, 31]

- **Commitment:** Many times, tardiness is plain lack of commitment and discipline.
- **Planning:** This ensures we aren't running at break-neck speed at the nth hour, setting us apart from our colleagues and competitors.
- **Buffer time:** Whether it is a project deadline, dialing for a conference call or leaving our desk for a meeting, work out the time required by factoring in a reasonable buffer to take care of all last minute delays.
- **Prioritize:** Regularly review our 'to-do' lists and prioritize: This helps in ensuring important agendas don't fall off.
- **Keep adequate backup:** It can be an extra shirt or an extra copy of that important presentation; it pays to think things thorough and also keeps stress levels low.
- **Punctuality** is a sign of responsibility and discipline. It is an essential component of professionalism and is a habit that can be developed. Shakespeare's "Better three hours too soon than a minute too late" sounds extreme but clearly shows that punctuality was a virtue even in that era.

- Summarizing the integrated approach as proposed by Mick Cairney (2007)[4] reveals:

  - There is a link between OC and workplace stress. An integrated approach, as proposed above, could bring improvements.
  - Stresses at workplace have been identified by research and can be targeted.
  - Establishing a positive culture is desirable for positive business outcomes.
  - Good leadership and management practices can deliver a positive culture, which can achieve higher safety and business values.
  - Releasing stresses through regular exercise, meditations and sound sleep is equally important.
  - Solutions must not only be sustainable but also sustained.

7.9.2 Lost performance at work (Presenteeism)[18]

7.9.2.1 Presenteeism

Apart from the workplace stress which has adverse impacts on the individuals who suffer from it and the organization to which they belong to, as described in the preceding paragraphs, there is another aspect of lost performance at work which is called 'Presenteeism', as described in the following sections with the help of a case-study. The case study presented herewith is by the Principal Health Advisor Dr. Rob McDonald belonging to a Multi National Company, Rio Tinto. This presentation entitles: 'Investing in the Health of our Workers' was presented at an international conference on safety and health on 14–16 November 2006 in Johannesburg, South Africa; organized by ICMM.

The presentation intended to give the following 'Core Messages'

- There is a clear link between a workers' health, safety and productivity;
- Poor health is costing significantly more than most businesses realize;
- Investing in employee health makes good business sense;
- Data capture is essential in implementing effective and sustainable wellness programmes.

7.9.2.2 Health Promotion Management (HPM) – What it is?[18]

Every organization has its ways and means to address the health-related issues and problems of their workforce (workers) as shown in Figure 7.10, Health Promotion Management (HPM) is an integrated management aim of health-related issues with the objectives as outlined below:

- The integrated management of health and injury risks, chronic illness, and disability to reduce employee's total health-related costs including medical expenditure, unnecessary absence from work, and lost performance at work (which is called 'Presenteeism)'
- **Presenteeism:** The problem of workers being on the job but because of medical conditions, not fully functioning. The health problems that result in presenteeism include such chronic or episodic seasonal ailments as: depression, back pain, arthritis, heart disease, high blood pressure, and gastrointestinal disorders.

Figure 7.11 shows parameters influencing productivity losses. Using HPM could benefit the organization in the following manner:

- Impairment at work (presenteeism) is the largest single component of productivity losses in the workplace;
- The extent and cost of productivity losses within a workplace are much larger than previously thought;
- The presence of modifiable risk factors is strongly correlated with reduced productivity;

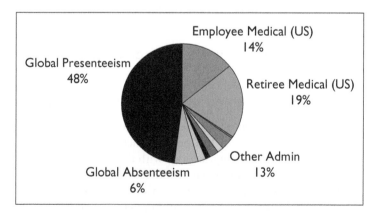

Figure 7.11 Parameters influencing productivity losses. (After McDonald, R. 2006).[18]

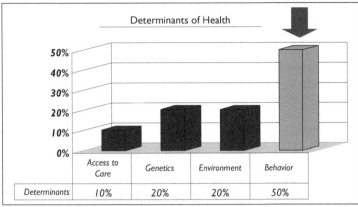

Mercer Human Resource Consulting

Figure 7.12 Parameters and determinants of health,; adapted from the US Department of Health and Human Services. (After McDonadl, R. 2006).[18]

> *Source:* Us department of health human service (1980) ten leading causes of death in United States Atlanta: center for disease control.

- Productivity-improvement projects can be designed and implemented for success and have a positive ROI.

7.9.2.3   *Health risks and behavior*[18]

Health risks have a close relationship with our life style, habits and behavior as illustrated in Figure 7.12.

7.9.2.4   *Developing a health profile for businesses – A case study*[18]

The Rio Tinto Health and Work Survey was conducted involving over 1300 employees (with 60% response rate) using the 'Harvard Productivity Questionnaire & Risk Factors' to assess the situation with the following objectives:

- Develop a health profile for businesses
- Identify diseases and risk factors for poor health that result in maximal productivity loss
- Assist participating businesses to develop their own wellness strategy
- Support the business case for a Group Wellness Strategy.

  Key Results – Risk factors were as under:

- 60% workers not active enough (*)
- 55% were overweight or obese (*)
- 20% at medium and 4% at high risk of significant psychological distress (#)
- 20% were at some risk from alcohol consumption patterns (NA)
- 18% reported fatigue or low energy (#)

- 15% smoked (#)
- 35% males and 19% of females aged over 40 had not seen a doctor for a routine check up in the last 12 months (NA).

(*) Equivalent to Australian Data
(#) Lower than Australian Data
(NA) Comparative data not available

Please refer to Figures 7.13 to 7.15 which are self-explanatory and so is Table 7.17.

The Business aspects of HPM – Implementation of HPM resulted Return on Investment as shown in Figure 7.16.

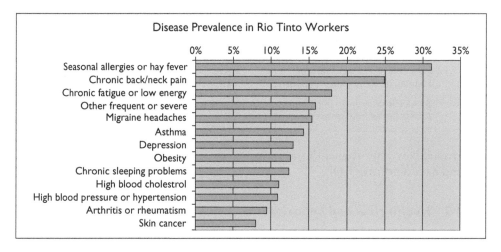

Figure 7.13 Types of diseases that could be prevalent amongst industrial workers. (After McDonald, R. 2006).[18]

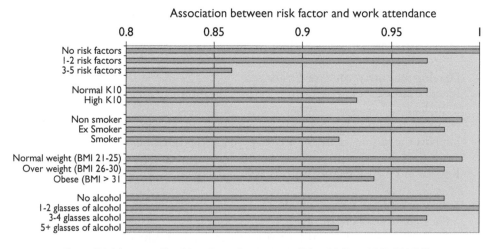

Figure 7.14 Impact of health-risk on absenteeism. (After McDonald, R. 2006).[18]

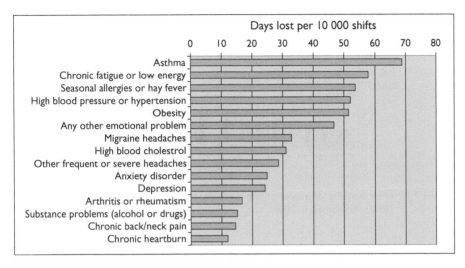

Figure 7.15 Days lost per 10 000 shifts on account of chronic or episodic seasonal ailments include: depression, back pain, arthritis, heart disease, high blood pressure, and gastrointestinal disorders. (After McDonald, R. 2006).[18]

Table 7.17   Impact of behavior on health. (After McDonald, R, 2006).[18]

Health risks and behaviour	
Health risk measure	High risk criteria
• Alcohol	More than 14 drinks/week
• Blood pressure	Systolic > 139 mmHg or diastolic >89 mmHg
• Body Mass Index	>27
• Cholesterol	>239 mg/dL
• Existing medical problem	Heart, cancer, diabetes, emphysema, stroke
• HDL	<35 mg/dL
• Illness days	>5 days per year
• Job satisfaction	Party or not satisfied
• Life satisfaction	Partly or not satisfied
• Perception of health	Fair or poor
• Physical activity	< one time per week
• Safety belt usage	Using safety belt <100% of time
• Smoking	Present smoker
• Stress	High
• Use of drugs for relaxation	Few times per month or more
Overall Risk Levels	
Low Risk	0 to 2 high risks
Medium Risk	3–4 high risks
High Risk	5 or more high risks

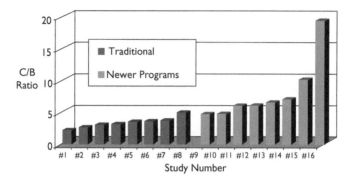

Figure 7.16 Cost benefits of implementing wellness programs. C/B: Cost benefit ratio.[18] (After McDonald, R. 2006).

- Operating in the top quartile for HPM saves $2562 per employee, when compared to the median (1998)
- Return On Investment (ROI) for HSPM $1.40–$13/$1 spent
- ROI traditional health promotion $1.00–$3.14/$1 spent.

Thus, HPM is one of the Best Practices having the following salient features:

- It is a Management Commitment towards wellness of all concerned
- It is an alignment between health, safety, production, HR and benefits, and overall business strategy of the organization
- Proper recording of data, measurement, reporting, evaluation and ROI studies are crucial for its success
- Incentives play an important role in its implementation and success
- Sharing and learning from this best practice by involving internal and external benchmarking could bring consistent improvements.

7.9.3 Occupational hygiene risk – Exposure assessment and control measures

7.9.3.1 Health-related variables influencing the working life of an industrial worker[2, 14, 15, 32]

Referring to Table 7.18 and figure 1.9, there is a definite relationship between the working life and health of any individual who is working in any industry. As per the demands of the job, one has to keep fit and healthy to perform his/her job effectively right from day one on joining till he/she leaves the job. And the following paragraphs describe the steps required to accomplish this.

Pre-placement assessment
- A pre-placement assessment involves a medical and/or physical assessment that determines an individual's capabilities and limitations with respect to the physical demands or requirements of a specific job.

Table 7.18 Working life of an industrial worker and the variables that influences his/her health during their working life span.

Wellness in workplace											
Employee joins			**Exposure to hazards**							**Employee exits**	
Medical Check up for fitness to the Job	Exposure to working culture, Systems	Undergoing Status Change Promotion, Transfer and Change in Responsibilities	Dust	Foul Gases	Radiations	Thermal stress	Noise	Physical Injury; Likewise other hazards could be included		Effects; Either of the below • Die • Injured • Sick • Medically Affected Exit • Employee Exits Sick • Employee Exits Healthy	
HR Related Issues			Fatigue							Outcome	

- Based on a job description that includes capabilities required of the employee, minimum standards of fitness as well as abnormalities that will prevent proper performance of the job.
- Applies to pre-employment, transfer to a new role, changed environment, or changed individual health status.

7.9.3.2   Periodic health surveillance based on exposure risk[15]

Assessing exposure to hazards and implementing control measures provides a sound footing to the preventive and proactive measures to the health of industrial workers or those concerned. It involves following steps:

- Workplace assessments are conducted and evaluated by competent professional hygienists using recognized standard methodology
- Initial observations, interviews with operational personnel and professional judgment lead to basic characterization and qualitative assessment
- Through quantitative assessment, statistical analysis and interpretation of data, a baseline exposure risk profile is established. The baseline exposure risk profile directs

    - Exposure control initiatives
    - Periodic review/re-evaluation
    - Health surveillance programmes.

The flow diagram (Figure 7. 17) illustrates following steps:

- Basic characterization
- Qualitative risk assessment and prioritization
- Exposure monitoring
- Statistical analysis including determination for adequacy of data

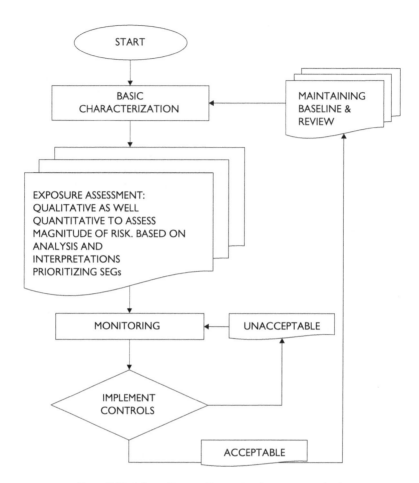

Figure 7.17 A flow diagram illustrating the steps involved.

- Interpretation
- Control measures – exposure risk management

Basic characterization[15]

In this phase hazards from various operations are listed, the locales are identified; category and number of peoples exposed are identified. A matrix as shown in Figure 7.19, of the Similar Exposure Groups (SEGs) is prepared.

Exposure Assessment[15]

Going through Figure 7.18, to identify a risk refer to the SEGs, for example: in the measurement operation 12 people have been involved and their exposure to Inhalable dust; Fluoride; HF and Noise is unacceptable and warrants effective measures;

Risk Assessment Criteria

ACCEPTABLE (A)	<50% OEL
As Low As Reasonably Practical (ALARP)	>/=50% to <100% OEL
UNACCEPTABLE (UA)	>/=OEL

SEGs	# People in SEG	Inhalable Dust	Res. Dust	SO$_2$	HF	Crystalline silica	Heat	Noise	Exposure Agent-X	Exposure Agent-Y
Measurement operator	12	12 UA	12 ALARP	12 ALARP	12 UA	A	12 ALARP	12 UA	12 ALARP	A
Crane driver	9	A	A	9 ALARP	9 ALARP	A	A	9 ALARP	A	9 ALARP
Anode butt transporter	10	10 ALARP	10 ALARP	10 ALARP	10 ALARP	A	A	10 UA	A	10 UA
Mech. Maintenance	5	5 ALARP	A	A	5 ALARP	A	A	5 ALARP	A	A
Supervisor	5	5 ALARP	A	5 ALARP	5 ALARP	A	A	A	A	5
Office staff	8	A	A	A	A	A	A	A	A	A

Figure 7.18 (Upper): Risk assessment criteria. (Lower): Illustration for the Qualitative Exposure Risk Assessment to various groups.

whereas exposure to SO$_2$ and Heat and exposure agent X (which could be any) is ALARP but needs to be cautious (ref. Sec. 10.); their exposure to the rest of the elements is acceptable.

In Figure 7.18, the Risk assessment criteria based on the occupational exposure limit (OEL) has been categorized as: **A** – Acceptable (Green) – Risk is low but should be monitored so as not to exceed; As Low As Reasonably Practical (**ALARP** Yellow) – Efforts must be made to achieve this, and **U** – Unacceptable (Red)*** – Risk is high, which needs measures to avoid exposure of all those involved including use of PPEs wherever required.

*** UNACCEPTABLE Exposure risk management includes:

• Immediate intervention where exposure is assessed as 'unacceptable'
• Interim measures may include the use of respiratory protective equipment under a well controlled programme
• 'Unacceptable' SEGs should be investigated to identify root causes of over-exposure
• Application of a hierarchy of risk control principles and action (Elimination, substitution and isolation; engineering controls and use of PPEs as illustrated in Figure 7.3) to reduce exposure to risk

- Where the estimate of the mean exposure exceeds 50% OEL, PPE is recommended, and an investigation of the cause of exposure should be conducted, and prioritized for control should be undertaken.

Considerations for implementation:

- A risk matrix should be developed for each department taking into consideration the locale, category or group of people exposed to risk/risks that have been identified
- Followed by assessment of the periodic medical exams (tests) and scheduling them for implementation
- Tracking the progress against the target set and reporting regularly to those responsible
- Medical exams (tests) results should be conveyed to the concerned employees
- An annual summary should be prepared to present to management to appraise trends and to review the follow-up actions.

Successful implementation and continuous improvement requires:

- Senior management's involvement/commitment
- Allocation of professional resources for the programme's implementation
- Programme objectives should be incorporated into the business plan
- The effective implementation includes

  - Involvement of the all sections/departments
  - Compliance with scheduled monitoring programme
  - Exposure reduction in 'unacceptable' SEGs
  - Compliance with risk based medical assessment programme
  - Compliance with PPE requirements
  - Regular audit and review for effectiveness.

7.10 NOTIFIED DISEASES AND PREVENTIVE MEASURES[7]

In order to work systematically and to safeguard against the occupational diseases, the following guidelines should be adhered to:

- Noise mapping should be made mandatory of various operations that generate noise along with personal noise dissymmetry of individual workmen exposed to noise level above 85 dBA.
- Vibration studies for any equipment that generates vibrations should be done before its introduction as per the ISO standards.
- All equipment before their introduction should be checked for their suitability in terms of 'Ergonomic fitness', as per the ISO standards. These checks should assess:

  - Working procedure
  - Working aids/tools
  - Working posture.

- Drinking water supplied to employees should be regularly tested, irrespective of its source, preferably after rainy seasons; samples of water should be collected from the points of consumption
- Initial medical examination (prior to offering job), as described in sec 7.9.3.1; is mandatory in any industrial setup irrespective of whether it is permanent, temporary or contractual. It should be strictly adhered to.
- Special tests should be included in the Periodical Medical Examination (PME) for employees exposed to specific health hazards;

a   For employees exposed to manganese, special emphasis should be given to behavioral and neurological disturbances such as speech defect, tremor, and impairment of equilibrium, adiadochokinesia H_2S and emotional changes.

b   For persons exposed to lead, PME should include blood lead analysis and delta aminolevulinic acid in urine, at least once in a year.

c   Employees engaged in food handling and preparation and handling of stemming material activities should undergo routine stool examination once in every six months and sputum for AFB and chest radiograph once in a year.

d   Employees engaged in driving Heavy Earth Moving Machines (HEMM) should undergo eye refraction tests at least once a year

e   Employees exposed to ionizing radiation should undergo blood count at least once a year.

The diseases, such as those listed below are usually recognized as occupational diseases in an industry where they could be caused:

a   All other types of Pneumoconiosis excluding Coal Workers Pneumoconiosis, Silicosis and Asbestosis. This includes Siderosis & Berillyosis

b   Noise induced hearing loss

c   Contact Dermatitis caused by direct contact with chemicals

d   Pathological manifestations due to radium or radioactive substances

There could be more in the above listing based on the type of industry, and due recognition should be given to them. No doubt there are rules, regulations, laws and standards, norms and practices that should be sufficient to guard against any occupational disease.

QUESTIONS

1   '5S' is a philosophy; list its phases, and outcomes when implemeneted.
2   Define: isotopes, radioactivity, Radon and Radon-daughters.
3   Describe the hazards of welding processes.
4   Does underground air differ from the atmospheric air? If, yes, in what manner?
5   During the regular production phase, what risks could industrial workers be exposed to?
6   Ergonomics is the process of making things user-friendly – Comment.

7 What practices would you propose for the good health of industrial workers?

8 Health Promotion Management (HPM) – what is it? How could it be made applicable in an industrial setup?

9 How useful to us is the application of ergonomic principles?

10 How is good house keeping the key to safety and good health?

11 How many elements are present in the human body and how do they contribute?

12 In addition to the regulatory provisions, what should effluent discharges into water-bodies not result in?

13 In the working life of an industrial worker, list the variables that influences his/her health.

14 List the fundamental principles of industrial hygiene.

15 List the heavy metals and their adverse health impacts when consumed as dust and fumes.

16 List the impacts of good ergonomics as well as poor ergonomics.

17 List the natural gases from strata and their impact when mixed with air.

18 List the problems of deep mining.

19 List the sources of Radiation, including Electromagnetic Radiation.

20 List the steps for managing industrial hygiene.

21 List the steps involved in 'Periodic health surveillance: based on exposure-risk'

22 List the places where aqueous effluents could be discharged?

23 Prepare guidelines for effective 'House keeping' of a shared workplace.

24 Summarize the integrated approach as proposed by Mick Cairney.

25 TLV-TWA – what does it stand for? Define this term.

26 What are 'Notified Diseases'?

27 What are sources of heat in underground mines and tunnels?

28 What are the logical steps in assessing workplace risk, and what are remedial measures to minimize them? You can draw a line diagram to answer this.

29 What are the ways and means to minimize the airborne radiations in underground mines?

30 What are workplace stresses, their adverse impacts and ways to avoid?

31 What is 'Organizational Culture' and what are its impacts?

32 What is 'Presenteeism' and what are its impacts?

33 Why should we work in 'Neutral' postures?

34 Write out the ultimate impacts of: Thermal stress; Noise; Dust; Radiations; Vibrations; Asbestos Fibers.

REFERENCES

1 Bailey, K.F. (2001) Industrial hygiene in mining. In: *Mine health and safety management*, Michael Karmis (edt.), Littleton, Colorado, SME, AIME. pp. 263–273. www.smenet.org.

2 Barnes, D. (2006) Tuberculosis control program. Pres. In: *International conference focusing on safety and health on 14–16 November 2006, Johannesburg, South Africa. London ICMM.*

3 Bridger, R.S. (1995) *Introduction to Ergonomics.* McGraw-Hill, Inc.

4 Cairney, M. (2007) Maintain a safer mining industry and organizational commitment. Pres. In: *'Mining 2020', International Mining Conference, 5–6 Sept. Sydney, Australia. Organizer: AIMEX.*

5 Crowl, D.A. & Louvar, J.F. (2002) *Chemical Process Safety.* New Jersey, Prentice Hall PTR. pp. 54–55, 74–95.

6 Dan MacLeod's Ergonomics Website, http://www.danmacleod.com/ [Accessed 2009].

7 Directorate General of Mines Safety, India; Circular on: Notified Diseases and preventive measures and interaction with authorities (2008).

8 Hethmon, T.A. & Doane, C.W. (2001) Health and safety management. In: *Mine health and safety management*, Michael Karmis (edt.). Littleton, Colorado, SME, AIME. pp. 17–32. www.smenet.org.

9 http://www.msha.gov [Accessed 2009].

10 http://keralites.net/ [Accessed 2009].

11 http://www.deir.qld.gov.au/workplace [Accessed 2009].

12 http://www.epa.gov/rpdweb00/glossary/termdef.html#decay [Accessed 2009].

13 http://www.epa.gov/rpdweb00/understand/health_effects.htm [Accessed 2009].

14 Jager, K.D. (2006) Wellness in the Workplace. Pres. In: *International conference focusing on safety and health on 14–16 November 2006, Johannesburg, South Africa.* London ICMM.

15 Kissane, L. (2006) Exposure Assessment and Risk-based Medicals. Pres. In: *International conference focusing on safety and health on 14–16 November 2006,* Johannesburg, South Africa. London ICMM.

16 Kroemer, K.H.E.; Kroemer, H.B. & Kroemer, K.E. (1994) *Ergonomics, how to design for easy and efficiency*, Prentice Hall.

17 Kupchella, C.E. & Hyland, M.C. (1993) *Environmental Science.* Prentice-Hall International Inc. pp. 507, 513.

18 McDonald, R. (2006) Health and Productivity Management. Pres. In: *International conference focusing on safety and health on 14–16 November 2006,* Johannesburg, South Africa. London ICMM. (Permission: Rio Tinto).

19 Mineral Information Institute (2001), *Web site,* Denver, Colorado. [Accessed 2001].

20 Pegg, M.J. (2000) Safety class-notes and course at Sultan Qaboos University, Oman.

21 Petroleum Development of Oman – HSE management (1996–2004).

22 Piscioneri, M. (2006) Hazard and Risk ID and Management Program. Pres. In: *International conference focusing on safety and health on 14–16 November 2006,* Johannesburg, South Africa. London ICMM. (Permission: BHP Billiton, Mitsubishi Alliance).

23 Pon, M. & Jones I.G. (2001) Other industrial hygiene concerns. *Mine health and safety management*, Michael Karmis (edt.), Littleton, Colorado, SME, AIME. pp. 307–334. www.smenet.org

24 Shell International Exploration and Production Company – HSE Management system (1996–2004).

25 Tatiya, R.R. (2000–04) *Course material for Petroleum and Natural Resources Engg; Chemical Engg. And Petroleum and Mineral Resources Engg.* Sultan Qaboos University, Oman.

26 Tatiya, R.R. (2005) *Surface and underground excavations – methods, techniques and equipment.* A.A. Balkema, and Taylor and Francis, Netherlands. pp. 7–10.

27 *The Ergonomics Kit for General Industry,* Second Edition, Taylor & Francis, 2006.

28 Times of India (Daily newspaper), 15 June 2007. WHO study on: Loss of healthy years based on low and high-income countries.

29 Vergne, J.N. (2000) *Hard Rock Miner's handbook.* McIntosh Redpath Engineering. p. 152.

30 Vutukuri, V.S. & Lama, R.D. (1986) *Environmental Engineering in Mines.* Cambridge University Press.

31 World Class Management (WCM) and Sharing Best Practices. Leaflets, displays, lectures, and talks on various topics – Essel Mining and Industries Limted, Aditya Birla group, India; (2004–2008; through interaction, and participation)

32 Wrigley, D. (2006) Silica Control in Southern Africa. Pres. In: *International conference focusing on safety and health on 14–16 November 2006,* Johannesburg, South Africa. London ICMM.

Chapter 8

Industrial safety

'Safety is not a commodity that can be bought but it is like a shrub which must be planted, cultivated, fertilized and pruned regularly through effective vocational training and education, consistent checks and supervision, and involvement of every one from top to bottom'. *This chapter attempts to highlight this theme.*

Keywords: Industrial Safety, Accidents, Loss Prevention, Training And Education, Inherently Safe, Hazards, Personal Protective Equipment, Unsafe Conditions, Unsafe Acts, Working Environment, Lifecycle, Layers of Protection, Accident Costs, Severity Rating, Injury, Fatality, Risk Matrix, Substandard Behavior. Simplify, Safety Awareness, Remedial Measures, Conceptual Planning, Consequences.

8.1 INTRODUCTION

Every industry involves resources to produce goods and services. The resources could be natural or man-made. Natural resources could be air, water and land; and man-made resources include man himself, machines, equipment and plant, materials and energy. In the process of running the plant, factory, mine, manufacturing unit, oil field or any other establishment, damage to any of these resources cannot be completely eliminated but efforts should be made to minimize it. (And also the causes of pollution, as described in chapters 3 and 4). The following three words are used to describe/define this damage.

Safety: Conditions that keep mind, body or property free from injuries, damage or destruction.

Accident: An unplanned, not thought about abnormal event or happening that results in loss of time, and may result in injuries or/and damage to property.

Loss – as given in Oxford dictionary: 'loss is diminution of ones possessions or advantage; detriment or disadvantage involved in being deprived of something, or resulting from change of condition'. All these parameters are tangible. 'Loss can also include failure to take advantage of something or failure to gain or obtain'.[16] Thus, loss is the result of interrupted practices or procedures that can lead to personal injury or property damage.[4]

In the present scenario, safety has been replaced by *loss prevention*. This term includes hazards identification, technical evaluation and design of new engineering features to prevent losses.

In this book a broader meaning is used for the word 'Safety', which is 'Protecting all input resources (as mentioned in preceding paragraph) from losses'. The losses could be loss of life, injuries, damage to property, harm to the environment (pollution) and to materials. They could be caused by hazards of various kinds as shown in figure 1.6. The most precious input-resource amongst them is the *'Human being' – our people/workers*. They must be saved at any cost whatsoever it may be. It is not possible to eliminate all losses but counter-measures are available to keep injuries less severe and losses smaller.

8.2   SAFETY ELEMENTS AND STRATEGIES

In Figure 8.1, the characters of the word 'SAFETY' defines a well-balanced safety program/ campaign that should be aimed at by every industrial setup to fulfill its objective towards safety. The outcome of safety is visible and can evoke emotions, if results are adverse.

8.3   SAFETY ELEMENTS

In the preceding paragraph the resources deployed to produce goods and services in any industry have been listed. The most precious resource amongst them is 'People'. People are the greatest asset in any industry. Utmost importance should be given to avoid their injury or illness (occupational disease) of any kind.

The word 'Health' has been used throughout the book to denote a healthy worker free from occupational diseases of any kind. The following three elements are responsible for industrial health and safety:

1   People/Industrial workers
2   Systems developed to run the show
3   The working environment.

8.3.1   People/Industrial workers

The key element is man himself, the people, the workers. Referring famous proverb: *It is the man behind the machine who really matters*; hence the quality of human

---

S – **S**ystem development & working systematically: keys for success of any safety campaign
A – **A**ttitude: Active participation, Awareness & Alertness are important to boost safety
F – **F**undamentals: clear and strong foundation for safety yield desired results
E – **E**xperience of Past and Expert Advice leads to right approach
T – **T**eam Work, Training and Timely Accomplishments are mandatory for success
Y – **Y**ou Means you, you – i.e. Participation By Everyone: From Top To Bottom is a must.

Figure 8.1 Letters of word 'SAFETY' defines a well-balanced safety program/campaign/strategy that should be aimed at by every industrial setup to fulfill its objective towards safety.

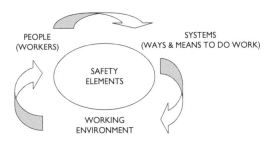

Figure 8.2 Safety elements.

Table 8.1 A comparison of qualities of men and machines.

People are better at	Machines are better at
• Detecting information (visual, auditory) from among background conditions • Handling unexpected occurrences • Reasoning inductively • Learning from experience • Creativity and originality • Flexibility and adjustment to change	• Quick response (human reaction time and response time is greater) • Sensitivity to stimuli (humans have limited range of sensitivity) • Precise repeated operations (e.g. automation, computing) • Deductive reasoning • Storing and processing large amount of data • Exerting force and power • Monitoring functions

(*After* Baruer, R.L. *Source: Mine health and safety management,* Michael Karmis (edt.), SME, AIME, Littleton, Colorado, 2001. Permission: www.smenet.org).

resources should be checked at the entry level during recruitment itself and thereafter. People vary in Age, Gender, Strength (Health), Skill [Unskilled (Raw), Semiskilled, Skilled], Capability, Knowledge, Experience, Expertise, Behavior (Maturity, Self-Discipline, Temperament), Attitude, Personality (Leadership, Initiative, Capabilities), Character (Values, Trustworthiness), Needs and Expectations.[9]

Again it is the man who has to the get the work done from a machine which is his faithful servant having qualities[3] as given in Table 8.1, and that too making best use of his own abilities.

Apart from the background, training aids are essential in order to do the operations in the right manner. Vocational training, refresher courses and application of behavioral science could help in inculcating safety habits amongst workers. Training needs are to be recognized using the *Why, Who, When, Where, What and How* concept to improve skill and knowledge.[7]

It has been published and established that 85% of accidents are caused by unsafe acts and 15% by unsafe conditions.[10] This percentage could be up to 90%; depending upon type of industry and prevalent safety culture within an industrial setup. Universally, people involved in safety management have accepted these data. Heinrich (1931) suggested 3Es; Education, Engineering and Enforcement to establish proactive safety culture. In fact the solution lies, in practicing truly the 4Es to establish proactive work culture including safety:

- Education
- Engineering
- Enforcement and
- Engagement.

EDUCATION[6]

This refers to the basic education that is required to carry out a function that is assigned to any individual, which is obtained from the Universities and Schools. On-the-job training then follows. As shown in Figure 8.3, people are responsible for accidents and losses of various kinds; 90% by unsafe acts and 10% by unsafe conditions. Research carried out on this aspect shows that *changes in human behavior with*

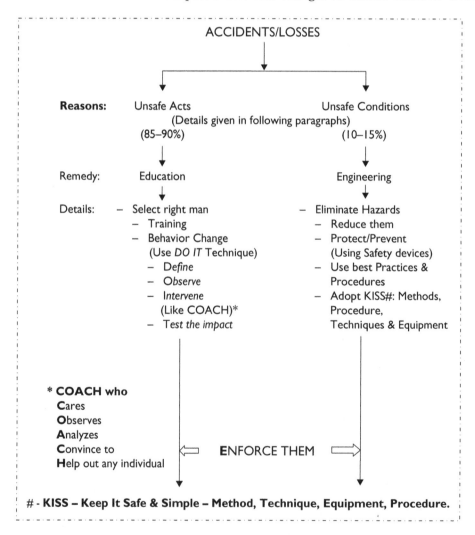

Figure 8.3 Strategies to minimize accidents and losses.[6, 9]

respect to safety could improve the situation. **DO IT** practice, as described below, has given positive results in changing behavior of industrial workers.

Define/Describe/Demonstrate
The safe practices that need to be inculcated should be first explained, demonstrated and their outcome in terms of benefits should be made known. It has been found that in most industries the unsafe practices that prevail include:

- Not using requisite PPEs or using them but not in the right manner
- Using inadequate and improper tools and appliances or not in the right manner
- Using improper, or inadequate guards/fencings
- Attending unauthorized operations/tasks
- Carrying out a job without requisite training on the job
- And likewise many practices.

Observe
Like a COACH, who Cares for you, Observe your performance, Analyzes it and ask you to Comply with it to Help you out from the situation.

Intervene
- After consistent watchfulness and observations; intervene to give him feedback, encouragement, suggest modifications, additions, alterations etc.

Test
- The change in safety behavior and if found OK Proceed further for the next change/improvement.

ENGINEERING
This aspect covers application of right methods, techniques, and equipment starting from the planning and design stage to execution stages. While designing, it is essential to know the capabilities of human beings. One size does not fit all. Care is taken as to how hazards can be eliminated or minimized in the design, layout, procedure, services and utilities. Running and operating the following sections/departments smoothly would facilitate proper working conditions.

- Communication and computing
- Energy, energy transfers and illumination
- Fire prevention and control
- Human resources – industrial relations
- Logistics and traffic control
- Maintenance, and designing for people – ergonomics
- Pollution control and ventilation
- R & D and renovations (extensive partnership with operators, manufacturers, academia and government)
- Rescue, recovery and emergency preparedness
- Safe handling, storage of raw materials and products
- Training and safety promotion (Refer Table 8.2)
- Workers/labor welfare, amenities, health and hygiene.

Thus, proper engineering helps in evolving systems.

Unsafe conditions:

- Inadequate guarding, equipment unguarded/not fenced
- Defective tools, appliances and equipment
- Improper design
- Improper layout of the working site/spot
- Defective safety wear or not using safety wear
- Inadequate ventilation – defective vision, air circulation and heat dispersion
- Inadequate illumination
- Improper roads, travel-ways, man-ways – where men and means transportation move.
- Inadequate infrastructures
- Lack of proper maintenance of equipment
- Use of defective or substandard material, consumables or spares
- Lack of proper communications
- Lack of standard practices and procedures
- Lack of means for health and hygiene – washrooms, toilets, drainage, sewage pipes etc.
- Odd working hours
- Adverse climatic conditions
- Faulty system

  - Material handling and disposal
  - Transportation
  - Movement horizontally or vertically
  - Drainage and pumping
  - Communication

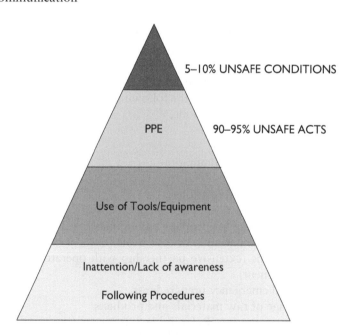

Figure 8.4 Unsafe acts and unsafe conditions are the reasons for accidents and losses of different types[12] (described in different sections including chapter 9).

- Ventilation
- Illumination
- Traffic control/logistics
- Not making available the following devices, equipment, appliances, mechanisms or means (use as applicable to your own setup/industry):

  - Electrical hoists and cranes
  - Hook, chains, eyebolts, slings and cables
  - Pressure vessels, and relief valves
  - Digesters, cookers and the like
  - Temperature control devices
  - Fire detection and extinguishing equipment and apparatus
  - Ladders, lifts and elevators
  - Proper crossovers.

Unsafe acts:

- Allowing unauthorized operations
- Allowing unauthorized operators/crews
- Lack of training and experience of working crews
- Improper use of safety appliances
- Over-confidence
- Inadequate and vague (not clear) instructions
- Improper posture while working – handling, loading, disposal
- Lack of coordination and team work
- Lack of knowledge and non observance of rules, regulations, standard practices and standing orders
- Not using Personal Protective Equipment (PPE) where warranted – masks, respirators, aprons, gloves etc.
- Using faulty, worn and rejected tools and appliances
- Tampering with materials, devices etc.
- Taking unsafe positions
- Lack of supervision or proper instructions and follow ups; it includes lack of:

  - Proper allocation
  - On the job training and instructions
  - Day to day observations and guidance for the correct procedure
  - Immediate corrections
  - Follow-up
  - Setting a good example by demonstration
  - Disciplinary action – when required.

ENFORCEMENT

- Education and training as illustrated in Figure 8.3
- Engineering concepts as illustrated in Figure 8.3
- Rules, regulations and statutory provisions – regulatory framework that prevails in the country where the industry is located.
- Safety devices and controls
- Best practices

The Engagement Pyramid

12) This last year, I have had opportunities at work to learn and grow.
11) In the last six months, someone at work has talked to me about my progress.

10) I have a best friend at work.
9) My associates or fellow employees are committed to doing quality work.
8) The mission or purpose of my company makes me feel my job is important.
7) At work, my opinions seem to count.

6) There is someone at work who encourages my development.
5) My supervisor, or someone at work, seems to care about me as a person.
4) In the last seven days, I have received recognition or praise for doing good work.
3) At work, I have the opportunity to do what I do best every day.

2) I have the materials and equipment I need to do my work right.
1) I know what is expected of me at work.

Figure 8.5 Ways to engage any individual fully to get the maximum for the benefit of both: the Employer as well as the Employee. A good-good situation.[5]

- Empowering approach that says "if I take responsibility of my safety, I can reduce workplace hazards. This will benefit my co-workers, my family, my life and me".[9]
- Incorporating HSE as a critical business activity on a par with other critical activities such as production and productivity.

8.3.2   Systems developed to run the show

A system incorporates the following features

- It has set Practices, Procedures and Norms to follow
- It follows rules, regulation and by-laws – statutory provisions
- It has certain inputs and yields outputs
- It interacts with other agencies
- It requires periodical check ups, maintenance and renovations.

The foundation is laid by the goal, mission and values of the company. Leadership plays an important role in establishing safety culture by giving it due importance. For example HSE is given a reasonable slot in corporate level meetings. Empowerment helps in self-motivation.

The systems that need to be established[9] for a successful HSE program include:

- Training and placement (right man to the right job)
- Auditing, checks, inspection by the statutory bodies and independent agencies
- Recognition of, and incentives for, safe workers
- Policies and procedures that establish best practices
- Effective communication
- Accountability and measurement of losses.

Search your own body to find out how many systems need to function properly to make you fit. And what could happen if any one of them doesn't function properly. And so is the case if systems related to safety, as listed above, fail. Incidents, injuries, illness, inefficiency, insurance-charges, investment–returns (not as expected), irregularities are the yardsticks to assess safety-systems' effectiveness and failures. Failures indicate that something is wrong with the system, and it is not simply human error.

8.3.3   The working environment

Peoples' attitude, behavior and sense of empowerment towards safety campaigns plays an important role. People are responsible for making systems and running them effectively. And the combination of these two aspects creates a working environment. Unsafe acts and unsafe conditions are the result of improper working environment. As described above (section 8.3.1) if unsafe acts could be controlled; 90% of accidents/losses could be prevented, and similarly proper working conditions can take care of the remaining 10% of accidents/losses.

The mining industry, which is the oldest industry in USA, has experienced many disasters, and the incident rates also have been higher; being a rough, tough and haz¬ardous industry; today it is considered amongst the safest industries as reflected by the industry's fatality and injury rate. The measures that have been taken include:

- Improved methods and techniques
- Increased automation that limits interaction between man, machine and material
- Education and intensive training
- Treating HSE as a critical business activity
- Greater recognition given to protecting the industry's greatest asset *'Its People'*.

In the USA during 1997 alone more than 5000 people lost their lives in work-related incidents and more than 3.8 million suffered disabling injuries and illness resulting in a direct cost to US industry of nearly $128 billions with estimated direct and indirect costs of $600 billions.[9]

Today, in addition to the concept of 4Es of safety (ref. Sec. 8.3.1), attention is paid to:

- Empowering workers
- Including/recognizing HSE as a critical business activity
- Adopting progressive labor practices
- Ergonomically (section 7.8) based designs and layouts – machines are made for the man. One size doesn't fit all.

8.4   STRATEGIES[10]

What strategies should be followed that would prevent losses of all kinds and result in a safe environment at the workplace? Incidents, accidents and losses are the result of hazards of various kinds, as illustrated in Figure 1.6. Safety strategies, based on a literature survey, can be classified into four categories:

- Inherent
- Passive
- Active
- Procedural.

The inherent ones don't need any human intervention, just incorporate them and forget. For example, a lift will not start if its door is not closed (inter-locking), and this is how the hazard of falling from a lift when it is in motion is removed. Passive ones need some human intervention. The hazard remains but it is under control. The active ones need periodic testing and maintenance, but a hazard could materialize if maintenance is not carried out properly. The procedural ones depend upon training the people concerned and hoping that they will do as trained.

In most countries speeding automobiles is the major cause of road accidents. The possible solutions are:

- Inherent – Design automobiles that they would not run beyond certain prescribed speeds (within practical limits).
- Passive – Install and maintain speed breakers/bumps (strategic locations, good design, warning signals and regular painting).
- Active – Synchronous road signals, police patrolling and radars; heavy fines for offenders.
- Procedural – Proper education to drivers.

Looking into the above scenario, one would agree that inherent safety is the best but others are equally important and should be enforced as much as possible. Inherent safe design works out to be the costliest; as such how much should be its share would depend upon the judgment of those concerned.

A well-known strategy that is followed in industries is:

- Eliminate hazards
- Minimize them, if they cannot be eliminated
- Substitute or moderate them with less hazardous materials/operations
- Protect through safeguards (safety devices), warning and pre-indicator mechanisms (devices)
- Use protection, barriers and guards to isolate them
- Follow best practices and procedures to accomplish the tasks.

  - Train thoroughly those concerned
  - Exercise strict controls and supervision
  - Periodical checkups, testing and calibration of the instruments
  - Provide adequate tools and appliances
  - Provide best working conditions
  - Obey behavior discipline

Inherent Safer Design Strategies:[6, 8, 11, 17] Pl. refer sec. 6.14 to 6.14.5 that describes these aspects.

8.5   LIFECYCLE APPROACH[18]

A lifecycle approach, as shown in Figure 8.4 (also refer to Figure 10.12), should be followed to incorporate following features:[18]

- Input of latest technological Research and Development
- Inherent safety in the system
- Sustainable development

- Eco-friendliness (least pollution)
- Safer and cost-effective designs.

Management has a responsibility to give recognition and encouragement to those who incorporate the above features in the system.

Total Life Cycle Costs: There is increasing interest in this concept in the environmental area, with recognition of the need to incorporate waste treatment, waste disposal, regulatory compliance, potential liability for environmental damage and other long-term environmental costs, in project economics evaluations. Some factors that should be considered are:

- Capital cost of safety and environment equipment
- Capital cost of passive barriers (for example; containment dikes, vacant land to provide spacing, required by codes, regulations and insurers)
- Operating and maintenance cost of safety instruments and interlocks, fire protection systems, personal protective equipment, and other safety equipment.
- Increased maintenance costs of process equipment due to safety requirements (for example: safety permits, equipment cleaning and purging (cleaning of impurities), personal protective equipment, training and restricted access to process areas).
- Operator's safety training costs
- Regulatory compliance costs
- Insurance costs
- Potential property damage, product loss and business interruption costs if an accident occurs
- Potential liability if an accident occurs.
- An inherently safe approach provides an opportunity to reduce or eliminate many of these long-term economic costs.

Note: All materials and processes have hazards, and it is not realistic or practical to propose that we can completely eliminate all of them.

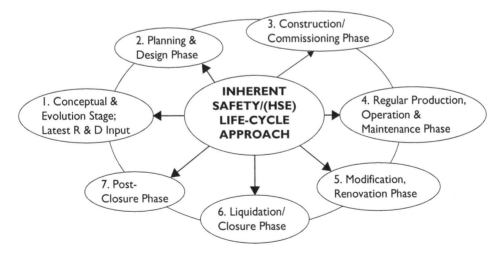

Figure 8.6 Incorporating Inherent Safety features at various phases of the lifecycle of any industrial setup.

8.6   LAYERS OF PROTECTION

One traditional approach of risk management is to control the hazard by providing layers of protection between it and the people, property and surrounding environment to be protected. These layers of protection may include:

- Operation supervision, control systems, alarms, interlocks etc.
- Physical protection devices (relief devices, dykes) and
- Emergency response systems (plant emergency response, community emergency response).

The layers of protection concept include examples which might be found in a chemical plant. It has significant disadvantages:

- The layers of protection are expensive to build, maintain through out the life of the process. Factors include initial capital expenses, operating costs, safety training cost, maintenance cost and diversion of the scarce and valuable technical resources (human and others) into maintenance and operation of layers of protection.
- The hazard remains, and some combination of failures of the layers of protection may result in an accident. Since no layer of protection can be perfect, there is always some risk that an incident will occur.
- Because the hazard is still present, there is always a danger that its potential impacts could be realized by some unanticipated route or mechanism. Accidents can occur by mechanisms that were unanticipated or poorly understood.

For these reasons, the inherently safer approach should be an essential aspect of any safety program. If the hazard can be eliminated or reduced, the extensive layers of protection to control those hazards will not be required.

There can be much discussion about whether a particular safety feature in a plant is 'inherent' or not, as different people can view the same thing in different ways.

8.7   ACCIDENTS

An accident is an abnormal event or happening whether it causes injuries, damage or not. When it does not cause any harm, or it is a 'Near-Miss', it is known as an 'Incident'.

Disaster: As already defined in sec. 6.1.2 this means a major accident or natural event or natural calamity involving loss of lives (human and other creatures), property and resources. It could be a natural or manmade disaster. The definition differs from country to country.

8.7.1   Accident – a three-step process

An *accident* is a three-step process:[11]

- **Initiation:** the event that starts the accident
- **Propagation:** the events that maintain or expand accidents
- **Termination:** the events that stop the accident or diminish it in size.

The initiation can be diminished, for example in a chemical industry;[9] through effective training, maintenance, process design and providing up-to-the-mark grounding, bonding, inerting, explosion-proof electrics, guide rails and guarding wherever required. And propagation could be diminished by reducing inventories of flammable material, providing effective mechanisms for quick transfer in emergency, by providing adequate space in the layout and using nonflammable construction materials. Quick termination of an accident could be achieved through effective fire fighting, relief and sprinkler systems and also by the installation of check and emergency shut-off valves.

Inherently safer strategy can impact or influence the accident process at any of the three stages. The most effective strategy will prevent initiation of accident. Inherently safe design can also reduce the potential of propagation of an accident, or provide an early termination of the accident sequence before there are major impacts on property, people or the environment.

A safety problem eliminated by use of inherent safety will cost $1 to fix it at the research stage; $10 at the process flow-sheet stage, $100 at the final design stage, $1000 at production stage and $10,000 at post incidental stage.[11] Thus, an early application of inherent safety pays dividends. Inputs from research scientists, business, engineering, environment and safety personnel are essential when deciding the life cycle of an industrial setup. Baseline assessment (existing scenario), impact assessment on HSE, and management plan (remedial measures) to deal with impacts during production and post closure phases, should be established. Impact assessment should cover consumers, producers, local inhabitants, flora and fauna.

Accidents cannot be totally eliminated due to the fact that plant, logistics, operations and maintenance are designed, constructed, operated, and maintained by human beings and human beings are not perfect. All accidents can be traced back and one would find human failure at their origin, which could be poor judgment, forgetfulness, ignorance, incapacitation, alcohol or drug dependence, fatigue etc.

8.7.2 Accidents/Incident analysis

The purpose of incident reporting and analysis is to:[14]

- Learn from mistakes
- Prevent re-occurrence
- Increase level of safety awareness
- Demonstrate commitment to continuous improvements

Causes

Substandard acts, substandard conditions and their underlying causes such as individual and work factors. Individual refers to inadequate knowledge, skill, motivation, capability, strength, and attitude; as described in sec. 8.3.1, etc.

Work factors – inadequate training, supervision, working conditions, design, layout, inspection and including those described in sec. 8.3. Underlying causes include:

- Improper, inadequate or not using PPE
- Inadequate tools, appliances and their improper use

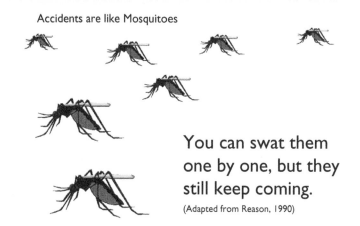

Accidents are like Mosquitoes

You can swat them
one by one, but they
still keep coming.
(Adapted from Reason, 1990)

Its best to drain the swamps
in which they breed

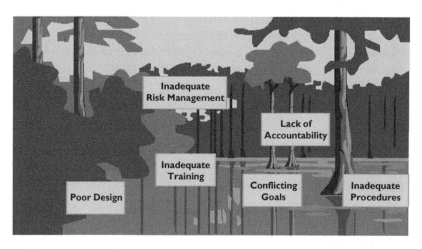

Figure 8.7 Accidents are like mosquitoes, it is best to drain the swamps in which they breed.[19]

Figure 8.8 Obsolete equipment (portable type); huge emissions; and poorly maintained transporting equipment (fixed type) illustrating poor working conditions in an industrial setup.[2]

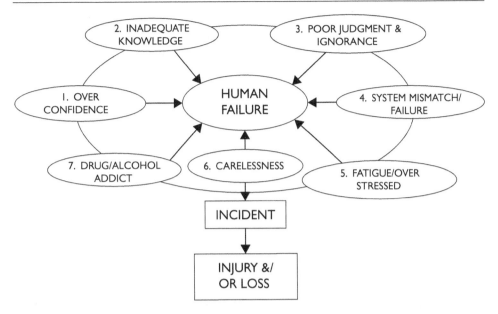

Figure 8.9 Human failure is responsible for most incidents.[2]

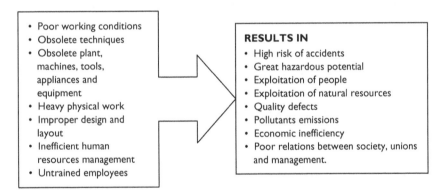

Figure 8.10 Improper practices and their consequences leading to non-sustainable development.

- Unsafe postures/positions while working
- Removing guards, fencing, or taking unauthorized entry
- Overexertion
- Unauthorized operation.

8.7.3 Accident-related calculations[1, 15]

Accidents are calculated as:

Frequency Rate: Number of injuries per 200,000 man-hours worked.#
Severity Rate: Number of lost days per 200,000 man-hours worked.#

– Definitions and calculations differ from one organization to other.

Frequency deals with likelihood of occurrence/hazard that will lead to an undesired event, incident, or accident. Severity deals with extent of damage or harm. Hazards with high frequency and high severity need the greatest attention.

Frequency rate, as per the prevalent practices, is usually calculated for the following losses:

- Lost time due to injury/illness
- First aid or minor injury/illness
- Reportable/Recordable injury/illness (to government/statutory authorities)
- Property damage
- All accidents
- Near-miss.

Severity rate, as per the prevalent practices, is usually calculated for the following losses:

- Accident severity rate
- Property damage (monetary losses) severity rate

Thus, frequency rate = Number of losses, as listed above, × 200,000 divided by total number of employee hours worked.

Employee hours = number of employees × hours worked per year.

8.7.4   Degree (type) of injuries

Fatal – Causing death
Serious – Causing forced absence from work
Minor Injury – Injury without loss of workdays.

8.7.5   Causes of accidents

Causes would differ from one project to another or one industry to another. Listed below are some common areas, or heads, to describe them: (also refer to sec. 10.5.1)

- Haulage – over speed, improper turns & gradient, inadequate safety fittings or their failures.
- Machines & Equipment (Hardware) – Failures due improper maintenance & operation.
- Structural Failures – Inadequate structures, design defects etc.
- Slips or Falls by Workers – Slippery roads, inadequate width of workings & roads.
- Material Handling – toxic, nontoxic, hazardous materials such as explosives, tools and appliances.
- Falling or Sliding of Material – absence of fencing, barricades, guards etc.

- Improper supervision – incompetent, negligent, overstressed, miscommunication and coordination.
- Miscellaneous – not covered above.

In the chemical industry many incidents have happened not because safety system engineering was lacking but safe procedures and preventive strategies were not followed.[8,11,16] for example:

- Failure to follow safety procedures caused release of UF_6 at Kerr-McGee, Oklahoma.
- The control systems were poorly maintained and not operated at the time of tragic accident at Bhopal, India.
- Temporary unsafe connections resulted in the accident at Flix borough, England.

8.7.6 Accident costs

- Injury, loss of body parts, disability; the greatest loss having practically no substitute. Any amount of compensation is just a token.
- Absence from duty, delays, loss of time.
- Loss of morale, loss of efficiency of crew/workers.
- Loss of material, property, equipment.
- Cost of treatment, cost of production loss, overtime payment.
- Cost of replacement, clean ups, repair, standby etc.
- Fines or penalties by the Government/safety authorities.
- Cost of legal assistance.
- Cost of compensation.
- High premium by insurance companies.

8.7.7 Remedial measures

The following steps are helpful in minimizing accidents:

- Conceptual planning, detailed design and evaluation
- Compliance with design specifications
- Safe working conditions – lighting, ventilation, sanitation
- Safe equipment – fittings, design, maintenance
- Safety wear, detectors and warning mechanisms
- Precautions and measures against fires and explosions
- Training, education and refresher courses
- Emergency measures
- Welfare amenities and medical check ups
- Legislation – rules, regulations, code of practices
- Accident analysis and preventive measures
- Risk analysis.

Some of the items listed above have been already discussed in the preceding paragraphs and sections, the remaining are discussed below.

8.8 CONCEPTUAL PLANNING, DETAILED DESIGN AND EVALUATION

Current trend divides any project into four phases:

- Planning and design
- Construction
- Operation
- Post operation period.

The planning and design phase is considered to be the most risky, as major decisions are taken during this phase. During this phase the conceptual model is first prepared and it is then evaluated using different alternatives and schemes. The best one is selected for engineering details. During this phase detailed drawings, equipment specifications, resources required and budget are forecast. Any deficiency could result in problems and delays during the construction phase. It could even become a cause of failures during the operational lifetime. It is important to plan and take into account the cost of ultimate closure of the industry as per the prevalent legislation of the land. This is known as 'Life Cycle' approach, as described in the preceding sections. During the design phase modern techniques concerning safety of plants, factories and industrial establishments, such as those outlined below and as described in sec. 6.13, could be considered:

- Failure scenarios
- HAZOP (Systemic identification of hazards and operability problems) and HAZAN (Systemic analysis of hazards and their potential consequences). Applications of these techniques are almost mandatory in chemical industries
- Fault tree analysis
- Event tree analysis.

It is important have a close liaison and collaborative efforts between designers, manufacturers, engineers, and all those who are involved with the project. This could minimize risks. There is considerable scope for improvement in standards of health and safety in the prevalent technologies applied to various industries worldwide. Some countries have more developed regimes than others. All can improve further. We should all share knowledge, guidance and good practice in these matters.

8.9 TRAINING AND EDUCATION[7]

Referring to Figure 8.3, in any organization the most important task is to appoint the best-fitted person for a job, and make use of this human resource, which is most valuable. He/she may be raw (fresher), semi-skilled, skilled or even highly skilled, or an executive of a company. With time, the methods, techniques, equipment, procedures, laws, regulations, policies, and job-related parameters change. To make oneself capable of performing the job in the right and efficient manner, every one needs some degree of training and education. It is an ongoing process, which starts from day one (when joining) till last day (when leaving).

For a company, to extract the maximum from its employees, it becomes almost mandatory to train and educate them. Education means updating the knowledge that is required to run the organization in the best possible manner. As stated, training needs should be understood using the well-established concept of: *Why, Who, When, Where, What and How*. Also refer to sec. 9.13.

WHY: Germain and Arnold (2001)[7] mentioned the following benefits for effective safety and health education and training:

- Reduced accidents, personal harm, property damage, and related losses
- Increased awareness of the value of tools, appliances, materials, supplies and facilities
- Decreased downtime and delays
- Improved morale and motivation
- Reduced mistakes and waste
- Optimum performance, productivity and profitability.

If you think education and training are expensive try ignorance and inefficiency (Germain & Arnold, 2001).

WHO: As stated above everyone needs training and education. Subject Matter Experts (SMEs) could judge the magnitude (how much) and direction (what exactly). SMEs could be managers themselves or external consultants/experts who can carry out an independent evaluation of training needs and make suggestions to management accordingly.

WHEN: This starts with induction, orientation and on-the-job training in most cases for a newly joined employee. Induction introduces the company's structure, organizational setup, policies, mission, goals, objectives, prevailing facilities, procedures, introduction to service providers such as: computing, maintenance, finance and accounts, medical, housing, transport, recreational (clubs, community centers), welfare establishments, external agencies, such as banks, schools, hospitals, post and telegraph services, public transport etc.

Orientation – rotation amongst cross-functional department/sections within the industrial setup/organization, and also in many cases other units of the same organization located at other sites.

On the job training, as per the needs of the individual then follows. Apart from this there are some statutory requirements to provide training periodically. Refresher courses, attending workshops, symposiums, seminars and conferences aid in updating knowledge, and may help in overcoming the inefficiencies, reducing mistakes and minimizing wastes and losses.

WHERE: Classroom and on-the-job training and education are equally important, as both have their specific merits and limitations. Classroom training with the help of modern aids such as audio, video, handouts and study material, drills and demonstrations are the effective ways of learning; particularly when experts and trained instructors are used for this purpose. Similarly on-the-job training by experienced supervisors and engineers proves to be very effective and useful.

WHAT: As mentioned by Germaine and Arnold (2001), it should cover Total Accident Control Training (TACT – Table 8.2). The areas that should be covered include:

- Occupational Health, Safety and Environment (HSE) – Routine
- Occupational Health, Safety and Environment (HSE) – Specific
- Leadership and management specific to HSE.

Leadership and management specific to HSE should ensure control so as to:

- Prevent accidents and
- Minimize losses.

HOW: The training and education imparted must be effective and it should be well received by the recipient so that best practices are followed at the work site. As mentioned, modern means/aids are useful in delivering the subject matter. The following steps are useful apart from the practice of 'DO IT' described in sec. 8.3.1.

- Motivate the trainee by explaining its advantages to him as well as to the organization. Make him comfortable and at ease to receive whatever is delivered. The subject matter should be easy to understand and very simple and straightforward.
- Demonstrate its operation physically, wherever practical. Also give practical examples. Stress the key points.
- Test what has been learned including the key points explained.
- Check – while on the job, that whatever is required is followed exactly, or whether it requires further explanation.

There will be excellent guidelines available from the Safety/HSE related government/regulatory bodies in your own country. For example: OSHA's 7 steps guideline model broadly covers:

1   Determine if training is needed: many performance problems can be solved by:[7]

    a   Hiring the right man to do the right job
    b   Improved design, layout and maintenance of the workplace
    c   Use of ergonomically improved/designed tools, appliances and facilities
    d   Job aids such as task procedures, flowcharts, decision tables, trouble-shooting guidelines, hotline etc.
    e   Effective communication and administration.

2   Identify training needs – as described in section 9.13
3   Identify goals and objectives
4   Develop learning activities
5   Conduct the training
6   Evaluate program effectiveness
7   Follow up.

Table 8.2 A Three-Day Supervisory Program – Total Accident Control Training (TACT).[7]

First day

1 INTRODUCTION – getting acquainted; program explanation; small group discussion exercise to spotlight key areas of concern for participants.

2 PEOPLE, PROPERTY & PROFITS – business and social reasons for a more professional job of supervisory management than ever before; basic concepts of risk management and Total Accident Control (TAC).

3 PROFESSIONAL SUPERVISORY MANAGEMENT – marks of a "pro", "management/leadership work" vs. "operating work", guiding principles; benefits.

4 PROBLEM CAUSES, EFFECTS & CONTROLS – supervisory management as a key to control; difference between "symptoms" and basic causes; three stages in managing controls; twenty elements for successful accident control.

5 SUPERVISORY INVESTIGATION – a practical, professional approach to accident/incident investigation; how effective investigation saves time and reduces losses; how to measure the quality of investigation; tips on interviewing, analyzing, documenting and following-up.

Second day

6 EFFECTIVE SUPERVISORY INSPECTIONS – kinds of inspections; why and how to inventory "critical parts," the importance of housekeeping; an effective hazard classification system; tips on using inspection reports to get more employee involvement and management action.

7 GROUP COMMUNICATION SKILLS – why supervisors should have group TAC meetings with workers; how to give a good talk (the 5P formula); when meetings should be held and how to make them most effective.

8 MEASUREMENT AS A MANAGEMENT TOOL – three key types of measurement (consequences – causes – controls); how to make best use of TAC measurements; values and benefits.

9 PERSONAL SUPERVISORY COMMUNICATION SKILLS – the supervisor's role in pre-job orientation, how to give key point tips for efficiency, safety and productivity; the "Motivate – Tell & Show – Test – Check" technique for proper job instruction; effective coaching guidelines.

10 THE SUPERVISOR'S ROLE IN DAMAGE CONTROL – bridging the gap between injury control and TAC; uncovering and reducing huge hidden costs; protecting people, preserving property and promoting profits.

Third day

11 MAINTAINING EFFECTIVE DISCIPLINE – tips on obtaining proper use of personal protective equipment, compliance with rules and regulations and better handling of "problem cases", comparing punitive and positive discipline.

12 CRITICAL TASK PROCEDURES – how to use the three-column (steps – potential problem – controls) analysis worksheet; how to save time effort and money with the "improvement check"; how to write the procedure from the worksheet; eight ways to put procedures to work.

(Continued)

Table 8.2 (Continued)

13   EMERGENCY PREPAREDNESS – doing a needs analysis; developing, implementing and monitoring a system, the use of emergency teams; key point tips on emergency response and emergency care.

14   GENERAL PROMOTION – variety is the spice of promotion; how to get the most benefit for the least expenditure, the place of "gimmicks"; double-barreled contests; employee involvement and supervisory leadership example.

15   MOTIVATING TAC PERFORMANCE – five basic guidelines for understanding motivation; six practical principles for managing motivation, basic aspects of a performance management and motivation system for supervisors.

The length of the program, the number of topics covered, what they are and the amount of time devoted to each can be changed in many ways … to meet the needs of the sponsor and the participants.

(After Germain, G.L. and Arnold, R. Mine health and safety management, Michael Karmis (edt.), SME, AIME, Littleton, Colorado, 2001. Permission: www.smenet.org)

8.10   PERSONAL PROTECTIVE EQUIPMENT (PPE)

PPE is defined as equipment designed to be worn by personnel to protect themselves against work related hazards, which may endanger their health or safety. The list of items below covers most of them.

- Head Protection
- Eye and Face Protection
- Body Protection
- Fall Protection
- Hearing Protection
- Hand and Arm Protection

  Respiratory Protective Equipment (RPE)

- Dust Masks
- Air Purifying Respirators (Filter Masks)
- Self Contained Breathing Apparatus (SCBA)
- Supplied Air Breathing Apparatus (SABA)
- Emergency Escape Breathing Apparatus (Escape Set)

It is important to use them in the right way and then maintain them. Replace them if damaged, or at the end of life.

8.11   RISK ANALYSIS[17, 18, 22]

Any industrial establishment involves huge investment and input of resources – man, machine and equipment for long durations. One traditional approach to risk management is to control the hazard by providing layers of protection between it and the people, property and surrounding environment to be protected. These layers of protection may include:

• *Operational supervision, control systems, alarms, interlocks etc.*
• *Physical protection devices (relief devices, dykes, pillars, barriers) and*
• *Emergency response systems (plant/project emergency response, community emergency response)*

In any project/industry, the modern techniques such as: 'Risk Matrix' (Tables 10.6 & 10.7 and Figure 10.3) could be applied to identify potential hazards. The illustration given below is in practice at one of the leading oil and gas companies in the Gulf region (Tables 8.3 to 8.5 and Figure 8.5).

Impact on the environment

1   The severity classification of Environmental Incidents is less straight forward. Environmental Incidents can be divided into two main types:

• Incidental releases of solid or liquid to either soil or water. The severity of these incidents can be classified quantitatively, using three elements – quantity, toxicity and sensitivity. These three elements are used to calculate a number referred to as the Environmental Incident Severity Rating Index (EISRI). The value of this EISRI is used to determine the actual severity and the potential risk rating of the incident.
• All other incidents such as exceeding statutory limits, releases of gases, harm to wildlife, complaints, nuisance due to odours or noise, etc., which cannot be classified according to quantity, toxicity and sensitivity. The potential consequence of these incidents must be classified in a more qualitative manner.

2   Severity of the impact on the environment is therefore determined either quantitatively by calculating the Environmental Incident Severity Rating Index EISRI or qualitatively by comparing attributes affected by the incident with those described below. The main attributes of the environment which may be affected by an incident are visual quality; chemical quality (air, soil, water, living resources); biological quality (diversity); noise level or smell.

Severity or consequence assessment

The vertical axis of the matrix classifies the severity or consequence of the incident. These classifications of consequence are consistent whether referring to actual or potential incident outcomes. They address harm to people, damage to assets, impact to the environment, and damage to the reputation of the company.

Probability of happening again

The horizontal axis of the HSE Risk Matrix in Figure 8.11, measures the likelihood of a similar incident, not the activity, happening again, with a reasonable potential severity, assuming that nothing is done to prevent it. This measure is implicitly intended to account for the frequency of the activity underway at the time of the incident *plus*, the number of people or things exposed to a given hazard during the activity *plus*, the likelihood of an incident being triggered in respect of the hazard controls and safeguards in place *plus*, any other aspect which has relevance in respect of the specific circumstances of the incident.

Consequence				Probability of Happening Again (If Nothing Is Done To Prevent It)				
				v. low	low	medium	high	v. high
				A	B	C	D	E
Severity Rating	P — People (Injury)	A — Assets (Damage or Loss)	E — Environment (Total Effect)	R — Reputation (Impact)	Never heard of in (**) Industry	Heard of incident in (**) industry	Incident has occurred in **	Happens >5 times per year in **	Happens >5 times per year at location
0	No injury	No damage	No effect	No impact			NEAR MISS		
1	Slight (eg. FAC)	Slight (<US$1k)	Slight effect	Slight impact			LOW		
2	Minor (RWC, MTC)	Minor (US$1-10k)	Minor effect	Limited impact					
3	Major (LTI, PPD)	Considerable (US$10-100k)	Localised effect	Considerable impact			MEDIUM		
4	Single Fatality/ PTD	Major (US$100k-1M)	Major effect	National impact				HIGH	
5	Multiple Fatalities/ PTD	Extensive (>US$1M)	Massive effect	International impact					

Figure 8.11 Risk analysis.
** The industry type – Chemical, oil/petroleum, mineral/mining, or any other[17, 21]

Table 8.3 Severity rating details – Damage to an asset or loss of production.

Severity rating	Type of damage to an asset or loss of production
1	Slight damage – no disruption of the production and negligible repair cost (i.e. less than US$ 1,000*).
2	Minor damage – some remedial work required which may include ordering parts. Production may be disrupted briefly and the repair cost is between US$ 1,000 and $9,999*).
3	Localised damage – partial shut down of facilities, production can continue with little deferment, estimated repair cost between US$10,000 and $99,999*
4	Major damage – substantial shutdown of facilities, the production will be deferred for some time, estimated repair cost between US$100,000 and $999,999*
5	Extensive damage – major shutdown or total loss of facilities, the production will be deferred for some time, estimated repair cost US$1,000,000* or more.

* Limit to be fixed based on policy of company. It could be fixed in US$ or in currency of your country.

Table 8.4 Severity rating details – Harm to people.

Severity rating	Type of harm to people
1	*Slight injury or health effects* (including first aid case (FA) and medical treatment case (MTC)) – Not affecting work performance or causing disability.
2	*Minor injury or health effects* (Lost Time Injury (LTI)) – Affecting work performance, such as restriction to activities (Restricted Work Case (RWC)) or a need to take a few days to fully recover (Lost Workday Case (LWC)). Limited health effects, which are reversible, e.g. skin irritation, food poisoning.
3	*Major injury or health effects* (including Permanent Partial Disability (PPD)) – Affecting work performance in the longer term, such as a prolonged absence from work. Irreversible health damages without loss of life, e.g. noise induced hearing loss, chronic back injuries.
4	*Single fatality (SF)* – From an accident or occupational illness (poisoning, cancer).
5	*Multiple fatalities (MF)* – From an accident or occupational illness (poisoning, cancer). The response of the Company shall be at the maximum level.

Table 8.5 Severity rating details: Environmental impacts.

Severity rating	Environmental impacts
1	*Slight effect* – An adverse effect on any attribute of the environment is observable or measurable above normal background levels (allowable threshold limits), is of short duration, confined to the Company site and with no complaints from third parties or governmental concern. For example: quoting norms fixed in Oil Field – Halon and CFC release <50 kg. Gas leak <1,000 cm^2. Remedial action cost less than US$1,000. EISRI less than 50.
2	*Minor effect* – Contamination. Damage sufficiently large to attack the environment. Exceeding the prescribed limits (statutory) or norms. Single complaint. No permanent effect on the environment. For example: quoting norms fixed in Oil Field – Halon and CFC release 50–100 kg. Gas leak of 1,000 cm^2 and greater. Investigation, monitoring or clean-up costs between US$1,000 and $9,999. EISRI between 50 and 4,999.
3	*Localized effect* – Environmental quality in the vicinity of operations becomes substandard or unfit over a limited area for one or more purposes including supporting normal wildlife population; interference with other users causes loss of earnings, complaints or claims. Repeated cases of exceeding permit requirement or internally prescribed standard. Losses exceeds those described S.N. (2).
4	*Major effect* – Severe environmental damage. The company is required to take extensive measures to restore the contaminated environment to its original state. Extended means exceeding the prescribed limits.
5	*Massive effect* – Persistent severe environmental damage or severe nuisance extending over a large area. In terms of commercial or recreational use or nature conservancy, a major economic loss for the company. Constant excess of the prescribed limits.

The user is constrained to use only one of five distinct probability classes. Either the incident:

'A' has never occurred in the Oil and Gas Exploration and Production industry. i.e. *Very low chance of re-occurring.*

'B' has occurred in the Oil and Gas Exploration and Production Industry but not in *our industry* i.e. *Low chance of re-occurring.*

'C' has occurred in *our industry* less than 5 times per year. i.e. *Medium chance of re-occurring.*

'D' Happens 5 or more but times per year in *our industry*, but not in any one location i.e . . . . *High chance of re-occurring.*

'E' Happens 5 or more times per year in one location such as an Operational Area, Drilling or Service Rig or Seismic Prospect (for example in case of an Oil Industry) i.e. *Very high chance of re-occurring.*

Obviously, most companies' incidents will fall into the C or D classes with the occasional E class potential. A and B classes will be very rare but they are included here for completeness.

8.12   CASE STUDY: WITHOUT A 'SUGAR' COAT!: BRITISH SUGAR[5]

Synopsis: In most manufacturing industries, active initiative for recognizing risks and averting the consequences is required. At times, it takes a tragedy to bring the necessity home. Until the year 2003, British Sugar Group, which produces more than half of UK's sugar requirements, had an excellent safety record. It had won several safety accolades. However, in the year 2004, the company suffered three fatalities, which made a profound impact on its safety thinking. The fatalities came from simple incidents. One involved a dispatch clerk, who was struck and killed by a large mechanical loading shovel while walking through a warehouse. At another plant, an employee was killed following a catastrophic failure of a boiler tube while at another, a subcontract worker died after falling from a height of 30 feet without a safety harness.

British Sugar took these tragedies to heart. Instead of simply addressing specific risks, it decided to tackle the entire safety culture. The company's operations director, safety manager along with industry safety experts began consulting widely with the workforce. Health and safety responsibilities were assigned to all directors and annual targets incorporated into their performance objectives. Each member from the top management team had to champion individual initiatives and conduct two behavioral safety audits annually. The entire workforce underwent behavioral change initiatives and training. All injuries/incidents had to be reported directly to the heads of teams immediately to ensure immediate investigations.

By the end of the year 2005, British Sugar recorded a 63 per cent reduction in major incidents and 43 per cent reduction in lost time injuries. In the words of the company's CEO, working safely is not an added option; it is a condition of employment. Employees need to always remember that however important the job at hand is, it should not prevent them from sparing two minutes to ensure safety. The first step towards this is to observe and challenge negligent/unsafe behaviors.

8.13   SUBSTANDARD BEHAVIOUR AND WORKPLACE ACCIDENTS, AND WAYS TO AVOID[12, 19]

The essence of accident prevention is improvement in the behaviour of those involved in performing any operation. In an analysis for a mineral-based industry it was found, as shown in Figure 8.13, that 700,000 substandard behaviours could results in 600- near misses; 30 incidents causing property damage; 10 minor accidents and 1 serious accident; on an average basis.[12] These ratios (proportions) may vary from one industry to another but the underlying theme is that it is the working culture and behaviour of the people that can bring the radical changes. A number of ways and means has been proposed in this chapter to accomplish this. The concept of 4Es, Engineering, Education, Enforcement and Engagement if implemented effectively could bring the desired changes.

Establishing the root causes of incidents leading to losses and injuries is the first step to bringing improvements, as illustrated in Figures 8.7, 8.9 and 8.12.

QUESTIONS

1   Define accidents. List the main reasons for accidents. List the costs of accidents.
2   Define the terms: Severity Rate, Frequency Rate.
3   How could safety education be imparted to factory/industrial workers?
4   In general, what are the main reasons or causes of accidents in the industry or stream you belong to: Petroleum, Chemical, Civil, Mechanical, Mining, or any other you belong to. What are the risks involved in accidents? Also list the safety wear you will recommend for a worker working in your stream.
5   List the types of layers that are usually provided between hazards that may arise by any process and the people, property and environment surrounding it.
6   How helpful in accident prevention is establishing the root causes of incidents leading to losses?
7   Describe how substandard behaviour and workplace accidents are interrelated.
8   What is the modern technique to assess risk?
9   Work out a 'Risk Matrix' for the industry you belong to; which could be; Petroleum, Chemical, Civil, Mechanical, Mining, or any other Petroleum, Chemical, Civil, Mechanical, Mining, or any other.
10   Propose a 'Three-Day -Total Accident Control Training program' for Shop-floor workers and supervisors in the industry you belong to.
11   In the absence of proper education and training what could result? Do you think these are expensive?
12   What logic you would follow to determine the training needs of your subordinates?

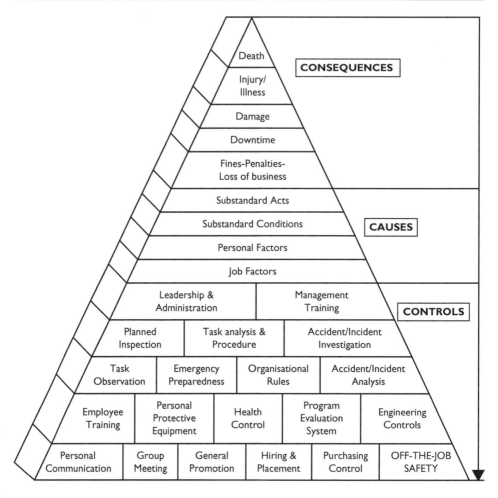

Figure 8.12 Summary: Incident/accident causes, consequences and controls ('3Cs')[7].
(*After* Germain, G.L. and Arnold, R. Germain, G.L. and Arnold, R *Mine health and safety management*, Michael Karmis (edt.), SME, AIME, Littleton, Colorado, 2001. Permission: www.smenet.org).

Figure 8.13 In essence it is the change in behavior that could allow any individual to leave his family happy and return happier.[12, 19]

13 List the phases of any industrial project.
14 List the degree (types) of injuries which can occur.
15 Name some improper practices; what could be their impacts to the growth of the industry you belong to?

REFERENCES

1 Arnold, R.M. (2001) Measurement techniques in safety management. In: *Mine health and safety management*, Michael Karmis (edt.). Littleton, Colorado, SME, AIME. pp. 51–57.

2 Baldermann, G. (2006) Occupational Health & Safety – An Element of Effective Business Management. *International conference focusing on safety and health on 14–16 November, Johannesburg, South Africa*. London ICMM.

3 Baruer, R.L. (2001) Engineering for health and safety. In: *Mine health and safety management*, Michael Karmis (edt.). Littleton, Colorado, SME, AIME. pp. 83–86.

4 Bird, F.E. & Gemain, G.L. (1986) *Practical loss control leadership*. Loganville, GA: Institute Publishing.

5 Case study – British Sugar, Aditya Birla Pub; *Sharing Best Practices, The Manager Mentor*, Sr. No: MBP EJ 09/07 SBP 02; Sept. 2007.

6 Geller, E.S; Carter, N; DePasquale, J; Pettinger, C. & Williams, J. (2001) Application of behaviour science to improve mine safety. In: *Mine health and safety management*, Michael Karmis (edt.). Littleton, Colorado, SME, AIME. pp. 67–78.

7 Germain, G.L. & Arnold, R. (2001) Management strategy and system for education and training. In: *Mine health and safety management*, Michael Karmis (edt.). Littleton, Colorado, SME, AIME. pp. 128–144.

8 Gupta, J.P. (2000) Safety course at Sultan Qaboos University, Oman.

9 Hethmon, T.A. & Doane, C.W. (2001) Health and safety management. In: *Mine health and safety management*, Michael Karmis (edt.). Littleton, Colorado, SME, AIME. pp. 17–32.

10 HSE Books, Sudbury, 1992; 1994; 1996; 2000; 2001.

11 *Inherently safer chemical processes*. Daniel A Crowl (edt.). New York, 1996; pp. 19–23.

12 Jansen, J. (2006) First Steps on the Journey to Zero Harm. *International conference focusing on safety and health on 14–16 November, Johannesburg, South Africa*. London ICMM. (Permission: Lonmin).

13 Kletz, T.A. (1985) Eliminating potential process hazards. *Chemical Engineering*, April 1.

14 Martin, D.K. (2001) Incident reporting and analysis. In: *Mine health and safety management*, Michael Karmis (edt.). Littleton, Colorado, SME, AIME. pp. 183-193.

15 Metzgar, C.R. (2001) Causes and effects of loss. In: *Mine health and safety management*, Michael Karmis (edt.). Littleton, Colorado, SME, AIME. pp. 39–45.

16 Pegg, M.J. (2000–03): Safety class-notes and course at Sultan Qaboos University, Oman.

17 Petroleum Development of Oman (PDO) – Lecture notes based on HSE practices at PDO (1996–2004)

18 Fthenakis, Vasils M (edt.), (1998) Preface in: *Prevention and control of accidental releases of hazardous gases.* pp. 1–5.
19 Sevren, T. (2006) Safe production. *International conference focusing on safety and health on 14–16 November, Johannesburg, South Africa.* London ICMM.
20 Tatiya, R.R (2005). *Civil excavations and tunnelling.* London, Thomas Telford. pp. 290–310.
21 Tatiya, R.R. (1996–2004) Class-notes, HSE for Petroleum, Chemical and Mining Engineering, Sultan Qaboos University, Oman.
22 Ridley, John & Channing, John (edts.). (1999) *Workplace safety.* Butterworth Heinmann, pp. 30–80.

Chapter 9

Loss prevention

"Maximum productivity, safety and recovery with minimum cost at the desired rate of production should be aimed at to achieve the optimum results. Success lies in: KISS (Keep It Safe and Simple) – Process/Method, Technique, Procedure and Equipment. This scenario would minimize losses."

Keywords: Loss Prevention, Input Resources, Efficient Systems, Legal Compliance, Human Resources Management, Managing Plant, Autonomous Maintenance System, Wastage, Financial Losses, Quality Management, Benchmarking & Standardization, Effective Training, Competency and Awareness, Effective Communication, World Class Management, Precision in Operation.

9.1 INTRODUCTION

9.1.1 Aims and objectives of an industrial set-up

Production is the bread and butter of those concerned (right from shop floor worker to the topmost executive) in any industrial setup. Productivity brings excellence to the production, but neither can be achieved if safety is jeopardized; pollution is paramount and workers' health is not looked after. Thus, a thorough balance is required amongst three critical business activities: Production, Productivity and HSE. How could this be achieved? There is no simple and straightforward answer but it is certain that by minimizing losses of various kinds we could achieve it. To understand the various types of losses and their causes, we should understand some basics.

9.1.2 Input resources

The following are the input resources needed to accomplish any industrial operation or manufacturing process:

- Human Resources – Manpower
- Plant and machinery: Equipment and machines, tools and appliances
- Raw Materials – natural (such as minerals, fossil fuels, flora) and man-made materials which are innumerable

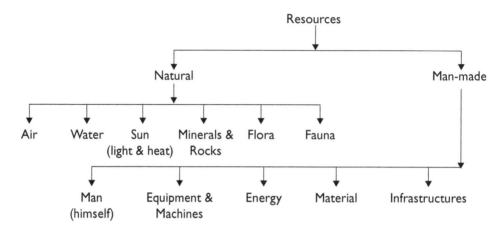

Figure 9.1 Resources – classification. Success lies in effectively utilizing these resources.

- Energy including the natural resources air and water.
- Infrastructures – logistics (rail, road, transport, sea and air-link), means of communication and Information Technology (IT)

Figure 9.1 is a line-diagram showing a classification of resources. Success lies in utilizing them through effective measures of loss prevention.

9.2   LOSS PREVENTION[10, 17]

Thus there is link between all these inputs in order to accomplish production from any industrial set-up, factory, mill or manufacturing unit. Any mismatch amongst them could jeopardize production or the ultimate goal/objective of any company. Any hindrance in their functioning could lead to losses, inefficiencies, defects, delays, customers' complaints, stockholders' dissatisfaction, complaints by communities and dissatisfaction amongst those concerned – the workers and managers and executives.

9.3   LOSS PREVENTION STRATEGY

Under its aims/objectives and mission, a company's policy is formulated. As shown in Figure 9.2 there are three pillars of equal strength for 'Loss prevention'. These are:

9.3.1   Content employees

Competent employees could be a result of effective human resources management. Table 9.2 describes how to engage any individual fully and then get the maximum for the benefit of both: the employer as well as the employee – A good-good situation. Should we not strive for it?

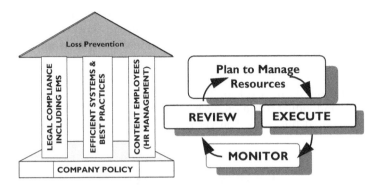

Figure 9.2 (Left) Three Pillars of equal strength for loss prevention which should be an integral part of any company's policy. (Right) In order to run any operation or system effectively, follow these steps: Plan, Execute, Monitor and Review.

9.3.2 Efficient systems

Inculcating habits to follow best practices: best practices are the result of effective systems that are developed to run any business. In this chapter some of the best practices that have been described are:

- Autonomous Maintenance System (AMS)
- Quality Management System (QMS)
- Environment Management Systems (EMS)
- Effective Training, Competency and Awareness
- Effective Communication
- Precision in Operations
- Emergency Preparedness and Response
- There could be few others also.

9.3.3 Legal compliance[10, 17]

Apart from Environmental Management Systems (EMS), every industry, institute, business or organization should work under a legal framework. A company can have its specific codes and practices too. All these are well thought out steps and well documented, and which are in the interest of all: the company owner, stockholders, employees and the public. Any non-compliance could cause damage to the environment, safety and wellbeing of people, which in turn could result in revenue losses and impacts on the goodwill and reputation of the company.

9.4 HUMAN RESOURCES (HR) – MANPOWER – HR MANAGEMENT

It is the man behind the machine who really matters; hence the quality of human resources must be checked right from the entry level. This means recruiting the right man for the right job. Equally important is inculcating habits for safe and best practices through vocational training, refresher courses and on-the-job training (based on the individuals' training needs). Allowing him/her to attend workshops, seminars and symposiums, which are essential for the professional growth of any individual; it should be an established practice.

After recruitment; efficiency lies in extracting the maximum out of any individual. An empty mind is the devil's workshop; as such the first thing is to allocate the best-fitted job to any individual based on his/her background. Table 9.2 describes how to engage any individual fully and then get the maximum for the benefit of both: the employer as well as the employee.

9.4.1 Some basics of man-management

Referring to Table 9.1, the letters of the word 'Manage' (the literal meaning) itself depicts ways to manage any industrial set-up.

Table 9.2 details some basics of man-management

9.4.2 Some basics of leadership[4, 15]

Apart from effective management, good leadership is equally important.

Leadership can be defined as one's ability to get others to willingly follow. Every organization needs leaders at every level. Leaders can be found and nurtured (David Hakala, 2008)[4] if we look for the following character traits:[4]

1 A leader with **vision** has a clear, vivid picture of where to go, as well as a firm grasp on what success looks like and how to achieve it.
2 **Integrity** is the integration of outward actions and inner values.
3 **Dedication** means spending whatever time or energy is necessary to accomplish the task at hand.
4 **Magnanimity** means giving credit where it is due.
5 Leaders with **humility** recognize that they are no better or worse than other members of the team.
6 **Openness** means being able to listen to new ideas, even if they do not conform to the usual way of thinking.

Table 9.1 The letters of the word 'Manage' (the literal meaning) itself depicts ways to manage any industrial set-up.

M	**M**anaging any industrial set-up by giving equal weighting to critical business activities: Production, Productivity, HSE and Social responsibilities *by*
A	**A**lways following laid-out Standards, Norms, Rules & Industrial Benchmarks *and*
N	**N**urturing Best-practice and a Reliable & Trustworthy Workforce (employees) *through*
A	**A**dministrating them effectively and efficiently, *also*
G	**G**earing up for the Abnormalities in Marketing & Sales, Failures and Emergencies; Security and Secrecy; *and expect to*
E	**E**arn Profit, Goodwill, Excellence and Reputation.

Table 9.2 Some basics of man-management.[17]

Man-Management (Human Efficiency) Engaged & Well Managed Employees = Happiness Amongst All (A Key To Achieve Set-Goals)	
Basic needs	1 Clear understanding: What he is expected to do – it should be made known clearly by the manager so that the employee can choose his best way. He should be made aware of the company's goals, mission and targets, norms and procedures.
	2 Equipping with necessary resources & empowering – manpower assistance, tools, appliances, office space and any other items that could help him and his team and ultimately the company to achieve the set goals in the best possible manner. A manager should ask his employee as how this specific resource could help him as well as the company. Empowering with responsibilities works equally well. It provides opportunity for self-development to the individual, and also enables him to show his worth.
Management Support (Work Culture)	3 Proper working environment at the workplace – the employee should feel that he has enough work to do and he has to fulfill the target given to him and he receives consistent feedback and encouragement for his work. The task is reviewed periodically. All required assistance and resources are made available to accomplish the work. The supervisor cares for the growth of his subordinate. Recognition is given for the excellence as and when any individual achieves it.
Team work	4 Team work: Equally important is to make a team, like in a game, to win. It should have workers (players) with different expertise and experience, sharing views and ideas with each other, and helping each other to carry out any operation in the best possible manner. Teams also need expert advice from internal as well as external agencies/sources. Equally important is the discipline and observance of a set rules, regulations, standing orders in the event of emergency, and set norms and procedures. At work, an individual's opinion and best ideas should be welcomed and recognized.
Growth & Self Development	5 A human being is very sensitive; he/she has expectations in addition to meeting his basic needs. Work satisfaction, incremental growth, professional development and a healthy working environment are some of the parameters that should be taken care of at every level. It benefits any organization. Information and transparency are vital components of trust. Across organizations, employees feel the need to be informed on the issues which are likely to impact their working conditions and life outside work. It helps not only to build confidence among employees about the organization and their future, but also to feel connected to organizational goals. This ultimately brings 'Organizational Commitment (OC) described in sec. 7.9.1.
But no room for workplace stress and frustration	If it is so, then something is wrong in HR Management, as described above. Please refer to Figure 7.9 and sec. 7.9.1.

7 **Creativity** is the ability to think differently, to get outside of the box that constrains solutions.

8 **Fairness** means dealing with others consistently and justly.

9 **Assertiveness** is not the same as aggressiveness. Rather, it is the ability to clearly state what one expects so that there will be no misunderstandings.

10 A **sense of humor** is vital to relieve tension and boredom, as well as to defuse hostility. Effective leaders know how to use humor to energize followers.

The difference between management and leadership has been very logically descibed by Stephen R. Covey in his book; The Seven Habits of Highly Effective People (1989) in the following text:[15]
The Leader is the one who climbs the tallest tree, suveys the entire situation, and yells, 'Wrong Jungle'. Leadership deals with the top line: What are the things I want to accomplish?

Management is Doing Things Right; Leadership is Doing the Right Things. For example: management's efficiency is in climbing the ladder of success; leadership determines whether the ladder is leaning against the right wall.

9.5   MANAGING PLANT, EQUIPMENT, MACHINES, TOOLS AND APPLIANCES

Repeating the proverb 'it is the man behind the machine which really matters'; yes, man has to manage the machines, plant, equipment, tools and appliances that are used to get the work done. In this regard the following 4 aspects should be addressed:

- Proper equipment selection
- Efficient equipment utilization
- Effective equipment maintenance for ensuring maximum availability
- Diagnosing problems.

9.5.1   Proper equipment selection[18]

Before selecting any equipment, consideration and a thorough analysis of the following factors should be given, as a *"Proper match of equipment, methods, techniques and layouts brings optimum results."*

- Environment Factors (*Noise and Vibrations; Exhaust gases; Dust; Fog and Fumes*)
- Accident Factors (*Equipment's overall design; Danger to third parties*)
- Ergonomic Factors (*Ergonomic design; Possibility of social contact; comforts it provides to the operator*)
- Technical Factors (*verify the technical needs*)
- Economic Factors (*Costs: capital, energy, maintenance, wages; capacity*)

A flowchart given in Figure 9.3 could be used as a guideline to select any equipment.

9.5.2   Efficient utilization[10, 17, 23]

In this era equipment to cater for large scale production rates are available but they cost huge sums of money. Efficiency lies in utilizing them every minute, else they prove to be white elephants. Two terms are used to measure equipment efficiency:

- Equipment availability &
- Equipment utilization.

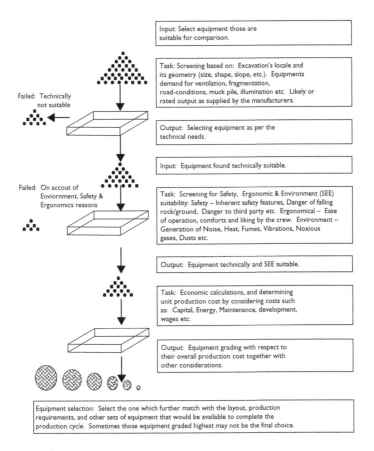

Input: Select equipment those are suitable for comparison.

Failed: Technically not suitable

Task: Screening based on: Excavation's locale and its geometry (size, shape, slope, etc.). Equipments demand for ventilation, fragmentation, road-conditions, muck pile, illumination etc. Likely or rated output as supplied by the manufacturers.

Output: Selecting equipment as per the technical needs.

Input: Equipment found technically suitable.

Failed: On accout of Enviornment, Safety & Ergonomics reasons

Task: Screening for Safety, Ergonomic & Environment (SEE) suitability: Safety – Inherent safety features, Danger of falling rock/ground, Danger to third party etc. Ergonomical – Ease of operation, comforts and liking by the crew. Environment – Generation of Noise, Heat, Fumes, Vibrations, Noxious gases, Dusts etc.

Output: Equipment technically and SEE suitable.

Task: Economic calculations, and determining unit production cost by considering costs such as: Capital, Energy, Maintenance, development, wages etc.

Output: Equipment grading with respect to their overall production cost together with other considerations.

Equipment selection: Select the one which further match with the layout, production requirements, and other sets of equipment that would be available to complete the production cycle. Sometimes those equipment graded highest may not be the final choice.

Figure 9.3 Flowchart to select equipment. This illustration is to select equipment in an excavation, tunnelling or mining industry. Concept is equally applicable to select any other equipment for the industry you belong to.[18]

The higher these factors are the better would be the output/unit time. In fact equipment efficiency and operator's competency play equal roles to achieve better outputs. Frequent breakdowns, reduction in speed or rated output often leads to losses.

Equipment utilization

Equipment over all efficiency (%) = Percentage Availability × Percentage Utilization

% Availability = [(Duration the equipment is made available for the purpose of carrying out useful work or Production)/(Working hours per day)] × 100

% Utilization = [(Duration the equipment utilized for the purpose of carrying out useful work or Production)/(Duration equipment available)] × 100

OEE: Stands for Overall Equipment Effectiveness. It is a comprehensive measure of the equipment's performance, which takes into account the following:

- Availability (A measure of its time utilization. Arrived at by subtracting the down time whether due to planned reasons or unplanned failures).
- Performance (A measure of its effectiveness in performance/underrated operation. Can be assessed by comparison with rated speed, loads) etc.
- Quality Rate (Measure of the total production quality compared with the acceptable quality).

Multiplying all the above three performance measures produces the OEE of equipment.

OEE = Availability × Performance Rate × Quality Rate   (9.1)

9.5.3 Effective maintenance[10, 17, 23]

AUTONOMOUS MAINTENANCE SYSTEM (AMS)
It includes following steps (please refer also to Table 9.4):

Step 1: Bringing equipment to its ideal state
- It is important to know equipment and plant fully by using 5 senses: Eyes, Nose, Ears, Touch and Feel.
- Perform *initial cleaning* to remove dirt, dust and grime.
- Expose all abnormalities including the latent defects; rectify minor flaws and abnormalities (refer Table 9.4).
- Bring equipment to its ideal state.

Step 2: Eliminate Sources of Contamination & Improve Accessibility
- Prevent leaking, spilling, spraying or scattering of: Products, Powders, Lubricants, Hydraulic fluids and other process materials, if any.
- Improve accessibility of places that are hard to: i) Clean, Check ii) Lubricate, Tighten iii) Operate, Adjust and carry out these operations within the least possible time.
- Stop/rectify sources of contamination, spills & leaks.

Step 3: Establish Cleaning & Inspection Standards
- Maintain requirements of Cleaning, Lubrication, Small fixes i.e. tightening.
- Introduce extensive visual controls & standardize gains made through improvements.
- Set the Standard and allocate responsibilities for complying with them.

Step 4: Assign Ownership (Responsibility) of the 'Equipment'
Allocate competent operator and maintenance crew for the sets of equipment who will inspect and maintain it in perfect condition. Provide them with needs-based training as and when required.

Step 5: Assign Ownership (Responsibility) of the 'Process'
Allocate a competent operator/supervisor for the process who will inspect it thoroughly. Provide them with needs-based training as and when required.

Table 9.3(*a*) Governing parameters for equipment availability and utilization.[17,23]

Equipment utilization – governing factors		Details	Remedial measures
Working conditions/ environment	Civil and steel structures	Unevenness of floors, ramps and roads. Presence of cracks and worn out structures. Presence of oil spills, waterlogged and slippery areas.	Presence of accident-prone areas/working spots are abnormalities that should be rectified promptly.
	Improper steps and landing platforms		
	Rotating machinery	Too Steep, Irregular, Torn Anti-Slip Covering, Corrosion, Missing Handrails etc. Displaced, Fallen Off or Broken Covers, No Safety or Emergency Stop Devices, Missing Guards etc.	
	Lifting gears	Unsafe Wires/Hooks/Brakes and Other Parts of Cranes & Hoists, Absence of Fencing etc.	
Absence of safety culture	Unsafe conditions	Presence of explosive substances and solvents. Presence of loose electric cables and wires. No Visual Displays/Caution Boards. Excessive pollution: Presence of excessive dust and noise. Emission of noxious and toxic gases such as NO_x, SO_x, CO_2, CO beyond permissible limits.	This ultimately brings dissatisfaction amongst workers, society and the stockholders. Due attention should be paid by all from the top executive to the shop-floor worker to this critical business activity.
	Unsafe acts	Not using PPEs, or their improper use.	
	Inadequate safety equipment	Inadequate/Improper Safety Equipment, and their dislocation and poor condition.	
Equipment health	Maintenance	Well-maintained equipment performs well.	Success lies in adherence to the strict maintenance schedule. Sec. -, described autonomous maintenance which should be adhered to.

(Continued)

Table 9.3(a) (Continued)

Equipment Utilization – Governing factors	Details	Remedial measures
Aging	Plant and equipment age also plays an important role. Old and worn out equipment cannot perform to full efficiency and warrants periodic replacement of its components as suggested by the equipment manufacturer.	
Free from abnormalities	Rectifying equipment abnormalities, as described in sec. 9.5.3, makes it healthy to perform the designated operation.	
Governing factors – operator efficiency:		A satisfied workforce is an asset to the company, and therefore the clue lies in managing the operator effectively.
Skill	Do working crews have adequate qualifications, practical knowledge and experience to perform the operations? Equally important is adequate training (needs-based) and up to date knowledge which should be imparted through refresher courses. Refer to sec. on training for more details on this issue.	
Ergonomics	Equipment should fit the man and not that the man has to adjust himself according to the equipment's design. Modern sets of equipment are designed to ensure the comfort the operator. Refer to sec. on ergonomics for more details on this issue.	
Lack of discipline	Over confidence, not complying with rules, regulations and Standard Operational Practices/Procedures (SOP).	
Strain/Stress	Excessive effort, exertion, overtime, non-fulfillment of basic needs (expectations), personal grievances. Refer to section 7.9.1.2 on workplace stress	

		Time is money. Success lies in developing efficient systems and adhering to them strictly.
Procedural delays		
Lost continuity	Frequent occurrence of Flickers/Disruptions, Breach/Drop in Pressure/Flow/Temperature/Voltage etc., Restrictions/Choking of passage or screen, Scaling or Failure of lining. Presence of false operation or functions, Dead sensors, Multiple handling.	
Absence or mismatching technology	Equally important is the process technology which should be up-to-date and matching the equipment as per the design	
Mismatch of sequential operations	Lack of proper coordination amongst production, logistics and dispatch. Mismatch amongst various Unit-Operations in a production unit.	
Inefficient services	This includes lack of proper communication, transport, logistics, drainage, illumination, hygiene, housekeeping, first-aid, fire-fighting, rescue and emergency preparedness. Refer to sec. 8.3 also.	

Table 9.3(b) Diagnosis Technique: 5 W – 2 H Analysis. This technique could be useful for revealing abnormalities and their causes.[17, 23]

---

5 W – 2 H Analysis

Who? – It identifies individuals associated with the problem. Also Characterizes customers who are complaining. Which operators are having difficulty?

What? – Describe the problem adequately. Does the severity of the problem vary? Are operational definitions clear (e.g. defects)? Is the measurement system repeatable and accurate?

When? – Identifies the time the problem started and its prevalence in earlier time periods. Do all production shifts experience the same frequency of the problem? What time of year does the problem occur?

Where? – If a defect occurs on a part, where is the defect located? A location check sheet may help. What is the geographic distribution of customer complaints?

Why? – Any known explanation(s) contributing to the problem should be stated.

How? – In what mode or situation did the problem occurred? What procedures were used?

How Many? – What is the extent of the problem?

Step 6: Systematize Autonomous Maintenance

Develop a competent team who could take care of plant and equipment. Standardize the procedures, maintenance and inspection schedules. Include trouble-shooting procedures and ways-out.

Step 7: Practice Full Self-Management

Self-development of the team is required to maintain the plant and equipment in their perfect condition ensuring their maximum availability and utilization.

9.5.4   Preventive maintenance[10, 17, 23]

Preventive maintenance is a schedule of planned maintenance actions aimed at the prevention of breakdowns and failures. The primary goal of preventive maintenance is to prevent the failure of equipment before it actually occurs. It is designed to preserve and enhance equipment reliability by replacing worn components before they actually fail. Preventive maintenance activities include equipment checks, partial or complete overhauls at specified periods, oil changes, lubrication and so on. In addition, workers can record equipment deterioration so that they know to replace or repair worn parts before they cause system failure. Recent technological advances in tools for inspection and diagnosis have enabled even more accurate and effective equipment maintenance. The ideal preventive maintenance program would prevent all equipment failure before it occurs.

Effective output is the function of equipment health, operator skill, operational delays, working environment and safety culture. The abnormalities, described in Table 9.4, could cause limited availability which has been described in Tables 9.3(a) and 9.3(b).

9.6 ABNORMALITIES[17, 23]

Table 9.4 Abnormalities of various types and suggested measures to deal with them. It could be used as a guide for the industry to which you are belonging to.[17,23]

Diagnosis of Abnormalities and Remedial Measures to overcome them

Abnormality	Type	Details	Remedial measures
Apparent Defects: Visible and invisible. Most important aspect is regular effective cleaning and using our 5 senses: ear, nose, eyes, feel and touch, which would reveal the abnormalities that need to be rectified.	Deposit & Accumulation	Dust, Dirt, Powder, Oil, Grease, Rust, Paint, Coolant, Soil, Ash, Sludge, Splash Stain, Sogginess, Leaks Solidified, q Debris, Jamming, Water accumulated in Air/Condensate Line etc.	• Know the equipment, its components & operational details fully. • Identify the abnormalities thorough inspection using senses for abnormal look, noise, vibrations, and smell/odor, if any. • Search for Invisible/visible defects. • Note abnormality, if any, and establish its root cause. If need arises apply: Why-Why, or 5 W–2 H Analysis or any other technique. • Remove them by applying appropriate measures; and where not feasible put a tag for the abnormality identified, and rectify it at the earliest.
	Damage & Deformation	Cracking, Crushing, Deformation, Chipping, Bending, Blasting, Bursting, Coating/Lining Corroding, Softened Surface etc.	
	Play & Gaps	Shaking, Falling Out, Tilting, Eccentricity, Wear, Distortion, Mis-aligned, Rattling Sound, Swaying, Locking, Backlash, Loose Contacts, Deviation	
	Sway & Slackness	Belts, Chains, Rollers/Spindles, Hangers, Hoods, Stays, Cables, Props, Fasteners, Ropeways, Bearings, Hinge & Linkage	
	Abnormal Phenomenon	Unusual Noise, Overheating, Vibration, Jerking, Resonating, Strange Smells, Discoloration, Incorrect & Improper Calibration, Improper Fluid Flow, Improper Operation	
	Adhesion & Restriction	Blocking, Hardening, Accumulation of Debris and Chemicals, Peeling, Malfunctioning, Choking, Sticky/Viscous Flow, Leaked Materials such as: Gas, Water, Steam, Oil, Air, Heat etc.	

(Continued)

Table 9.4 (Continued)

Diagnosis of Abnormalities and Remedial Measures to overcome them

Abnormality	Type	Details	Remedial measures
Basic Non-fulfilment *Ignoring these defects could be the main causes of bottlenecks preventing rated output or planned production from the plants.*	Lub./Coolant/ Priming	Insufficient, Unidentified & Unsuitable, Sub Grade, Hot, Carbonized, Gritty & Dirty, Foaming, Choked etc.	Following guidelines could be useful:
	Supply & Systems	Dirty/Damaged/Deformed/Choked Lubricant Inlets, Faulty/Longer/Haphazard Lubricant Pipes, Inadequate Pressure & Flow etc.	• Ensure that all meters operate correctly & are clearly marked with specified values.
	Inadequate Gauges	Dirty/Damaged/Leaking Glasses, Distant location, Improper type, Incorrect Scale & Readings, Low Visibility.	• Investigate any leaks of product, steam, water, oil, compressed air etc. • Hunt for scaling blockages inside chutes etc.
	Inadequate Tightening	Slackness in Spring & Fasteners. Missed/Crossed/ Crushed Threads. Too Long/Short, Corroded, Wrong/Damaged/Deformed Washers. Unsuitable/Over Tight/Improper Gasket. Mismatched Seats (Tong and Grooves).	• Adhere to routine, scheduled, preventive and predictive maintenance. • Ensure periodic overhaul, Checking, parts replacement. Pay attention to slight defects for example while cleaning check for bolts – are they loose? missing? Is length adequate and protrude 2–3 threads length from the nut? Are washers of suitable type used?
	Improper Measuring Instruments	Non-calibrated, Inappropriate Storage Condition, Faulty Calibration Method. Dirty, Damaged, Faulty Pressure/Temperature/ Volume/Weight/Flow/Power/Voltage/Current Meters, Integrators & Recorders.	
	Unmatched Equipment & Machines	Under/Over Size, Mismatched, Inadequate Control Devices, Sluggish, Mal-routed, Wrong/ Improper Feeders, Inadequate Stirring/Mixing/ Agitation/Crushing/Sizing/ Grading, Improper Material Flow and Movement Devices, Restriction/Fitting Obstruction Improper Installation, Induced Vibration, Frequent Monitoring, Improper Layout, Overhanging, Loose Base etc.	

Defective Design	Design/Shape/Main operating components of Nuts & Bolts, Equipment, Belts, Chains, Bearings, Gaskets, Glands, Hooks & Props etc.	• Note whether equipment is easy to clean, lubricate, inspect, operate & adjust (Identify hindrances such as large obstructive covers, ill positioned lubricators etc.)
Lack of Maintenance	No Maintenance Schedules & Records. Improper Maintenance Planning/Resources etc.	
Non-approachable Places *They are the hindrances to the smooth running of plant and equipment, and cause delays.*		
For Cleaning	Construction/Layout/Installation of Covers/ Footholds, Space around Machines and Equipment	
For Checking	Construction/Layout of Covers, Boxes to Lubricate, Position of Inlet, Construction, Shape, Height, Footholds, Instrument Position, Orientation of Main Panel/Meters/ Recorders, Operating Range Display, Material Level in Silo/Tankers/Wagons/Gear, Lubricant Overflow.	
For Tightening	Covers, Construction, Layout, Size, Footholds, Dusty/Fuming/Dark/Hotter Zone etc.	
For Operation	Machine Layout, Position of Valves, Switches and Levers, Footholds, Starters, Breakers etc.	
For Adjusting	Position of Pressure Gauges, Thermometers, Valves, Relays, Controllers, Flow Meters, Moisture Gauges, Vacuum Gauges, Level Sensors, Limit/Time Switches	• Keep the vicinity of the equipment in order and tidy.
Outside Contaminants from: *They interfere with the smooth running of plant and machinery. They could be also responsible for quality defects in some cases.*		• Remove unnecessary things and attachments, if any.
Raw Materials	Leaks, Spills, Spurts, Scatter, Overflow etc.	**Check while Cleaning Lubrication System**
Lubricants	Leaks, Seepage, Overage, Spent, Spoiled, Spills, Spurts, Scatter, Overflow etc.	• Storage: Is lubricant storage clean!
Gases	Leaking of Compressed Air, Gases, Steam, Exhaust Fumes, Oil Vapor etc.	
Liquids	Leaking, Spilt and Spurting Cold/Hot Water, Half-Finished Products, Cooling/Washing Water, Effluents, Wastewater, Oil etc.	

(Continued)

Table 9.4 (Continued)

Diagnosis of Abnormalities and Remedial Measures to overcome them

Abnormality	Type	Details	Remedial measures
	Scrap materials	Flashes, Cuttings, Packaging Materials, Metal Cutting, Sand, Gritty Matter, Spent/Set/Over-aged Resin, Rubber etc.	• Is lubricant container always capped? • Lubricant Inlets: Are grease/lubrication points kept clean? • Are the inlets labeled with correct type/quantity? • Level Gauges: Are the gauge glasses/indicators kept clean? • Is the correct oil level clearly marked?
	Other Carriers	Contaminants brought by People/Fork-Lifter/Trucks etc. Infiltration through Cracks in Buildings/Hoppers/Chutes/Bins/Silos etc. Pollutants ingress by Wind/Storm.	
Materials Obsolete & Non-Urgent *There have been a number of accidents when due attention has not been paid to guidelines, given in column 4.*	Machinery & Equipment	Pumps, Fans, Blowers, Compressor, Columns, Tanks, Agitators, Vessels, Elevators, Conveyors, Crushers, Pulverizers, Ball Mills, Heat Exchangers, Coolers, Filters, Kiln, Dryers, Transformers, Press, Cells etc.	These are auxiliary machines/equipment/tools/appliances which play vital role in the smoothl running of plant and equipment. The following guidelines are useful: • Their proper selection and matching with the system must be ensured. • Equally important is their regular maintenance and timely replacement. • Substandard quality of material used for their construction often leads to failures and proves to be problematic and could cause losses of different types. • Used and worn-out items should not be used as replacement. • Outdated and mismatched items are often not reliable, and should not be used.
	Interface Connection & Piping	Pipes, Hoses, Ducts, Valves, Dampers, Chutes etc. Measuring & Control Instruments, Temperature/Pressure Gauges, Voltmeters/Ammeters, Electronic Ears, Sound/Gas Emission Monitoring, Relays Switches, Regulators, Instruments/ etc.	
	Electrical Equipment	Wiring, Piping, Power Leads, Switches, Plugs, Sockets, Fuses, Stampings, Insulators, Circuit Breakers etc.	
	Jigs & Tools	General Tools, Cutting Tools, Jigs, Moulds, Dies, Frames, Fixers, Pulley Blocks, Jacks, etc.	

9.7   CLASSIFICATION – LOSSES[17, 23]

9.7.1   Direct losses in various forms or types

1 Production-related losses
2 Lack of effective utilization of time
3 Human Efficiency (Effectiveness) – Man-Management
4 Poor Recovery (Yield)
5 Imbalance amongst production, productivity and HSE
6 Equipment and plant performance
7 Use of substandard materials, auxiliary machines & spares
8 Inadequate internal as well as external infrastructures
9 Quality Defects
10 Frequent interruptions, inconsistencies & shutdowns
11 Defective design & layouts; and lack of planning
12 Energy losses & excessive material consumption
13 Lack of supervision and accountability
14 Lack of tools, techniques (up to date technology) and renovations
15 Lack of attention to abnormalities
16 Excessive waste generation.

9.7.2   Indirect losses

- Affects adversely the goodwill and reputation of the company
- Reduces talent retention
- Increase/rise in indirect costs
- High insurance premiums
- These increase frustration and lower morale amongst working crews/workers.

9.7.3   Losses in a manufacturing plant – reasons and suggested measures to minimize them

Table 9.5 lists types of losses in a manufacturing plant or unit and their reasons, and suggests measures to minimize them. The net result should be zero: Breakdowns, Losses, Pollution, Customer complaints, Quality defects, Delays, Accidents and Wastage; AND IF NOT ZERO all these should be MINIMUM.

9.8   WASTAGE[17, 23]

Waste is a natural phenomenon, which is part and parcel of any process or system. To understand let us consider the human body. We take food and drinks which are converted into many useful products such as blood, bones, flesh and many more that keep us alive, and out of the total-intake, the useless components are rejected from the body in the form of stools, urine, perspiration, etc, etc. But imagine, if we overeat and drink, can we keep our body balanced? Rather there would be side effects and illness.

Table 9.5 Types of losses in a manufacturing plant or unit; their reasons and suggested measures to minimize them.[10, 17, 23]

Types of losses	Reasons & measures to minimize them
Production-related issues Increase in the cost of production results due to reasons given in col. 2	**Deviation from best practices or excellence to achieve planned (Rated Production).** To achieve targeted production, deviation from good practice, as outlined below could result in losses or reduced profits: • Quick handling/removal of the finished goods is essential for safety and productivity, and therefore methods allowing this feature should be preferred. • Concentrating the activities within a compact layout and deploying resources there can yield better productivity i.e. output/man/shift due to effective supervision and better coordination and minimum movement. • Proper match of equipment, methods, techniques and layouts brings optimum results. • Production rates: Higher output rates can reduce expenses on services, over-heads and fixed costs. However, this is governed by the designed capacity and market demand. • Minimize cost of production by keeping minimum inventory. Selecting safe, eco-friendly, technologically sound and economically viable equipment always pays. • Minimize the schedule required to achieve rated production i.e. pre-production or gestation period. • Maximize flexibility and adaptability at the manufacturing units. Less Production than targeted due to: • Shutdown or frequent stoppage • Frequent interruptions due to failure of equipment (keeping a standby ready, following maintenance schedules and practices strictly could minimize this problem) • Mismatch of Unit Operations. Production greater than targeted • Profitable if market is favourable; else • Results in blocking capital. Usually creates storage and handling problems • Concentrating on marketing and sales' promotion strategy could solve this problem.
Lack of effective utilization of Time***	Improper allocation, ambiguity in job profile and responsibility assigned, sequencing operations improperly, mismatch amongst equipment, layout and working conditions, and absence of clear instructions are some of the reasons for time lost and delays. In addition, the following guidelines are useful: • Punctuality always pays. • Every moment is precious*** and could be utilized to add value. • Time has a direct relation with speed; accomplishing any task at a faster speed means adding to productivity. • Waiting time should be minimised. • Undue delays and interruptions add costs and overall dissatisfaction amongst those involved.

(Continued)

Table 9.5 (Continued)

Types of losses	Reasons & measures to minimize them
	*** – **Time is precious as evident by the following facts:**
	1 *To realize the value of ONE YEAR – Ask a student who failed a Grade*
	2 *To realize the value of ONE MONTH – Ask a mother who gave birth to a premature baby*
	3 *To realize the value of ONE WEEK – Ask the Editor of a weekly newspaper*
	4 *To realize the value of ONE HOUR – Ask the lovers who are waiting to meet*
	5 *To realize the value of ONE MINUTE – Ask a person who missed a train*
	6 *To realize the value of ONE SECOND – Ask a person who just avoided an accident*
	7 *To realize the value of ONE MILLI SECOND – Ask a person who won a silver medal in the Olympics*
Human Efficiency (Effectiveness) – Man-Management	It is a measure of effectiveness of System, which includes personal efficiency; equipment or plant efficiency. Sec. 9.4 deals with this aspect in detail.
	• Maximize mechanization, wherever feasible
	• Maximize automation (deployment of remote controlled equipment)
	• Computer Aided Design (CAD) is found to be effective; use wherever appropriate.
	• Adverse working conditions add to inefficiencies. Use of robotics and automation are viable alternatives, if practicable in a given situation.
Poor Recovery (Yield)	Recovery is a vital factor that measures overall yield out of the input resources. **Process Recovery = (Process losses)/(Input);** Whereas, **Process loss = Input Quantity – Output Quantity.**
	Process losses are a function of applied method, technique and equipment to accomplish an operation or task; particularly when applied in conjunction with extraction of natural resources such as minerals: petroleum, natural gas, and solid minerals.
	• The better the recovery, the higher would be the financial gains.
	• Optimize recovery by minimizing dilution and contaminants. Contaminants are due to Leaks, Spills, Spurts, Scatter, Overflow etc. of Lubricants, Raw Materials, Gases, Liquids and Scrap. Contaminants are also brought by ingress of pollutants and wind storms. Refer to sec. 3.11 for more details.
Imbalance amongst production, productivity and HSE	Productivity is a function of effective utilization of available resources to accomplish a job. It is expressed as:
	Productivity = Output/Man/Shift. (9.2)
	Thus, it is a measure of amount of work per unit time by a man; and also by a machine or equipment.
	• Carry out scientific studies to reduce cycle time for various unit operations.

(Continued)

Table 9.5 (Continued)

Types of losses	Reasons & measures to minimize them
	• Maximize natural supports such as gravity, wind direction, surface terrain and topography to boost productivity. Equally import is the safety of man (directly involved and third–party), equipment and process. • Failing which, it directly affects production targets, productivity and ultimately the costs. • Inadequate Safety results accident. An accident has indirect costs, it degrades morale of people involved and also brings bad reputation to the company. Refer to sec. 8.7 for details. • Pollution has an adverse impact on workers' health that reduces their efficiency.
Equipment and plants performance	As described in sec. 9.5, equipment plays a most important role in prevention of losses and in cost reduction as well. Equipment availability and its effective utilization is the key for success. World Class Management (WCM) recommends Autonomous Maintenance System (AMS), as described in sec. 9.5.3; which aims at: • Developing a team who could take care of plant and equipment. The team is trained to diagnose abnormalities, troubles and deficiencies of the equipment, ways to remove them and to bring the equipment its ideal condition. • And then running it to its full efficiency regularly without interruptions.
Man-Management	Sec. 9.4 describes the role of human resources and human efficiency • When human resources are managed effectively the overall gain to the company is obvious. • Mismanaged companies ultimately land into losses, closures and industrial disputes and unrests.
Use of substandard materials, auxiliary machines & spares	Input material and its quality plays an important role in preventing losses. Substandard material has the following demerits: • It can damage the process, equipment and plant. • It can endanger safety not only of those involved but third parties. • Operational failures could be higher. • Thus, it could cause losses rather than savings. • It must be avoided. • Timely procurement of right-quality material, its proper storage, handling and issue system is equally essential to avoid losses.
Inadequate internal as well as external infrastructures	Any factory, mill, or mine could be compared to a busy city with a concern for water, light, power, communication, transportation, supplies, sewerage and construction. • As such any shortcoming or interruptions to these services results in losses.
Quality defects 	Quality of any product is as important as its Quantity. It is a function of so many parameters as described in sec. 9.10. Deviation from these parameters results in defects or a substandard product. In achieving a quality product, the following pays an important role:

(Continued)

Table 9.5 (Continued)

Types of losses	Reasons & measures to minimize them
Refer sec. 9.10 for more details.	• Preventing (or minimizing) mishandling, dilution, foreign matters and moisture. • Effective controls on grain size (in case of solid products) and viscosity (in case of liquid products) should be adhered to. • Strict supervision, laid-out tests and checks by the internal as well as external agencies should be mandatory. • Process control mechanisms must work effectively. Malfunction of process control meters and gauges should not be tolerated. • Variability/Non-Uniformity in the process often leads to quality defects, and must be removed. • Defective products are either rejected or reprocessed which are non-value adding activities adding costs. They are nonproductive and cause delays in delivery. This usually becomes a cause of customers' complaints and dissatisfaction. Product becomes non-reliable. • Producers' motive must be to improve the quality rather than leaving any scope for defects. Research and Development (R & D) should be part and parcel of quality improvement and control measures.
Frequent interruption, inconsistencies & Shutdown	Unplanned and unscheduled layoffs, lockouts, shutdowns are evils which must be prevented. • They bring non-value additions and unrest amongst those involved – the workers, management and stockholders. • They could be due to mismanagement, lack of proper planning and foresight resulting in huge losses. • They must be avoided. During planned or scheduled shutdown: • Ensure supplies' delivery, effective communications and supplier's presence, to avoid delays. • Proper planning including use of Critical Path Method (CPM) is helpful to monitor the progress. It should include resources: men, tools, appliances and material handling mechanism, which should be readily available. • Use experienced crews during shutdowns to do the right job within the time allocated.
Defective design & layouts; and lack of planning	It is always sensible to prepare different designs, alternatives and options during economic studies – feasibility and DPR stages, and select the best. • Spending a dollar during this phase could prevent losses of many dollars if design is defective and the equipment has been procured and plant has been set. • Input from the experienced experts and professionals during conceptual, feasibility, engineering studies or detailed project report (DPR) stages pays. • Eco-friendly design should be incorporated wherever practical. Such designs pay in the long run, and are sustainable. • Inadequate internal as well as external infrastructures also add to poor layouts.

(Continued)

Table 9.5 (Continued)

Types of losses	Reasons & measures to minimize them
Energy losses & excessive material consumption	*Energy & material consumption:* • Methods which involve higher consumption of materials and consumables, are often costlier; and should be avoided. • Another consumable item is energy/unit of output, which is required to carry out different unit operations and services such as ventilation, pumping, illumination etc. • Minimize power distribution and transmission losses • Minimize wastage of water. It would add to savings beside judicious use of this precious resource. • Pollution means more energy has been already spent than required. Least pollution and clean environment are the keys of success. • Also includes selection of an appropriate energy source: electric, diesel, compressed air, hydraulic or non-conventional (if available), which could bring overall benefit to the Company.
Lack of tools, techniques & renovations	• Lack of tools, techniques (up to date technology) and renovations is like a solder at the front without adequate and up-to-the-mark arms, ammunitions, weapons and know-how. And imagine if the enemy is equipped with all these fully. • 'Kaizen' is applied to remove or minimize bottlenecks in the process, techniques, equipment and machines by making use of existing resources. It should be encouraged. • Renovation is change in technology, method and equipment which can bring improvement in the system and results in increase in profits as well as a safe and clean environment. For growth, and to compete in the market, renovations are essential. Periodical renovations are almost mandatory in any industry to improve quality as well as output rates.
Lack of attention to abnormalities	Sec. 9.6, describes types of abnormalities that need to be given immediate attention and must be removed and rectified, failing which losses of various kinds could occur.
Wastage	Abnormal waste generation results in losses. Details are given in sec. 5.3.

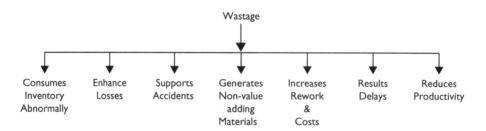

Figure 9.4 Impacts of wastage.

In any industrial process the same concept is also applicable when inputting certain ingredients; a final product is obtained together with foul gases, heat, noise, and also solid waste in some cases. Efficiency lies in converting more and more input material into useful products, and generating waste as little as possible as it has no value, rather its handling, disposal and storage costs money. When produced in excessive quantity, it degrades environment. The excessive production of waste (except in the case of mining and petroleum industries where it is a natural phenomenon); indicates a drawback in the technology, or systems. It is a non-value adding activity that adds to the cost of production, and hence it is a loss. The line-diagram, Figure, 9.4, lists the impacts of wastage.

Wastage is not confined to the various operations we carry out at an industrial setup. Think of reduction in wastage and savings in our routine activities. It also includes social cultures and customs with which we have to abide in one way or another. In this era we could save a lot by using modern techniques, but we would have to bring about a change in our thinking. We would have to be proactive. On our festivals and New

Structure of Cost-Loss Matrix:

Total Losses	Loss Category / Losses	Availability Loss			Performance Loss					Quality Loss				Management Loss			
		1	2	3	4	5	6	7	8	9	10	11	12	13	14	15	16
		Shutdown Losses	Production adjustment Losses	Equipment Failure Losses	Normal Production Losses	Abnormal Production Losses				Quality defect Losses	Reprocessing Losses			High Inventory Losses	Substandard Material Losses	Substandard tool & Equipment Losses	Man Management Losses
	Stoppage Days																
	Qty Loss (MT)																
	Loss in (***)																
A] Total Variable Cost	1 Raw Material 2 Utilities																
B] Total Fixed Expenses	1 Salary 2 Stores & Repair 3 Overhead																
Total [A + B]																	

Manufacturing Losses

Figure 9.5 Assessment of abnormalities and the resulting financial losses on various accounts.

*** – *Numerical figures in the currency to which the industrial unit belongs.*

Years' Eve, or any other occasion it is customary to send a card, but we can do the same by sending 'e-Cards', which could result in the following benefits:

1　Savings on paper, card and envelope costs
2　Savings on the costs of handling, labelling, printing, enveloping and dispatching
3　Savings on courier costs
4　Elimination of additional garbage at the receiver's end
5　Elimination of delayed delivery/incorrect delivery/non-delivery

Considering the size of your Company/Group, and the customer base, not only would the cost savings be huge but the environmental conservation impact would be substantial too.

Likewise there could be number of alternatives that could lead not only to savings but reducing pollution.

9.9　CASE-STUDY ILLUSTRATING COMPUTATION OF FINANCIAL LOSSES[17, 23]

The abnormalities and losses of various types results in financial losses, which could be assessed as illustrated in Figure 9.5. Figure 9.6 is a graphical illustration for ana-

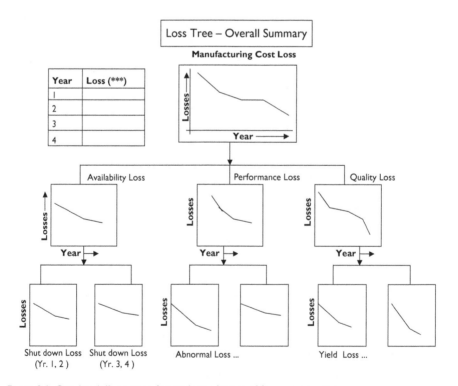

Figure 9.6 Graphical illustration for analyzing losses of four consecutive years.
　　*** Numerical figures could be in the currency to which the industrial unit belongs.

lyzing losses of four consecutive years in a manufacturing unit. It provides a basis for how various types of losses could be taken into account to assess the financial implications of an industrial setup.

There is a link amongst the inputs including human resources, plant and machinery, raw materials (natural such as minerals, fossil fuels, flora; and man-made materials which are innumerable); energy including the natural resources air and water; and infrastructures to accomplish production from any industrial set-up including the industry which one belongs to.

9.10   EFFECTIVE SYSTEMS – BEST PRACTICES

9.10.1   Quality Management System (QMS)[10, 17, 23]

This practice covers the following aspects:

* Quality Policy
* Adequate Resources
* Responsibilities and Authorities
* Training
* System Documentation
* Process Controls
* Document Control
* System Audits
* Management Review

9.10.1.1   Quality is a function of these factors

Composition, Bulk Density, Porosity, Warping, Wrinkles, Trueness, Surface Finish, Tensile Strength, Hardness, Conductivity, Insulation, Denier, Setting, Shelf Life, Color, Viscosity, and many more as applicable to the product or commodity in question. Quality defects are caused due to deviation from the laid-out standards, or agreed norms, practices and standards.

Quality defects result in: Non Value Adding Activities, such as: repeated work of Mixing/Drying/Blending, Reworking, Material Handling, Inspection, Recording or Logging, Unplanned Stocking, Storing and Sorting etc.

Such a scenario, in turn, adds costs. It delays delivery to customers, thereby often resulting in complaints, dissatisfaction and reduced reliability, and thereby could jeopardize goodwill. It reduces productivity, adds a quantum of work which in turn adds pollution, and it could jeopardize safety.

9.10.1.2   Reasons for quality defects[10, 17, 23]

* **Due to improper handling** – Dropping, Jolting, Collision, Vibration, Skidding/Slipping, Sudden Stoppage, Thermal Stress etc.
* **Due to Foreign Matter** – Inclusion, Infiltration, Entrapment of Dirt, Rain Impurity, Rust, Chips, Wire Scraps, Insects etc. *Clean operations* means not contaminating the product with foreign contaminants.

- **Due to Grain Size** – Crushing, Screening, Disintegration, Separation, Abnormalities in Screens, Centrifugal Separators etc
- **By Moisture** – Wide Variation, Absorption, Condensation, excessive Leakage.
- **Changed Viscosity due to:** Inadequate Warming, Heating, Compounding, Mixing, Evaporation etc.
- **Improper Process Controls** – Inappropriate Settings/Control of Parameters, Inadequate Standard Operating Procedures (SOP), SOP not Updated, Poor Techniques, Use of Crude Tools, Controlling with Trial and Error, Rough Handling of Machine/Equipment Measurement System, Use of Non-validated/Crude/Sluggish Measurement Systems.

Incapability of process functions and features results in quality defects. This includes: inadequate & inefficient Metering/Control Appliances; Constraints related to 'Unbalanced Inputs'; Technical Bottlenecks and limitations; Sluggish/Slow/Rough Meters & Methods; Scaling/Sludge Deposits; Operational Changes or Stoppages; Non adherence to standard practices and planned schedules; Variability/Non-Uniformity and many more reasons.

9.10.2 Six sigma[12]

This is a program aimed at the near elimination of defects from every product, process and transaction. The concept of '**Six Sigma**' (Figure 9.7) was introduced and popularized at Motorola in its quest to reduce defects of manufactured electronic products. (Motorola is known around the world for innovation and leadership in wireless and broadband communications). 'Six Sigma' technically means having no more than 3 or 4 defects per million opportunities in any process, product or service.[12] Thus, this is an effective tool to manage '**Uncertainty**'. The purpose of Six Sigma is to gain breakthrough knowledge on how to improve processes to do things better, faster and at lower cost.

9.10.3 Quality Control Tools (QC Tools)[10, 12]

These are proven techniques for systematic gathering of data, arranging them in order to interpret trends, and analysis and formulation of relationships between parameters. Basically QC tools necessitate arriving at facts based on systematic collection of actual data and its analysis. The commonly used QC tools are: Check sheets, Histograms, Scatter diagrams, Pareto analysis, Cause & Effect diagram, control chart.

9.10.4 Benchmarking & standardization

Amongst various best practices and systems, standardization of equipment, operations, procedures and practices would mean following the laid down norms, guidelines and instructions. Standards are laid down taking into consideration the best performances and industrial benchmarks. They are based on experience, scientific studies, debates and consensus arrived at amongst those involved and concerned. They are aimed at effective utilization of resources.

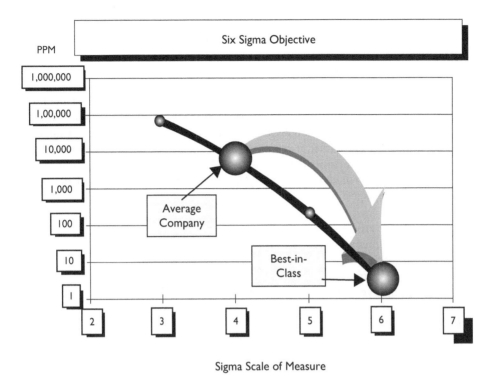

PPM

Six Sigma Objective

1,000,000

1,00,000

10,000

1,000

Average
Company

100

10

Best-in-
Class

1

2   3   4   5   6   7

Sigma Scale of Measure

Figure 9.7 Six Sigma Scale – Technically means having not more than 3 or 4 defects per million opportunities in any process, product or service. Industry with 6 Sigma on this scale is known to be the best. (After Severn, T. 2006).[12]

ISO stands for the International Organization for Standardization, located in Geneva, Switzerland. ISO promotes the development and implementation of voluntary international standards, both for particular products and for environmental management issues.

9.10.5   ISO 9000[3, 7, 11, 12, 13, 21, 22]

This is a family of standards for quality management systems. ISO 9000 is maintained by ISO, the International Organization for Standardization and is administered by accreditation and certification bodies. Some of the requirements in ISO 9001 (which is one of the standards in the ISO 9000 family) include

- a set of procedures that cover all key processes in the business;
- monitoring processes to ensure they are effective;
- keeping adequate records;

- checking output for defects, with appropriate and corrective action where necessary;
- regularly reviewing individual processes and the quality system itself for effectiveness; and
- facilitating continual improvement.

A company or organization that has been independently audited and certified to be in conformance with ISO 9001 may publicly state that it is "ISO 9001 certified" or "ISO 9001 registered". Certification to an ISO 9000 standard does not guarantee any quality of end products and services; rather, it certifies that formalized business processes are being applied. Indeed, some companies enter the ISO 9001 certification as a marketing tool.

According to *the Providence Business News*[3], implementing ISO 9001 often gives the following advantages:

1   Creates a more efficient, effective operation
2   Increases customer satisfaction and retention
3   Reduces audits
4   Enhances marketing
5   Improves employee motivation, awareness, and morale
6   Promotes international trade
7   Increases profit
8   Reduces waste and increases productivity

It is widely acknowledged that proper quality management improves business, often having a positive effect on investment, market share, sales growth, sales margins, competitive advantage, and avoidance of litigation.[11, 22]

However, a broad statistical study of 800 Spanish companies[3] found that ISO 9000 registration in itself creates little improvement because companies interested in it have usually already made some type of commitment to quality management and were performing just as well before registration.[22]

In today's service-sector driven economy, more and more companies are using ISO 9000 as a business tool. Through the use of properly stated quality objectives, customer satisfaction surveys and a well-defined continual improvement program companies are using ISO 9000 processes to increase their efficiency and profitability.

Problems: A common criticism of ISO 9001 is the amount of money, time and paperwork required for registration.

Even without certification, companies should utilize the ISO 9000 model as a benchmark to assess the adequacy of its quality programs.

9.10.6   Other models of standards[3, 7, 11, 12, 22]

- ISO 10006 Quality management – Guidelines to quality management in projects
- ISO 14000 – Environmental management principles, systems and supporting techniques
- ISO 14001 – Environmental Management Systems – Specifications with guidnace with use.
- ISO 14010 – Guidance for Environmental Auditing – General principles of environmental auditing

- ISO 14020/23 – Environmental Labeling
- ISO 19011 – Guidelines for quality management systems auditing and environmental management systems auditing
- ISO/TS 16949 – Quality management system requirements for automotive-related products suppliers

9.11 LEGAL COMPLIANCE INCLUDING ENVIRONMENT MANAGEMENT SYSTEMS (EMS)[10]

Compliance with legal requirements is a critical business activity and also one of the pillars that supports loss prevention (Figure 9.2).

While the requirements listed in the checklist in Table 9.6 relate directly to an organization's management of legal requirements, each of these elements can contribute to enhanced compliance (including communication, documentation and document control, records management, EMS audits, and management review).

Likewise such a checklist could be prepared for different aspects that need to be followed up and monitored, and for keeping the status up to date.

Environmental Management systems (EMS) cover the following aspects

- Environmental Policy
- Adequate Resources
- Responsibilities and Authorities
- Training
- System Documentation
- Operational Controls

Table 9.6 Checklist of legal, statutory & other compliances.[10]

Parameters/aspects to be considered	*Status*
Do you have an **existing process** for identifying applicable legal and other requirements?	
If yes, does that process need to be revised? In what way?	
Who needs to be involved in this process within your organization?	
And what should be their responsibilities?	
What **sources of information** do you use to identify applicable legal and other requirements?	
Are these sources adequate and effective? How **often do you review** these sources for possible changes?	
How do you ensure that you have **access** to legal and other requirements? (List any methods used, such as on-site library, use of web sites, commercial services, etc.)	
How do you **communicate information** on legal and other requirements to people within the organization who need such information?	
Who is **responsible** for analyzing new or modified legal requirements to determine how you might be affected?	
How will you keep information on legal and other requirements **up-to-date**?	
Your next step on legal and other requirements is to ...	

- Document Control
- System Audits
- Management Review

For standardization and benchmarking, models laid out by the International Organization for Standardization (ISO) could be used.

9.12   ISO 14000 AND ISO 14001[8, 11, 21, 22]

As descibed in preceding paragraphs, ISO 9000 is a family of standards for quality management systems. Similarly ISO 14000 refers to a series of voluntary standards in the environmental field under development by ISO. Included in the ISO 14000series are the ISO 14001 EMS Standard and other standards in fields such as environmental auditing, environmental performance evaluation, environmental labeling, and life-cycle assessment. The EMS and auditing standards are now final. The others are in various stages of development.

9.12.1   Procedure to develop these standards[8]

All the ISO standards are developed through a voluntary, consensus-based approach. Each member country of ISO develops its position on the standards and these positions are then negotiated with other member countries. Draft versions of the standards are sent out for formal written comment and each country casts its official vote on the drafts at the appropriate stage of the process. Within each country, various types of organizations can and do participate in the process including industry, govenment (Federal and State), and other interested parties, including various non-government organizations (NGOs). For example, EPA and States participated in the development of the ISO 14001standard and are now evaluating its usefulness through a variety of pilot projects.

9.12.2   ISO 14001standard[8]

The ISO 14001standard requires that a community or organization or industry put in place and implement a series of practices and procedures that, when taken together, result in an environmental management system. ISO 14001 is not a technical standard and as such does not in any way replace technical requirements embodied in statutes or regulations. It also does not set prescribed standards of performance for organizations. The major requirements of an EMS under ISO 14001 include:

- A policy statement which includes commitments to prevention of pollution, continual improvement of the EMS leading to improvements in overall environmental performance, and compliance with all applicable statutory and regulatory requirements.
- Identification of all aspects of the community or organization's activities, products, and services that could have a significant impact on the environment, including those which are not regulated

- Setting peformance objectives and targets for the management system which link back to the three comitments established in the community or organization's policy (i.e. prevention of pollution, continual impovement, and compliance)
- Implementing the EMS to meet these objectives. This includes activities like training of employees, establishing work instructions and practices, and establishing the actual metrics by which the objectives and targets will be measured.
- Establishing a program to periodically audit the operation of the EMS
- Checking and taking corrective and preventive actions when deviations from the EMS occur, including periodically evaluating the organization's compliance with applicable regulatory requirements.
- Undertaking periodic reviews of the EMS by top management to ensure its continuing performance and making adjustements to it, as necessary.

9.12.3 Potential benefits of an EMS based on ISO 14001[21,22]

- Improvements in overall environmental performance and compliance
- Provide a framework for using pollution prevention practices to meet EMS objectives
- Increased efficiency and potential cost savings when managing environmental obligagtions
- Promote predictability and consistency in managing environmental obligations
- More effective targeting of scarce environmental management resources
- Enhance public posture with outside stockholders.

The standard is flexible and does not require organizations to necessarily "retool" their existing activities. The standard establishes a management framework by which an organization's impacts on the environment can be systematically identified and reduced.

An EMS is a set of management processes and procedures that allows an organization to analyze and reduce the environmental impact of its activities.

First adopted by private industry, the EMS approach is increasingly common in the public sector.

9.13 EFFECTIVE TRAINING, COMPETENCY AND AWARENESS[10,16,17]

Training Needs (Refer also to sec. 8.9): Nobody is perfect; we lack many skills which are essential to do our job in the right way. Training is essential right from the shop-floor workman (employee) to the topmost executive. Training aims at preparing an employee to best fit the job assigned to him. Practice makes the man perfect, for which going through the right training is the first step. It becomes almost mandatory in the following circumstances:

- To train the new entrant/recruit for the assigned job/vocation
- On transfer to another location, or other section or other skill

- Changes in existing process, technology or method
- Introduction of new systems
- New regulations, policy or any other compliances affecting the Company's activities
- To remove deficiencies or weaknesses in any individual
- As prescribed in or directed by the prevalent rules, regulations and laws.
- In addition, the following could be reasons to access training needs:

  - Based on employee's own request
  - As per recommendation of the immediate supervisor
  - As recommended by the equipment or technology supplier
  - Based on audits, reviews and management directives.

Who should impart training?

  - Internal trainers (available from within the organization)
  - Expert trainers/Consultants from outside
  - Academic Institutes – Universities & Colleges
  - Technical/trade/business associations
  - Training consortia (teaming with other local companies)

Key Steps in Developing a Training Program

  - Assess training needs & requirements
  - Define training objectives (topics and subject matter)
  - Select suitable methods and materials***
  - Prepare training plan (who, what, when, where, how)
  - Conduct training program
  - Take feedback on effectiveness of the program
  - Track training (and maintain records)
  - Evaluate training effectiveness (During post training period)

*** A well planned training program uses the following aids – both sound and light. Use a combination of:[5]

  - Graphs, Charts, Models, Drawings, Illustrations
  - Posters, Displays, Photographs, Paintings, Process Flow-Diagrams
  - Transparencies, Slides
  - Videos, Computer-Assisted PowerPoint Presentations
  - Encourage: Questions, Discussions, Quizzes, News Items, and Case Studies.
  - Include Actual Machines, Tools, Appliances, Equipment, Working Models Wherever Practical.
  - Include Live Demonstrations, Field Visits and Delivery from Guest Faculties, wherever practical.
  - QC tools**

** The commonly used QC tools are, Check sheets, Histograms, Scatter diagrams, Pareto analysis, and Cause & Effect diagrams, control chart.

Training could be on-the-job; in the classroom; in the field; within the company; outside the Company or any combination of these.

Training provides skill, competence, confidence, motivation and opportunity for improvement. It helps in compliance and brings awareness amongst employees. If effectively imparted and well received, it is an investment in the right direction that pays; otherwise it results in reduced productivity and frustration.

Awareness comes not only through the classroom teaching or training taken in the manner described above; it also comes through proper practice and implementation, and through participation in group discussion, symposiums, seminars, conferences, workshops and other forms where technical issues are discussed and knowledge is shared.

Equally important are publicity and propaganda, as these are easy-to-conceive ways to impart knowledge. Exposure to posters, slogans, illustrations, films and videos on the theme of occupation health, safety and environment, best practices, industrial hazards, best industries and industrial setups are also important for keeping up-to-date in your field of specialization or vocation.

9.14 EFFECTIVE COMMUNICATION[10, 19]

Effective communications help us to:

- Motivate the workforce
- Gain acceptance for our plans and efforts
- Explain the company's programs, policies, tools, and initiatives and how it is related to the overall organizational vision
- Ensure understanding of roles and expectations
- Demonstrate management commitment
- Monitor and evaluate performance; and
- Identify potential system improvements.

The following are the salient features of an effective communication system:

- Effective **internal** communication requires mechanisms for information to flow top-down, bottom-up and across functional lines. Since employees are on the "front line" they can be an excellent source of information, issues, concerns and ideas.
- In communicating with employees, it is helpful to explain not only **what** they need to do but also **why** they need to do it.
- Use KISS – **K**eep **I**t **S**hort & **S**imple – which means the message should be simple, clear, concise, accurate and easy to understand (conceive). Convey it with respect and trust.
- Everyone wants to be heard. Communication is a two-way process.
- One size does not fit all. Understand your audience and communicate in the best way to reach it, and so that out of hundreds of messages received everyday, yours is accepted and conceived well.
- Communicate courageously. If you communicate openly and honestly, people will understand and will respect your courage.

- Good communication is a good investment. In the absence of good communication, the grapevine thrives. And the grapevine will leach the resources from your business – productivity, commitment, and reputation.

9.15 WORLD CLASS MANAGEMENT (WCM)[23]

The Best Practices such as WCM, which has been adapted by the Fortune-500 companies, or the renowned industrial houses world-over, could be a useful guide. This concept addresses 8 focused areas to achieve Visual Management & Control – QCDIP – Control on Quality as well as Quantity, Cost, Delivery (JIT-Just-in-time), Innovations and Productivity. This is what is known as 'QCDIP' as shown in Figure 9.8.

9.16 PRECISION IN OPERATIONS[16]

Precision brings accuracy, which is the key to reducing wastage. It saves time, energy, material and resources. And ultimately results in cost reduction. The illustration presented describes the mining of minerals. If done with accuracy this would result in clean mining with less dilution from other rock-materials. It would result in better recovery and reduced handling of waste rocks thereby reducing pollution.

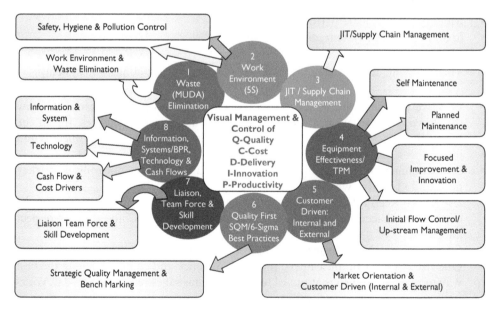

Figure 9.8 WCM concept addresses 8 focused areas to achieve Visual Management Control on Quality as well as Quantity, Cost, Delivery, Innovations and Productivity (QCDIP).[23]

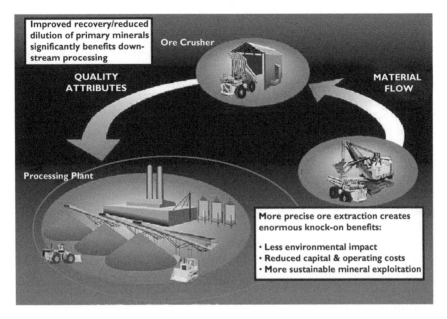

Figure 9.9 Enormous knock-on benefits due to precision in ore extraction technology as applicable to mining as well as processing operations.[16]

Clean and precise operations are required in any industry. They add to hygienic conditions and a smart culture and outlook. Figure 9.9 displays enormous knock-on benefits due to precision in ore extraction technology as applicable to mining as well as processing operations. The concept is equally applicable in any industry whatsoever it may be.

9.17   EMERGENCY PREPAREDNESS AND RESPONSE

This aspect should be part and parcel of best practice, as described in chapter 10.

9.18   WAY FORWARD[10, 17]

Every industry emits emissions in various forms, and much would depend on how these could be minimized taking into consideration the influencing agencies and prevalent culture in the country of operation. A Policy should be formulated on the following critical business activities:

*   Production Targets
*   Productivity
*   Occupational Health, Safety, Environment and Social welfare (ref. Sec. 10.1.1 also)

This would be influenced by the following agencies, which could be Government, Non-government, public, private or their combination:

- Politics (Government & other political parties), laws, lobbyists, media, public opinion and concerns.
- Economic status of the country of operation – Developed, Developing or Underdeveloped.
- Input of Science and Technology, R & D, Education
- Type of Industry, Insurers, Investors, Past History and Existing Goodwill

Prevention of losses is related to the processes that are used in an industrial set-up, and require consistent approach and efforts for improvements. Apart from the best practices listed and described in the preceding sections; the concept such as World Class Management (WCM), described above, could be very useful for the effective utilization of resources.

The flow diagram in Figure 9.10 could be used as a guide to achieve continual improvements that would lead to the outcome shown in Figure 10.13. The idea is to achieve an outcome that supports safety at the workplace together with sustainable development, which is beneficial socially, economically and ecologically to the present as well as future generations.

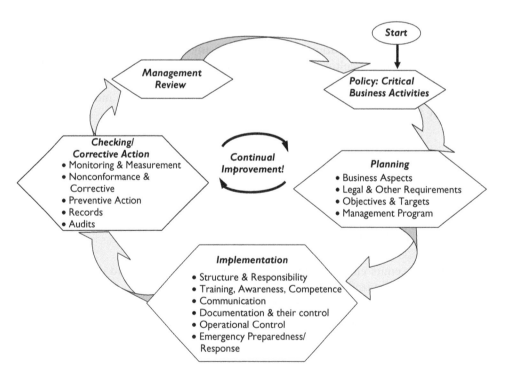

Figure 9.10 An effective model to accomplish continual improvements.

9.19 HEALTH, SAFETY AND LOSS PREVENTION (HSLP) MANAGEMENT SYSTEM AND ITS EFFECTIVENESS[14]

As proposed by Severn (2006)[14] using following relation Effectiveness in Accident Prevention (E) can be assessed:

$$E = Q \times A \tag{9.3}$$

E = Effectiveness in Accident Prevention
Q = Quality of HSLP System
A = Acceptance of HSLP System

Quality of HSLP System would depend upon how effective is the management in HSLP related issues, which have been described in the preceding sections. And acceptance would depend upon the behavior of those concerned, as sub-standard behaviors are the underlying causes of most workplace accidents. According to Severn (2006)[14] behavior can be Predetermined and/or Deliberate Behavior, which involves:

- Risk Taking
- Short Cuts
- Non Conformance

  - Time, Comfort
  - Convenience

- Conflicting Priority

Making It Work (which means practical ways to implement it) – Zero Tolerance to Risk Behavior

- The meaning of Zero Tolerance is:
  - 100% compliance with rules, procedures, standards
  - Not walking past an unsafe act or condition
  - Believing all incidents are preventable
  - Not operating defective equipment (defect is a condition beyond a standard, guideline, or regulation)
  - Applies to all levels, all areas of operation, and all employees ... equally applied to salaried as well as hourly
  - Applies to individual perceived hazards when rules, etc., are not available
  - The context from which managers, supervisors and employees make decisions.

Making It Work – Implementation Strategy

- Strong Kick Off (involve all concerned and appraise them about the aims and objectives)
- Calendar Test (periodic as well sudden check ups and inspections)
- Involvement (Right from top to bottom)
- Accountability Standards (fixing up the Benchmarks)
- Create and Maintain a Safety Presence (by inculcating safe habits)

- Communications (as described in sec. 9.14)
- Impromptu Discussion

Making It Work – The Complete System

- Involve everyone in the process. It is *teamwork.*

9.20   CASE STUDY – THREE PILLARS OF EQUAL STRENGTH FOR LOSS PREVENTION[9]

This case study describes the organizational changes that were brought about by 'Lonmin' who undertook the task of bringing changes in the safety culture of a platinum producing company having operations at several places.

Lonmin[12] followed the concept which was written by John P. Kotter published as *"Leading Change"* in the March 1995 issue of the *Harvard Business Review*, where he listed the most common errors made when people attempt to change organizations. *Kotter suggested an eight-stage process of creating major change, as mentioned below:*

1 Establish a Sense of Urgency
2 Create a Guiding Coalition
3 Develop a Vision and Strategy
4 Communicate the Change Vision
5 Empower Broad-Based Action
6 Generate Short-Term Wins
7 Consolidate Gains and Produce More Change
8 Anchor New Approaches in the culture.

- The first four steps help defrost a hardened status quo.
- Phase five to seven introduce new practices.
- The last stage grounds the changes in the corporate culture and helps make them stick.

Figure 9.11, illustrates the safety status of the company before Lonmin[12] took over. Looking into it, the situation called for the radical changes.

The program started with creating an understanding that the company was in trouble and the current ways of doing things were no longer working. This enabled them to establish a 'Sense of Urgency'. (How much urgency is enough? More than most people think! Increasing urgency demands that you remove sources of complacency).

Within 2 days after Lonmin taking over the operations, a serious accident occurred.

A decision was taken to suspend operations to:

- Investigate the accident,
- Conduct clean-up of working places, and
- Fix sub-standard working conditions.
- At the same time three 1-day workshops were conducted to:
  - Introduce the Lonmin management team to all employees, Communicate the Lonmin vision and values,

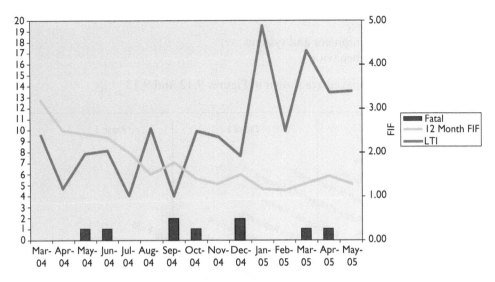

Figure 9.11 Injury Frequency before Lonmin Takeover. (After Jansen, 2006).[9]
FIF – Fatal Injury Frequency; LTI – Loss Time Injury.

- Make it clear that the current level of performance is not acceptable, and
- Give employees the opportunity to voice their opinions.

Two levels of management were removed, which forced more junior managers to accept accountability for their areas of responsibility. This assault on complacency created some anxiety. Uncertainty was managed by giving support and direction to junior managers. It reduced anxiety and allowed creativity. The concept of 'zero Harm' was passed on using:

- Direct communication with the workforce
- Regular meetings with those concerned
- Weekly meetings with Production Supervisors
- Regular meetings with safety representatives
- Other channels of communication used included: Monthly newsletters, Notice boards, Big screen TV and Electronic Boards.

REPITITION of all the above because ideas sink in only after they have been heard many times.

Those concerned were made aware that:

- We are committed to zero harm to people and the environment
- We are committed to integrity, honesty and trust
- We are committed ethical people who do what we say
- We have transparency, openness, honest communication and free sharing of information
- We have respect for each other

Changes were brought about in three phases (Figure 9.12):

1   Focus on compliance
2   Focus on plant, equipment and systems
3   Focus on peoples' behavior.

And the results achieved are shown in Figures 9.12 and 9.13.

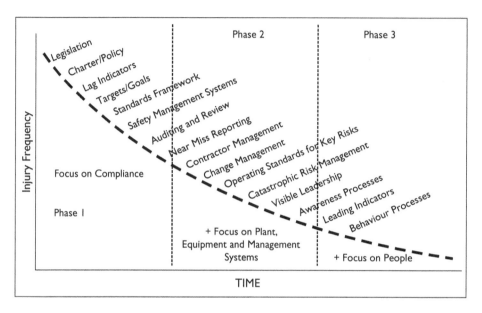

Figure 9.12 Impact of three phased change on injury rate with respect to time. (After Jansen, 2006).[9]

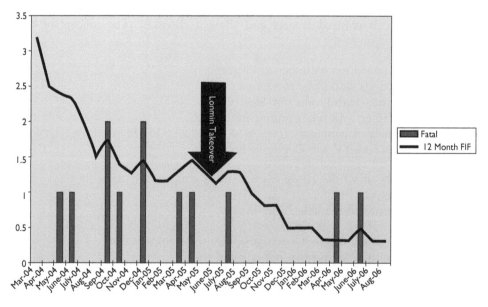

Figure 9.13 Fatal injury frequency after implementation of 3 phase – eight stage program
by Lonmin. (After Jansen, 2006).[9]

Thus, based on this case study; attention to the Three Pillars of equal strength for loss prevention: Legal Compliances, Effective Systems and Content Employees, as described in section 9.2 could bring the radical changes to any industrial setup.

QUESTIONS

1   AMS: what is it? List the steps that are required to follow it.
2   Are 'Benchmarking & Standardization' essential to run any industrial setup?
3   As per Stephen R. Covey what is the difference between management and leadership? List the qualities a leader should possess.
4   Classify losses of various types.
5   Classify the resources used to accomplish any industrial operation.
6   What does WCM stand for? List the focused areas that are addressed by WCM. What does (QCDIP) stand for?
7   How does 'Minimizing losses of various kinds' help in achieving the aims and objectives of an industrial setup?
8   How is 'Precision' helpful in bringing excellence to any industrial setup?
9   How can you ensure your subordinates (human resources) are reporting to you?
10   How is effective communications helpful in running any industry? How could they be achieved?
11   How would you calculate equipment Overall Efficiency (%)?
12   How you can achieve human efficiency?
13   Illustrate by means of a line diagram the impacts of 'wastage'.
14   In which circumstances does training become mandatory?
15   List the 'Best practices' which should be implemented to run an industrial setup.
16   List the 'Three Pillars of equal strength' for loss prevention.
17   List the aspects that are covered by a Quality Management System (QMS).
18   List the aspects that should be considered in managing machines, plant, equipment, tools and appliances.
19   List the eight-stage process of creating major change as suggested by Kotter to bring about organizational changes.
20   List the governing parameters for equipment availability and utilization.
21   List the potential benefits of an EMS based on ISO 14001.
22   List the steps that should be followed to run any operation or system effectively.
23   List the type of abnormalities which are usually found in industries in general. How useful is the assessment of abnormalities and the resulting financial losses ?
24   List the types of losses in a manufacturing plant or unit, and their reasons. Suggest measures to minimize them.
25   Propose an effective model to accomplish continual improvements.
26   Quality is a function of what factors? List the reasons for 'Quality Defects'.
27   What does 'Six Sigma' mean technically? What is its purpose?
28   **What does EMS stand for? What aspects does an EMS cover?**
29   What is ISO 9000? What advantages could be expected by implementing it.
30   What is the significance of 'Diagnosis Technique – 5 W–2 H'?

REFERENCES

1 A Small Business Guide to ISO 14000. (1995) Canadian Standards Association, Competing Leaner, Keener and Greener.

2 Carter. P. (2007) The mine of the future. Pres. In: *Symposium on Mining 2020, 5–6 Sept. Sydney, Australia.*

3 Clifford, S.: So many standards to follow, so little payoff. *Inc Magazine,* May 2005. (6).

4 Hakala, D. Through: http://www.hrworld.com/features/find-future-leaders-020508/March 19, 2008.

5 Germain, G.L. & Arnold, R. (2001) Management strategy and system for education and training. In: *Mine health and safety management,* Michael Karmis (edt.). Littleton, Colorado, SME, AIME. pp. 128–144.

6 http://www.epa.gov/owm/iso14001/isofaq.htm

7 Institute of Quality Assurance, *Quality Systems in the Small Firm: a Guide to the Use of the ISO 9000 Series,* March 1995.

8 International Organization for Standardization, *ISO 14001 (1996) Environmental Management Systems – Specification with Guidance for Use.*

9 Jansen, J.: First Steps on the Journey to Zero Harm. *International conference focusing on safety and health on 14–16 November, Johannesburg, South Africa.* London ICMM. (Permission: Lonmin).

10 Philip J. Stapleton, Principal, Margaret A. Glover, Principal: *Environmental Management Systems: An Implementation Guide for, Small and Medium-Sized Organizations.* S. Petie Davis, Project Manager. NSF ISR; 789 N. Dixboro Road; Ann Arbor, MI 48158; 1-888-NSF-9000.

11 *Providence Business News,* August 28, 2000: Reasons Why Companies Should Have ISO Certification. (4).

12 Seddon, J.: *A Brief History of ISO 9000: Where did we go wrong? The Case Against ISO 9000,* 2nd ed. Chap.1; Oak Tree Press. November 2000.

13 Seddon, J.: The 'quality' you can't feel. *The Observer,* Sunday November 19, 2000 (7).

14 Severn, T. (2006) Work culture in loss prevention. Pres. in: *International conference focusing on safety and health on 14–16 November, Johannesburg, South Africa.* London ICMM.

15 Stephen R. Covey (1989) *The Seven Habits of Highly effective People.*

16 Tatiya, R.R. (2009) Exploration to exploitation in the Indian mining sector – proposed strategy and milestones. *International conference on developments in exploration to exploitation in the mining sector; Associate of Mining Engineers; 10–12 Feb. Jodhpur, India.*

17 Tatiya, R.R. (2009) 'Loss Prevention' – A need of hour in the Indian Mining Sector. *National conference on latest trends in equipment, technology and management in mineral sector; Indian Mining and Engineering Journal, 11–12 May, Bhuwneswar, India.*

18 Tatiya, R.R. (2005) *Surface and underground excavations – methods, techniques and equipment.* A.A. Balkema, and Taylor and Francis, Netherlands. pp. 140–141.

19 *Ten principles for leadership communication* by Hewitt Associates, Guidelines to executives (2009) – Aditya Birla Group.

20 United Nations Environment Programme (UNEP), the International Chamber of Commerce (ICC), and the International Federation of Consulting Engineers (FIDIC). *Environmental Management System Training Resource Kit*. Version 1.0, December 1995.

21 Voehl, Frank; Jackson & Ashton, *ISO 9000: An Implementation Guide For Small and Mid-Sized Businesses*, St. Lucie Press, 1994.

22 Wade, Jim.: Is ISO 9000 really a standard? *ISO Management Systems* – May–June 2002 (3).

23 World Class Management (WCM) – Leaflets, displays, lectures, and talks on various topics – Essel Mining and Industries Limted, Aditya Birla gropu, India; (2004–2008; through interaction, and participation).

Chapter 10

HSE management system

"Whole-heartedly working by each and every member of a team brings about team-spirit; and it is the team-spirit that could achieve victory."

Keywords: HSE Management, Leadership and Commitment, Policy and Objectives, Hazards and Effects Management, Planning and Procedures, Implementation and Monitoring, Audit, Occupational Health & Physique, Environment Management, Pollution, Sustainable Development, Risk Analysis, Emergency Measures, Mass Balance Equation.

10.1 INTRODUCTION

In any industrial setup or organization, *Human beings/people* are the most valuable asset in running it efficiently. Care is taken right from recruitment and thereafter to develop them to prove them the right man for the right job. And how sad it would be if any of them met with a fatal accident, or were seriously injured and became disabled; and it would be equally serious if any one of them suffered from any chronic diseases by inhaling polluted air or emissions of any kind which were injurious to health. This, in turn could compel them to remain absent or not perform up to the mark. Thus, incidents (which consist of near-misses or accidents) and pollution, as outlined in the preceding chapters/sections, are the two main enemies/threats against which any industry formulates its policy, organizes itself, allocates resources, assigns responsibilities, decides the standards to be achieved, lays down procedures and documents them. This is what is known as an HSE management system, and for it to function effectively, these aspects must be implemented, monitored, audited and reviewed regularly. Corrective action is taken in cases of non-compliance. Thus, an efficient HSE management is a structured, well planned, approach that concentrates on the continuous improvement of those activities which have greatest impacts on the people and the environment. This chapter under the umbrella 'HSE Management System' attempts to cover the following aspects:

1 Leadership and Commitment
2 Policy and Strategic Objectives
3 Organization, Responsibilities, Resources, Standards and Documents

4 Hazards and Effects Management
5 Planning and Procedures
6 Implementation and Monitoring including Emergency Measures/Preparedness
7 Audit
8 Review
9 Management Commitment
10 Management: Occupational Hazards (Health & Physique)
11 Environment – pollution control
12 Environmental management
13 Sustainable Development.

10.1.1 HSE – A critical business activity

Producing goods and services, as planned, brings in the bread-and-butter, and if it is achieved efficiently, it would multiply the profits; but accidents adversely affect the production, costs and productivity of any industrial setup. Production spoils reputation and demoralizes workers, and many a time it results in court inquiries and disputes. It could damage the environment and is of public concern. Ultimately it results in a deviation from the laid out objectives and goals of a company (Ref. sec. 1.6; sec. 8.7; 9.9). The logical approach is a thorough balance between production, productivity and safety, giving equal weighting to each of these components, like the three sides of an equilateral triangle (Figure 1.8), together with giving due attention to the welfare of the people who are in and around the place of operation. This results in fulfillment of objectives though sustainable development (Figure 10.1).

Thus, do you recognize that Occupational Health, Safety and Environment (HSE) together with social welfare are amongst the critical business activities in achieving the set goals and objectives of any company? Do you wish to build your reputation amongst the best in the 'Chemical, Petroleum, Mining, or any other Industry which your company belongs to, in the areas of HSE? This means all the employees and corporate level management including the owner (persons from top to bottom) have joint responsibility for willingly carrying out HSE operations in the most cost effective manner practicable.

Figure 10.1 Interrelationship of critical business activities as shown in Figure 1.8, with addition of Social-Welfare with HSE to achieve sustainable development.

10.1.2 Vision

Your company should aim at being recognized as one of the best on HSE management and operations in the region/zone/country.

10.2 HSE LEADERSHIP AND COMMITMENT[11, 15, 16, 24]

An example is a world-class company such as Shell International Exploration and Production Company, whose HSE Management system requires:

- Top/senior management should provide *strong and visible leadership* to promote a culture in which all employees share a commitment to HSE.
- Top/senior management should be *proactive in target setting*.
- Top/senior management should show *informed involvement* in HSE issues, to *demonstrate leadership and commitment*.

In fact senior management should provide *strong and visible leadership* to promote a culture in which all the employees share commitment to HSE. Figure 10.2 depicts the elements of HSE leadership and commitment.

10.2.1 Visibility

Senior management/managers can set a personal example in their day-to-day work by:

- Putting HSE matters as a priority while listing agendas of meetings.
- Participating in the review of performance against all HSE planning and targets.
- Providing immediate and visible response and involvement in case of an incident/ accident or disruption to normal business.
- Communicating implementation of HSE-related issues in business decisions and in communications with stockholders.
- Seeking internal and external advice/expert opinion on HSE issues.
- Recognizing achievements.

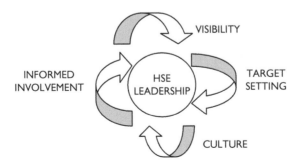

Figure 10.2 Elements of HSE leadership and commitment.

10.2.2 Target setting

Senior management should be proactive in target setting. Senior managers can demonstrate this by:

- Holding joint discussions with those concerned: Managers, Engineers, Supervisors and Workers Representatives in setting targets and deciding expectations.
- Making sure that in appraisals, HSE performance by any individual is given due recognition.

10.2.3 Culture

Efforts should be made to create and sustain a culture in which everybody shares the commitment to HSE. Such a culture often results in:

- The slogan: '*Safety First & Production Must*' becomes *reality.*
- *Reporting unsafe acts and conditions* immediately *becomes an established practice.*
- *Receiving due recognition and reward on HSE*-related issues become *a reality.*
- *Empowerment to stop unsafe work* becomes a *practice.*

10.2.4 Informed involvement

Senior managers can demonstrate this by:

- Regularly reviewing progress made in developing the HSE management system.
- Allocating resources and expertise to meet HSE targets (e.g. finance manpower, technology, skills and training etc.).
- Undergoing relevant training themselves.
- Making oneself and others aware of high priority areas and issues.
- Consistently taking follow up actions wherever required.
- Key messages[11]
  - If you don't set a good example by working safely, your safety and your team's safety is at risk.
  - If you don't set a good example by following procedures, then you can't expect others in your team to.
  - Refer figure 10.8 the degree of change that could be brought about based on the involvement of senior management.

10.2.5 Accountabilities[1, 11, 12, 15, 16, 24]

Successful HSE Management reflects teamwork (Figure 10.3) involving every one from top to bottom and demonstrating responsibilities as outlined below:

As an effective corporate level executive/senior manager
- Protect health and safety of staff, contractor's employees and other persons affected by the company's activities
- Protect the environment and minimise pollution

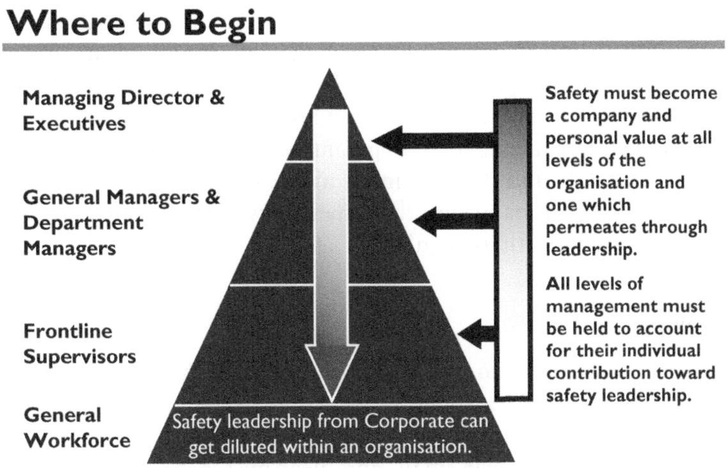

Where to Begin

Managing Director & Executives

General Managers & Department Managers

Frontline Supervisors

General Workforce

Safety leadership from Corporate can get diluted within an organisation.

Safety must become a company and personal value at all levels of the organisation and one which permeates through leadership.

All levels of management must be held to account for their individual contribution toward safety leadership.

Figure 10.3 Successful HSE Management reflects teamwork involving everyone from top to bottom (After Latus, M. 2006).[11]

- Use materials and energy safely and efficiently
- Manage HSE as any other critical business activity
- Reduce long term liabilities
- Promote HSE Issues
- Conduct workplace HSE inspections personally (whenever feasible)
- Develop pro-active HSE targets
- Allocate resources to implement the HSE management system
- Prepare an annual HSE letter of assurance.

As an effective manager

- Make sure those workplaces, tools and equipment used at the working site, which could be a factory, mine, construction site or any other establishment, are in accordance with prevalent regulations.
- Set a good example for your team by always following safe procedures and norms and abide by the rules and regulations.
- Make sure that subordinates including contractor's staff are competent enough to carry out work safely and they are made aware of the possible hazards.
- Periodically check and make sure that emergency arrangements are intact.
- Conduct HSE-related meetings regularly and giving due importance to incident/ accident analysis and steps taken for preventing any reoccurrence.
- Conduct workplace HSE inspection personally.
- Develop pro-active HSE targets.
- Allocate resources to implement the HSE management system.
- Prepare an annual HSE progress report.

As an effective supervisor

- Make sure that workplace, tools, equipment used by the subordinates are in safe condition and work does not commence if they are found unsafe.
- Set a good example for your team by always working safely and following set procedures and norms and abide by the rules and regulations.
- Make sure that subordinates including contractor's staff are competent enough to carry out work safely and they are made aware of the possible hazards.
- Periodically check and make sure that emergency arrangements are intact.
- Attend HSE related meetings regularly.

As an employee

- Learn the right way/procedure to the do job and follow it
- Use safety wear required for the job
- Notify supervisor of unsafe acts and conditions, if you see any
- Report any incident, however small, and 'near-misses' to your supervisor
- Maintain equipment, tools and appliances in safe working condition. Report any defect immediately
- Know about rules and regulations, and follow them
- Attend meetings, talks and discussion related to HSE matters and share your views
- Know what events are considered as an 'emergency' and what you are supposed to do during such occasions
- Take care of yourself and those around you.

As an effective HSE Advisor

- Lead by your own example and always work safely. Provide leadership by personal commitment towards HSE. (Figure 10.4)

Safety Leadership Model – Personal Commitment

☐ **I understand and actively promote our safety and health management standards.**

☐ **I actively promote the vision of an injury free workplace.**

☐ **I have high standards of safe performance.**

☐ **I look out for, and manage, hazards around my workplace.**

☐ **I achieve and set challenging safety goals and objectives.**

☐ **I actively participate in safety and encourage others.**

☐ **I recognise good safety practices and acknowledge good performance.**

☐ **I openly and frequently discuss safety with my colleagues.**

Figure 10.4 Safety leadership model – personal commitment. (After Latus, M. 2006).[11]

10.2.6   Checklist

Know your current status!!! Refer to Tables 10.1 and 10.2.

Table 10.1 Checklist to know your current status with regard to HSE Management – Leadership & Commitment.[24]

Critical aspects	Current status?
Leadership & Commitment	How does management demonstrate its leadership on HSE issues?
• Visibility	Does your senior management provide strong and visible commitment to HSE?
• Target setting	Are they involved in setting targets for 'Results' as well as 'Improvement of HSE' for those concerned including for themselves?
• Industrial culture	How would you describe the 'Industrial Culture'* at your establishment? And how does thecompany's culture contribute towards achieving a commitment to HSE?
• Involvement	How does the 'Line Management' demonstrate their involvement in activities such as resource allocation, high priority to HSE, follow-up actions and undergoing training?

* for any type of industry, which could be Chemical, Petroleum, Mining, processing or any other.

Table 10.2 Checklist to know your current status concerning organization, responsibilities, resources, standards and documentation with regard to HSE Management.[24]

Critical aspects	Current status?
Organization	How does the structure of your company support HSE management?
Responsibilities	What are the responsibilities of managers, supervisors and staff including contractor's representatives towards HSE management?
Resources	Who allocates responsibilities for effective HSE management?
– Awareness, training and education	When, where and who should undergo training including refresher courses? What about contractors' staff? Are they being trained?
– HSE appliances, tools and materials	Are they of requisite standards and sufficient in numbers?
Documents, reports, records etc.	Are they maintained as per standard formats and practices, and counterchecked by those responsible? Are statutory compliances taken care of and there is no violation?
Long pending HSE issues	Is any priority attached to them?
Communication	How does your company involve those concerned in communicating HSE matters?

10.3   HSE POLICY[13, 15, 17, 19]

To understand this aspect, we quote the HSE policy of an oil company in Gulf region; it describes HSE policy as an integral part of its business policy. It states: "The company will endeavour to conduct its business in such a way as to protect the health and safety of its employees, its contractors' employees, other persons affected by its activities, as well as to protect the environment, minimise pollution and seek improvement in the efficient use of resources"[15] which could be natural* or man-made**. (* – Natural resources include: air, water, minerals in its all the three forms – solid, liquid and gases; land, flora and fauna; ** – man-made resources include: man-himself, machines and equipment, material and energy).

Top/senior management/managers formulate business policy and their active participation is essential for successful HSE management. A systematic approach to HSE management would focus on the following issues:

- Deliver compliance with the laws of the land with regard to safety, environment, occupational health and welfare
- Achieve continuous improvement in HSE performance
- Set objectives and targets
- Measure, appraise and publicly report HSE performance
- Require outsource agencies such as contractors, sub-contractors to abide by the set policy of the company
- Include HSE performance in the appraisal and reward of all those concerned (from top to bottom)
- Allocate resources and expertise to meet HSE targets (e.g. finance, manpower, technology, skills and training etc.)

The impact of formulating a Policy and its effective implementation results in continuous improvement in the management of the risks to the health and safety of employees, contractors' employees and others affected by the industry's operations. It also results in a continuous reduction in the adverse impacts on the environment. *This scenario leads to sustainable development, which is an important aspiration for any country.*

10.4   ORGANIZATION, RESPONSIBILITIES, RESOURCES, STANDARDS & DOCUMENTS[10, 11, 15, 16]

All employees of any industrial setup, both individually and collectively, are responsible for HSE performance. This could be achieved through:

- A proper *organizational set up* that defines *Roles and Responsibilities* of each individual
- *Resources* required to accomplish it
    - *Human – their competence and training needs*
    - *Physical- labs., office space, equipment, appliances, software, hardware etc.*
    - *Finance*

- What *standards* it should met – ISO system, as formulated by the company itself, or those specified by the statutory (legal) bodies/institutes
- *Planning and scheduling* of HSE-related activities and their proper follow-up and *documentation*
- *Training and education* – to have awareness and effective implementation.

10.4.1 Training needs[16]

HSE awareness amongst all is a must. Training with regard to HSE-related competence is almost mandatory for each individual right from top executive to the shop floor worker in any industry. It is equally applicable to the workers/people engaged as service providers, contractors, and sub-contractors. The steps given below should be followed to train them.

- Consider the degree of HSE knowledge required by the individual to perform the job
- Evaluate the individual's HSE-related knowledge
- Establish shortcomings and training needs
- Impart training
- Record training imparted
- Obtain feedback of its effectiveness from the individual himself, his co-workers and supervisors.

10.4.2 Resources required

Allocation of human, physical and financial resources should be part of the policy formulated. Resources should be sufficient and top executive/Managing Director should ensure it. Line-managers should ensure that they are utilized judiciously.

10.4.3 Roles and responsibilities[13, 17, 19]

In a leading Oil and Gas Company, the bullets listed below illustrate how HSE management is focused:

- Board of shareholders – Review and endorse the company's strategic objectives
- Managing Director's Committee – endorse company's HSE policy and subsequent revisions
- Director's meeting – facilitates two way communication of HSE issues, including issues considered by HSE steering Committee and HSE Implementation Committee
- HSE Managers meeting – hold regular meetings with their team leaders to incorporate HSE-related issues
- Roles and responsibilities of any individual include:
  - Promote and support safe work practices.
  - Be aware of relevant changes to HSE requirements and make sure you know what these mean to you and your team.
  - Provide input into HSE discussions.

- • Make sure you are aware of, and understand, your own HSE 'Tasks and Targets'.
- • Share your HSE experiences and knowledge with your team.

- • The various committees formulated to manage HSE effectively include:
  - • HSE Steering Committee
  - • HSE Implementation Committee
  - • Integrated Audit Committee
  - • Emergency Response Steering Committee – accountable to MD
  - • Incident Review Committee – for incident notification, investigation, reporting and follow-up.

In addition, the responsibilities right from top executive to an employee towards HSE have been described in the proceeding section 10.2.5.

10.5 HAZARDS AND EFFECTS MANAGEMENT[15, 23]

Various operations undertaken at any industrial establishment have the potential to damage/harm people and the environment. The degree could vary from the known hazardous industries such as Chemicals, Petroleum, Mining and Metallurgical to less hazardous process industries producing goods and services of many kinds.

A Hazards and Effects Management Process (HEMP), which is used in many industries, provides a structural approach to manage the hazards and the potential effects. It involves four steps (as shown in Figure 10.5):

1 Identification of hazards to people and environment
2 Assessment of the related risks
3 Implementing measures to control these risks
4 Recovery in the event of failure of these measures.

10.5.1 Steps in hazards and effects management process[15]

1 Identification of hazards to people and environment could be achieved:

- • Through Experience and Judgment
- • Using *checklists* For example, while considering *Potential sources that may endanger safety,* the following should be considered:
  - • Flammable and explosive substances.
  - • Chemicals and toxic substances.

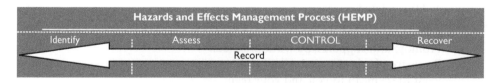

Figure 10.5 Steps in Hazards and Effects Management Process.

- Pressure systems.
- Differences in height.
- Confined spaces.
- Slips, trips and falls.
- Machinery with moving parts.
- Extreme temperatures.
- Electricity.
- Objects under induced stress.
- Transport and moving equipment.
- Security.
- Natural environment – weather.

- Referring code and standards.
- Using more structured review techniques.

When is HEMP required? This is shown in Table 10.3.

The planning and design phase is considered to be the most risky, as major decisions are taken during this phase. During this phase a conceptual model is first prepared, and then it is evaluated by using different alternatives and schemes. The best one is selected for engineering details. During this phase detailed drawings, equipment specifications, resources required and budget are forecasted. Any deficiency could result in problems and delays during the construction phase. It could even become a cause of failures during the operational lifetime. It is important to plan and take into account the cost of abandoning or closing any industrial setup (Post closure phase) as per the prevalent legislation of the land. This is known as the 'Life Cycle' approach. During the design phase, modern techniques concerning the safety of operations, such as those outline below and as described in sec. 6.13, could be considered:[7, 14]

- Failure scenarios
- HAZOP (Systemic identification of hazards and operability problems) and HAZAN (Systemic analysis of hazards and their potential consequences). Applications of these techniques are almost mandatory in chemical industries.
- Fault tree analysis
- Event tree analysis

Table 10.3 The life cycle phases requiring HEMP.

Phase	Focus

– Planning and design, construction and commissioning.
– Regular production phase, which could have the following operating conditions:
  • Normal operating conditions.
  • Abnormal operating conditions (including shutdown, maintenance, start up, and upset condition).
– Reasonably foreseeable accidents, incidents and/or emergency situations.
– Decommissioning, abandonment, dismantling and disposal.
– Post closure phase.
	Identification and assessment of hazards that may be avoided, reduced or eliminated.

It is important have close liaison and collaborative efforts between designers, manufacturers, engineers, and all those others who are concerned with the project. This could minimize risks. There is considerable scope for improvement in standards of health and safety in industries of all types worldwide. Some countries have more developed regimes than others. All can improve further. We should all share knowledge, guidance and good practice in these matters.

2   Assessment of related risks for each operation/job, tasks involved should be listed. Assessment of health, safety and environment hazards then follows.

10.5.2   Control of hazards and effects[7, 15, 16]

Table 10.4 details guidelines for control of hazards and effects. Table 10.5 lists measures used to reduce or limit the consequences arising from hazards.
Determine the harmful effects associated with each hazard. Consider the acute and chronic effects and routes of exposure, for example:

- Inhalation.
- Ingestion.
- Skin contact.

Table 10.4 Control of hazards and effects.

ALARP As Low as Reasonably Practicable	• Risks are said to be reduced to a level of ALARP, at a point where the time, trouble, difficulty and cost of risk reduction measures have been assessed, • But with further reduction measures considered to be unreasonable with respect to the additional risk reduction obtained.
Controls	• Includes risk reduction measures, • Preventative measures, • Barriers or mitigation measures.
Threats	Due to extremes, for examples: • Thermal (high temperature), • Chemical (corrosion), • Biological (bacteria), • Radiation (ultraviolet), • Condition (poor visibility), • Uncertainty (unknowns) or • Human factors (competence).
Incidents	• The release or near release of a hazard, which exceeds defined limits. • These are unplanned events or a chain of events, which could cause injury, illness and/or damage and loss to assets, the environment or third parties.
HSE critical activities	• Activities that are undertaken to provide or maintain controls for major hazards.
Top event	• The 'release' of a hazard.
Major hazard	• Any hazard giving rise to "high" HSE risks.

Table 10.5 Measures used to reduce or limit the consequences arising from hazards.[7, 15, 16]

Controls	Use	Description and examples
Prevention measures	To reduce the likelihood of hazards or To prevent or avoid the release of hazards	Examples include: Guards or shields (coatings, inhibitors, shutdowns), Separation (time and space), reduction in inventory, Control of energy release (lower speeds, safety valves, different fuel sources) Administrative (procedures, warning, training, drills)
Mitigation measures	To reduce or limit the consequences arising from a hazardous event or effect	Active system To detect and abate incidents, for example: gas, fire, and smoke detectors Passive systems Intended to guarantee the primary functions, for example fire and blast wall, isolation, separation, drain systems etc. Operational (non physical) Intended for emergency management for example: contingency plans, procedures, training, drills
Recovery measures	Includes top events	All technical measures, operational and organizational measures which can: Reduce the intensity/severity of the hazardous event to develop further Provide life saving capabilities should the 'top event' develop further.

10.5.3   HEMP Tools – Risk analysis[20]

Any industrial setup involves huge investment and input of resources – man, machine and equipment for long durations. One traditional approach to risk management is to control the hazard by providing layers of protection between it and the people, property and surrounding environment to be protected, as described in section 8.6.

In such projects, the modern techniques such as: 'Risk Matrix' (Tables 10.6 & 10.7 and Figure 10.3) could be applied to identify potential hazards.[20]

10.5.4   Recovery measures[3]

• Those measures aimed at reinstating or returning the situation to normal operating conditions.
• Recovery from high risk and emergency scenarios shall be in place and subject to testing and review.

Table 10.6 Probability ranking.

Category	Definition
A	Common or frequent occurrence (1 yr.)
B	Is known to occur or it has happened (1:10 yrs.)
C	Could occur or heard of happening (1:100 yrs.)
D	Not likely to occur (1:1000 yrs.)
E	Practically impossible (<10,000 yrs.)

Table 10.7 Consequence ranking.

	People	Equipment	Production loss	Environment
1	Fatality	$10 m	2 weeks	Major
2	Disabled	$1–10 m	1 day–2 weeks	Serious
3	Major LTI (Weeks)	$100 m	1 shift	Moderate
4	LTI	$10 k	2 hrs.	Minor
5	Minor	$1 k	<1 hr.	Insignificant

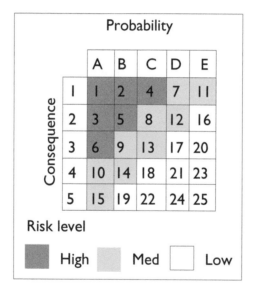

Figure 10.6 Risk ranking (matrix) – combined consequence and probability consequence ranking. Also refer to illustration given in section 8.11 regarding risk matrix as practiced at a leading Oil and Gas Company in the Gulf region (Tables 8.3 to 8.5 and Figure 8.11).

- In addition, recovery procedures should be regularly updated in light of incidents, following analysis of drills and testing, and kept in line with industry best practice.
- Performance against all recovery procedures should be recorded and formally reviewed periodically.
- Parties responsible for implementing recovery procedures shall be competent and clearly understand their responsibilities.

Table 10.8 HEMP and its details; based on the practices used at an Oil Exploration & Production Company in the Gulf Countries. (WHIMS – THESIS these are Software.) [15,22]

HEMP stage	Objective of stage	Result of HEMP	Source of information	Tools & techniques available
Identification of hazards and effects	To identify the hazardous events (consequences) to be avoided, to identify the hazards, threats and escalation factors which might contribute to that hazardous event.	List of potential hazardous events List of potential hazards List of potential threats List of potential Escalation Factors.	Individual knowledge Group knowledge Contractor knowledge Established activity documentation (business reports, inspections, audits, incident investigations etc.) Specialist reports (HAZAN, HAZID, SAFOP etc.) Specialist tools QRA, HAZID, etc.) HSE Cases Task Hazard Control Sheets	Brainstorming Hazard Identification. (HAZID) Health Risk Assessment (HRA) HRA and exposure evaluation for chemical agents Human Factors Environmental Assessment (EA) Soil and Groundwater Guides Job Hazard Analysis.
Evaluation of risk	To assess the risks to health, safety and environment from the identified hazardous events by consideration of; 1 The likelihood of occurrence; 2 The severity of exposure. Risks are to be assessed against established and demonstrably effective controls (barriers).	For each hazardous event, hazard, threat, escalation factor, an evaluation of: • The probability of occurrence • The severity of exposure List of appropriate standards which are in place and enforced List of hazardous events for which standards are not in place or not acceptable.	HSE Risk Matrix Incident Reports HSE Info Tool QRA Datasheets Country's Legislation HSE Standard Manual International Standards (BSI, API, ISO etc.)	Most of the above techniques, plus: HSE Risk Matrix Quantitative Risk Assessment. QRA Historical records Consequence Analysis as in: Physical Effects Modeling Environmental Dispersion Models Monitoring water quality Oil Spill Trajectory Models Risk Assessment Models for Contaminated Groundwater models

(Continued)

Table 10.8 (Continued)

HEMP stage	Objective of stage	Result of HEMP	Source of information	Tools & techniques available
Recording of hazards and effects	To record the results of the identification and evaluation stage for those hazardous events where risk is significant so that; 1 The information can be communicated to others; 2 A record exists when challenged by inspection or audit.	Hazard Registers Activity Specification Sheets Task Specification Sheets Task Hazard Control Sheets.	THESIS software (TSE/1) WHIMS (BOS/7)	THESIS software WHIMS software HSE-MS HSE Case Specifications
Comparison with Objectives and Performance Criteria	To establish the gap between the risk as evaluated in previous stages and acceptable risk as defined in Company standards.	A list of unacceptable risks ranked against severity of exposure and likelihood of occurrence.		HSE Risk Matrix
Risk Reduction Criteria	To reduce the residual risk to as low as reasonably practicable	New standards to be in place and enforced	HSE Risk Matrix HSE-MS Guidelines	HSE-MS

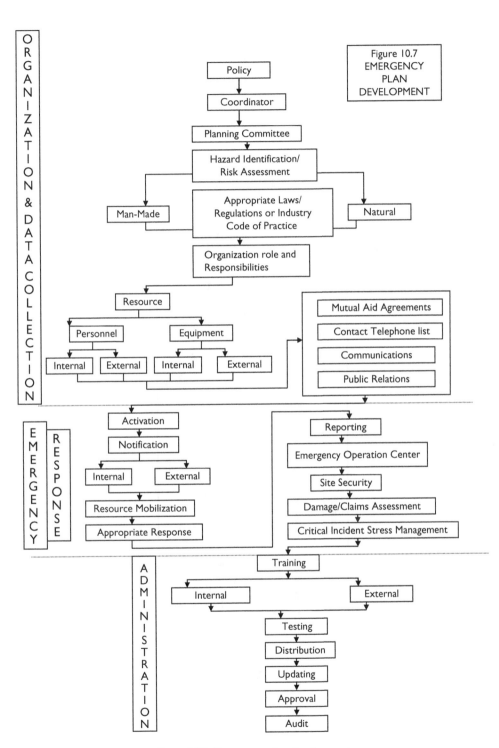

Figure 10.7 Emergency plan development.

Table 10.9 What-if check list – A risk analysis.

Reviewer			Operation Type			Location			
What if? (1)	Cause	Consequences	Risk control measures	Monitoring measures	Likelihood	Severity	Risk	ERP Recommondations
An Event Or Situation That May Arise	What could have potentially caused the event situation in column (1)	What Could Have Potentially Occurred As A Result Of The Event/ Situation (1) (Escalation)	What Measures Are In Place To Prevent, Mitigate And Recover An Incident/ Accident	What Active/ Reactive Monitoring Measures Are In Place? Are These Measures Are Adequate & Reliable?	The Probability Of Frequency That The Event/ Situation In Column I Will Occur Based On Hazard Matrix Definitions.	The Severity Of The Consequence Based On The Hazard Matrix Definitions	Taken From The Hazard Matrix Risk = Consequence × Likelihood	Any Additional Information Or Investigation Required? Are There Practicable Changes Which Could Reduce Risk Or Eliminate Hazards?

10.6 PLANNING AND PROCEDURES[2]

Managing HSE risk and improving HSE performance requires **careful planning** at all levels in any industrial setup. This could be achieved in the following manner:

- Management sets **objectives and targets,** and develops plans to meet them. Everyone working for the company has HSE action items that need to be completed if the Company is to achieve its objectives.
- Implementing HSE Plans also relies on **controlling day-to-day activities** through operating procedures, maintaining the safe operation of equipment, appropriate supervision and inspection, and the safe management of change.
- Finally, planning also includes **emergency response plans** (Figure 10.7), which enable a quick and effective response to unplanned incidents.
- In summary, effective planning will ensure that you know 'where you want to go' (setting objectives and targets), and 'how you will get there' (putting a plan in place). **TARGETS** should be:
  - Quantified and measurable wherever practicable.
  - Realistic and achievable and have identified time scales.
  - Clearly and unambiguously documented.
  - Communicated to all relevant staff and contractors.
  - Reviewed regularly to ensure their continuing relevance and suitability.
  - A mechanism for delivering continuous improvement in HSE performance.

10.6.1 Emergency measures/preparedness[3]

Table 10.9 gives the checklists that should be prepared for the project in hand, to assess the likely hazards or events that could cause problems, and remedial measures that will have to be taken. Figures 1.1 and 1.6 outline and briefly reviews types of hazards – natural or man-made. An emergency management system is an integral part of industrial safety. Standing orders (Emergency Development Plan, Figure 10.7)[3] should be prepared and workers should be trained to act as per these orders in the event of an emergency. It requires coordination, liaison and co-operation between different agencies. It requires enforcing scheduled training, refresher courses and demonstration and mock practices to those concerned – rescue and recovery crews, trained staff and others.

10.7 IMPLEMENTATION AND MONITORING[1, 14–16]

The successful implementation of any company's HSE requires embedding HSE into its business, that is, into the Company culture, responsibilities, and line ownership. An important part of this is the effective management of HSE risks. This involves:

- Setting performance measures and indicators used to monitor how HSE risk management is achieved.
- Establishing a monitoring programme to measure HSE performance against targets.

- Addressing non-compliance within the Company's HSE requirements and ensuring that corrective action is taken.
- Reacting to incidents to make sure that they are notified, analyzed, reported and followed-up.

In fact, full implementation of the HSE Management System, and our safety, depends on staff at all levels of the organization doing what the management system states they should be doing.

One must take ownership of his own safety. This means following work place procedures, understanding the hazards associated with the work and understanding how the procedures help to avoid these hazards.

10.8   HSE AUDIT[1, 15, 16]

HSE audit is a structured, independent examination aiming at:

- Providing management and external stockholders with a systematic, independent way to assess the effectiveness and implementation of the HSE Management System.
- Checking that it is working correctly by highlighting potential problem areas.
- Checking the compliance of the relevant legislative requirements.
- Providing outcome, which is used to identify changes needed to improve HSE Management System.
- To review recommendations of pervious audit and their compliance.

A team with a leader carries out the audit. The team could be internal, external, or both (as is a usual practice). The size of team could be 3 to 5 members depending upon the size of operation to be audited. The frequency of internal audit could be fortnightly, monthly, quarterly, half-yearly or yearly depending upon the policy of the company and nature of the job. Frequency of audit by external team/agency could vary from once in a year to a longer period. The team members should be well trained in the areas they intend to audit. They should have a thorough exposure and experience, and devotion to the duty that has been assigned to them. They should be aware of auditing techniques, elements to be looked into and ways to make audit a positive experience.[12]

Management should ensure that everyone is properly trained for the job they are performing. When employees question need for HSE audit, the answer is – it prepares safe working areas/conditions with least pollution for people. And failure to properly audit working places/spots where men-machines-equipment operate and certain procedures are followed places them (people) at risk.

Audit tools

- An audit-kit contains a set tools. This includes sample slide presentation; questionnaire; template and guidelines that can be customized and supplemented by the audit team-leader to suit individual requirements
- Inspection guide contains guidelines and tools for use in conducting HSE inspections.

A three-tiered approach, as is the practice at some companies, is as under:

- Level 1: Includes HSE audits conducted on behalf of the Company's Internal Audit Committee (IAC) as part of the Integrated Audit Plan. This includes independent audits carried out by external bodies, such as ISO 14001 certification audits.
- Level 2: Includes Asset Level HSE audits carried out on behalf of Asset Managers as part of their own Asset Level assurance processes.
- Level 3: Includes task verification and workplace inspection activities to supplement the formal HSE audit process.

10.9   REVIEW[1, 10, 15, 16]

An effective HSE Management System requires that the Operating Company's senior management review it and its performance, at appropriate intervals, to ensure its continuing suitability and effectiveness. For example: International Organization for Standardization (ISO) (ref. To Sec. 9.10.4) Standard for Environmental Management Systems (ISO14001) requires that the organization's top management shall, at intervals that it determines, review the environmental management system to ensure its continuing suitability, adequacy and effectiveness.

The reasons for review are as follows:

- Situations change and there is always a need to improve the HSE Management System. As such the HSE Management System is reviewed from time to time by senior management to determine what needs to be improved, and how.
- Reviews are an important part of continually improving HSE performance.
- Components of review include:
  - The **possible need for changes** to the Company's HSE Policy and strategic objectives, in the light of changing circumstances, and the commitment to strive for continual improvement.
  - Resource allocation for implementation and maintenance of the HSE Management System.
  - Sites and/or situations, on the basis of evaluated hazards and risks and emergency planning.
  - The Letter of Representation, issued by the Managing Director, is the culmination of Company's HSE review processes.

10.10   MANAGEMENT COMMITMENT[18]

The link between leadership activities and a high performance safety culture is more than anecdotal. In working with more than 1,450 sites we have repeatedly seen (Severn, T. 2006)[18] that one of the biggest differences between sites that struggle, and sites that succeed, is leadership behavior.

Senior leaders can dramatically improve safety performance in their organizations by fostering a high performance culture. Going beyond vocal support, leader-

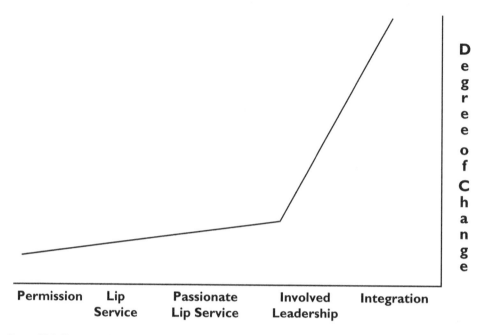

Figure 10.8 Degree of change that could be brought about based on the involvement of senior management. (After Severn, T. 2006).[18]

ship involvement uses a deliberate strategy to elicit the leadership behaviors needed to build a high performance culture in the organization.

Through what they choose to focus on, and their everyday actions, senior leaders can strengthen the organization's culture in ways that drive performance.[18]

Leaders in successful organizations don't just provide resources and verbal support; they influence the organizational climate through their actions. Figure 10.8 illustrates this aspect.

10.11 MANAGEMENT: OCCUPATIONAL HAZARDS (HEALTH & PHYSIQUE)[3, 4, 6]

Many industrial operations are rough, tough and hazardous. For example: working crews regularly inhaling dusts of various kinds for a long period, may result in the long run in silicosis, asbestosis, manganese poisoning and some other lung diseases. Dusts and toxic gases affect many parts of our body. Adverse working conditions give undue strain to our body. The idea is that one should not add any strain or disease due to his duties at the site. To ensure fitness, two tasks are necessary in conjunction with occupational health:

• Periodical check ups for fitness to do the occupation – in terms of vision, hearing, etc. (sec. 7.9; 7.10)

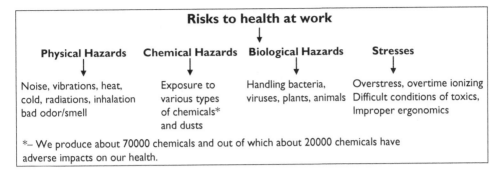

Figure 10.9 Classification of risks to health while working in an industrial set up.

- Danger of suffering from any disease that could be caused, while working in any industry, due to some of the reasons given below: [3, 4, 9]

  - Dermatitis – due to regular contact with and long exposures to cement material.
  - Skin and respiratory problems due to epoxy materials.
  - Impaired hearing – exposure to noise. It is a common health problem. Noise should be reduced at source to the extent that use of personal protective wear is at a minimum.
  - Heat strain/heat stroke due to working in an improper and inadequately ventilated working area including working at great depths.
  - Vibration syndrome due to continuous working with hand-held pneumatic or hydraulic tools. Low vibration tools and proper maintenance of the tools could minimize this.
  - Musculoskeletal injury due to handling heavy loads repetitively.
  - Lungs diseases due to respirable silica dust.

The line diagram in Figure 10.9 summarizes risks to the health of an industrial worker. [3, 4, 6, 21]

10.12 ENVIRONMENT MANAGEMENT[22–24]

10.12.1 Why pollution?

Environment means surroundings. As described in preceeding sections and chapters, various Unit Operations carried out in any industry are bound to pollute air and water, and degrade land. To understand the basics of environment in relation to these industries, let us take the example of gold mining. Gold is mined if its grade/concentration is 5 gms/tonne. This means while mining 1 tonne (1000 kg) of gold ore, only 5 gms (on an average) would be of useful and rest 999.995 kg of rock that would be generated is waste. In recovering this 5 gms gold it has gone through mining (breaking rock into small fragments from in-situ), concentration (crushing, grinding into powder and separation from the rest of the rock-mass using chemicals), smelting and lastly refining and casting into bars or any other shape. One can imagine how much energy it required, materials of different kinds it consumed, foul gases it produced, and land it required to dispose of the wastes generated. This is the reason mining is blamed for pollution. But minerals are our basic needs we cannot do away with them. One should not forget that more than 75% of power is generated using fossil fuels (coal, oil and gas). Automobiles are run by fossil fuels. Fossil fuels have a greenhouse effect. Exhaust from automobiles is responsible for bad ozone. Industrial pollutants are causing acid rain, for example in a recent survey it was found that in China it is affecting its one- third of the land.

Refer sec. 1.5.1 to understand what has happened in the past that compelled society to suffer on account of ill health, aesthetic and cultural pleasures, and economic opportunities.

10.12.2   Mass balance system/equation

Referring to the mass balance equation as described in section 1.5.1, what approach we should we follow? ADD or NOT ADD? And up to what extent? This has called for an awakening, which is known as: 'Sustainable Development'. The United Nations World Commission on Environment and Development defined the term 'sustainable development' in 1987 for Ecological Sustainable Development[17] as "Development that meets the needs of the present generation without compromising the ability of future generations to meet their own needs".

While considering Potential sources that may have *adverse impacts on environment*, the following should be considered:

- Emissions to the air.
- Discharges to water.
- Spills and leaks.
- Waste management.
- Use of energy, materials and resources.
- Environmental noise and vibration.
- 'Land take' and land degradation.
- Beneficial and adverse impacts:

  - On natural resources of air, water, land.
  - On human health.
  - On flora and fauna.
  - On cultural and heritage resources.

10.12.3   Environmental degradation in an industrial setup

Figure 10.10 describes types of pollution caused by an industrial setup. It also includes operations, activities materials or reasons, which are responsible for causing industrial

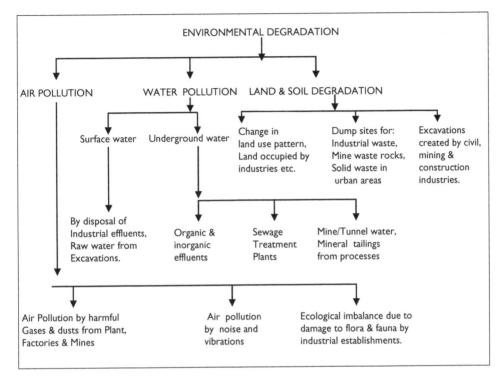

Figure 10.10 Types of pollution caused by an industrial setup.

pollution damaging air, water and land regimes. Figure 10.10 also details air pollutants responsible for making it harmful.

10.12.4   Main sources of pollution in air, water and land environments

Pl. refer figure 3.1 which classifies Air Pollutants.

10.13   ENVIRONMENTAL MANAGEMENT[22-24]

It is a mechanism that involves the following three steps

- Base line i.e. existing or pre-establishment scenario.
- Environmental Impact Assissent (EIA)
- Environnental Management Plan (EMP)

Table 10.10 briefly summarises steps that should be followed during all the three phases, bulleted above.

Table 10.10 Summary of aspects to be looked into while preparing the 'Management Plan'. [3,6,9]

I Base-line information	II Impact assessment (EIA)	III Management plan (EMP)
1 Ground/Land Environment	1 Likely Impacts	– Diversion plans of existing land use pattern (if any) together with mechanism, schedule and cost implications.

1.1 Surface topography
      Plain land
      Hilly terrain
	1.1 Surface topography
	      Magnitude, location.
		– Excavation/Mining sequence ensuring minimum land disturbance at any one time

1.2 Soil
      Type (quality)
      Thickness
	1.2 Soil
	      Quantity
	      Extent
		– Waste rocks and materials dumping schemes with location and procedure ensuring minimum horizontal spread. Provision for bunds around dump-yards to prevent silt flow to the surrounding areas.

1.3 Land use pattern
      Industrial
      Residential
      Grazing & Forest
      Occupied by infrastructures
      Barren land
      Others – not covered above
	1.3 Land use pattern
	      Magnitude & location of disturbance. Land occupied by waste rock dumps, effluents from the process plant.
		– Stacking different wastes, including mineral wastes, separately and keeping a record for future reference.

1.4 Aesthetic – Existing natural beauty	1.4 Impacts on the existing natural beauty.	– Land reclamation and beautification schemes. Useful soils' reuse schemes.
2 Water Environment	2 Impacts to water environment	– Water flow diversion if required, with costs, schedule and mechanisms.
2.1 Existing water bodies and their proximity to the project	2.1 Diversion of water-bodies, if any.	– Provision for effluent discharge
2.2 Position of water Table Potable water availability (Quality & Quantity)	2.2 Water pollution – Quality, quantity (Resultant water Table lowering).	– Provision for raw water treatment (including acid water, if any) before its discharge.
2.3 Sewerage water disposal	2.3 Sewerage water – quantity.	– Assessment of pumping requirements including standby arrangements to avoid any flooding or inundation.
2.4 Industrial water disposal	2.4 Effluents – types and quantity	– Water surveys, quality monitoring and plans to meeting potable water requirements.

2.5 Rain
      Average rain
      Rainy months
      Natural drainage system
	2.5 Impacts of rain water to the workings. Chances of flooding. Water pumping requirements. Impacts to existing drainage system. Chances of acid water, acid rain, if any.	

3. Air	3 Air pollution assessment	– Provision for effective ventilation – fan types, location. Meeting water-gauge and air quantity requirements.
3.1 Existing climate	3.1 Assessing any rise in temperature, humidity, if any.	
3.2 Air quality (existing industrial and general pollution)	3.2 Pollutants – types and quantity.	– Periodic air surveys. – Noise containment measures.
3.3 Noise – magnitude and sources.	3.3 Noise sources and quantitative assissent.	– Dust sampling, analysis and suppression measures.

(Continued)

Table 10.10 Continued.

I Base-line information	II Impact assessment (EIA)	III Management plan (EMP)
3.4 Dust – air borne dust (quantity and quality)	3.4 Likely dusts generation – sources, quality and quantity.	– Measures to maintain temperature and humidity of the surroundings within allowable limits.
4 Seismic – existing (natural) Induced by industrial activities	4 Assessing vibrations due to pneumatic machines, and as a result of blasting operations.	– Measures to design blasts enabling peak particle velocity to be within allowable limits, so that damage to the surrounding structures is at a minimum. Scheduling blasts by considering minimum disturbance to the workers and local inhabitants.
5 Bio-Environment	5 Impacts to the existing flora and fauna.	– Repeating vegetation plantation schemes: location, schedule, manner and costs.
5.1 Flora – vegetation type, tree density, crops –types,	5.1 Assessing magnitude of vegetation removal, deforestation.	– Possibility of relocating existing fauna, if any.
5.2 Fauna – birds, insects, animals (types and population)	5.2 Fauna likely to be affected by the project activities.	
6 Social Environment	6 Assessing likely impacts to the people and facilities within a radius of 5 km or so (as required by the prevalent regulations).	– Planning relocation of local inhabitants, facilities, if any, with schedules, costs and mechanisms. Assessment of compensation, if any.
6.1 Local inhabitants – population, trade and vocation	6.1 Relocation of existing inhabitants	– Diversion plans of existing infrastructures, facilities with their details and costs involved.
6.2 Places of worship & historical importance	6.2 Disturbance/demolition, shifting of places of public interest.	– Provisions to settle public disputes and grievances.
6.3 Existing facilities – education, medical, housing, power, roads, recreation, playgrounds, transport, communication	6.3 Diversion or damage to the existing facilities.	
7 Existing legislation, & labor laws		– Provision for employing competent persons to comply with the regulations, and carry out activities with safety and least harm to the environment.
7.1 Rules, regulations, laws regarding environment, health and safety.	7.1 Implications of enforcing existing laws – impacts on production, productivity and costs. Necessity of formulating code of conduct, or procedures, in absence of local laws.	– Provisions for periodical tests, checkups and inspections. – Provision for safety wear, training, and basic health and welfare facilities.
7.2 Trade unions – types of labor.	7.2 Assessing likely problems from local labor and trade unions in existence, if any.	

10.14 SUSTAINABLE DEVELOPMENT[20]

Apart from descriptions given in sections 1.7 and 1.7.1 on policy matter, the following aspects should also be considered:

- Many industrial operations consume a huge amount of fossil fuels that contributes to the generation of 'Green House Gases'. Improving energy efficiency in every operation is warranted, which should be strived for through continuous R and D efforts. These aspects have been covered in detail in chapters 2 and 3.
- New materials that will drive sustainability should be tried.
- Design engineers need to include sustainability as a purchasing and design criteria.
- The industry needs to participate in the development of new technologies and products. The collaborative efforts of designers, manufacturers, engineers, government and university research agencies could achieve this.
- Consideration of environmental impacts must be made during the design phase itself. Matching sustainable products and materials with construction requirements is an emerging area requiring the industries to develop expertise and/or provide additional training to those concerned.

Based on the best practices that are being adopted by some of the countries who recognized sustainable development to be the way forward; the following principles, if followed religiously, could achieve sustainable development.[20]

1   Integrating sustainable development considerations within the corporate decision-making process. Top/senior management should provide **strong and visible leadership** to promote a culture in which all employees share a commitment to Sustainable Development. Referring to Figure 10.12 in which ICMM has very recently released guidelines in this regard.

2   Implementing risk management strategies based on valid data and sound science. Adhering to the laid out procedures, guidelines, standing orders, rules and regulations to deal with industrial hazards including dealing with emergencies could achieve this. Modern techniques such as "Risk Matrix' should be applied to assess the risks involved (Ref. Sections 8.11 and 10.5.3).

3   Conducting businesses with ethical practices and sound systems of corporate governance. Bringing transparency to the system.

4   Seeking continual improvement in Occupational Health and Safety (OHS) performance (chapter 7). It includes implementing a management system focused on continual improvement of all aspects of operations that could have a significant impact on the health and safety of our own employees, those of contractors and the communities where we operate.

5   Upholding fundamental human rights and respecting cultures, customs and values in dealings with employees and others who are affected by our activities. Ensure fair remuneration and work conditions for all employees and ban forced, compulsory or child labor. It should ensure equipping an employee with necessary resources & empowerment.

6   Seeking continual improvement of our environmental performance based on a precautionary approach. The idea is to assess the positive and negative, the direct

and indirect, and the cumulative environmental impacts of new projects – from beginning through closure.

7 Contributing to conservation of biodiversity and integrated approaches to land use planning and management. This includes respecting legally designated protected areas and culturally sensitive sacred graves.

8 Facilitating and encouraging responsible product design, use, re-use, recycling and disposal of our products (Ref. Sec. 2.8.6). Conducting and supporting research and innovation that promotes the use of products and technologies that are safe and efficient in the use of energy, natural resources and other materials and encourage putting into use the positive aspects that emerge out of such R & D.

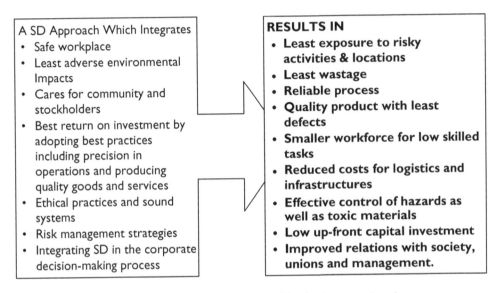

A SD Approach Which Integrates
- Safe workplace
- Least adverse environmental Impacts
- Cares for community and stockholders
- Best return on investment by adopting best practices including precision in operations and producing quality goods and services
- Ethical practices and sound systems
- Risk management strategies
- Integrating SD in the corporate decision-making process

RESULTS IN
- **Least exposure to risky activities & locations**
- **Least wastage**
- **Reliable process**
- **Quality product with least defects**
- **Smaller workforce for low skilled tasks**
- **Reduced costs for logistics and infrastructures**
- **Effective control of hazards as well as toxic materials**
- **Low up-front capital investment**
- **Improved relations with society, unions and management.**

Figure 10.11 Our ultimate aim is sustainable development, in whatever industry.

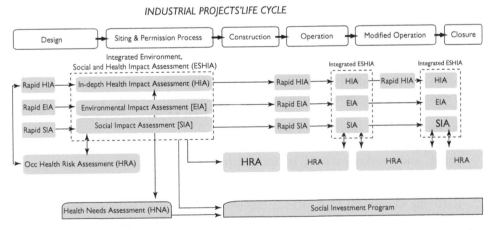

Figure 10.12 Integrated Environment, Social and Healthcare Impact Assessment during life cycle of an industry. (After ICMM, 2010).[5]

9 Contributing to the social, economic and institutional development of the communities in which we operate. Encouraging partnerships with governments and non-governmental organizations to ensure that programmes (such as community health, education, local business development) are well designed and effectively delivered.

10 Implementing effective communication, and independently verified reporting arrangements with our stockholders. We should regularly provide them information that is timely, accurate, relevant and independently verifiable.

It could be achieved by adopting the model as shown in Figure 10.12.[5] This model advocates an integrated Environment, Social and Healthcare Impact Assessment during life cycle of any industry. A good-good situation for everybody, particularly those concerned. Should we not strive for it?

10.15 CONCLUDING REMARKS[11]

Based on the foregoing discussion/description on the various subject matters of this book, the following aspects have emerged:

- Various operations undertaken at any industrial establishment have the potential to damage/harm people and the environment. The degree could vary from the known hazardous industries such as Chemicals, Petroleum, Mining and Metallurgical to less hazardous process industries which produce goods and services of many kinds.
- An effective HSE management requires involvement and commitment of everyone right from top to bottom. Experience suggests, as shown in Figures 10.3 and 10.13; that it is teamwork which can bring the best results.
- The Key-words shown in Figure 10.13 have the following meanings: Reactive – action in response (which is the natural human tendency); Dependent – one who depends upon others (it has an element of fear in that if not obeyed as instructed there could be disciplinary action against those concerned); Independent – not dependent (as described in Table 9.2; empowering with responsibilities works equally well. It provides opportunity to the individual for self-development, and also to show his worth); Interdependent – mutually dependent; illustrating the fact that it is *teamwork* which could bring the desired results out of the various options available to us.
- One must take ownership of his safety by following laid down procedures, methods and techniques.
- Safety checks, inspections and audits are the independent ways to assess the effectiveness and implementation of the HSE Management System.
- With the changes in scenario/situations from time to time, reviewing HSE Management System is mandatory for continuous improvements.
- Standing orders (Emergency Development Plan; Ref. Sec. 10.6) should be prepared, and workers should be trained to act in the event of an emergency. It requires coordination, liaison and co-operation between different agencies.

Referring Figures 10.11 & 10.12, our ultimate aim is sustainable development, may it be in whatever industry. A good-good situation for everybody, particularly those concerned. The solution lies in 'Teamwork' as shown in Figure 10.13. Should we not strive for it?

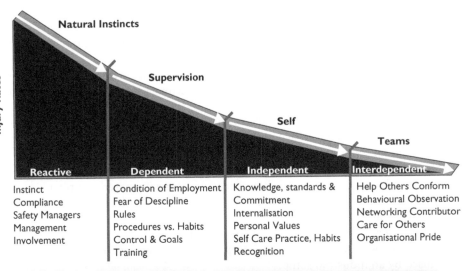

Reactive	Dependent	Independent	Interdependent
Instinct	Condition of Employment	Knowledge, standards &	Help Others Conform
Compliance	Fear of Descipline	Commitment	Behavioural Observation
Safety Managers	Rules	Internalisation	Networking Contributor
Management	Procedures vs. Habits	Personal Values	Care for Others
Involvement	Control & Goals	Self Care Practice, Habits	Organisational Pride
	Training	Recognition	

Figure 10.13 (Upper) It is teamwork that could minimize losses, which could approach zero-illness, zero-injury, zero-incident, zero-waste, zero-emissions and zero-rework. (After Latus, M. 2006)[11] (Lower Photo: Winning the world cup (Billy). Courtesy: BHP Billiton Mitsubishi Alliance)[2]

QUESTIONS

1   What does 'Informed involvement' signify?
2   Based on your profile, what should be your Role and Responsibilities with respect to HSE-related issues?
3   Do you agree – successful HSE Management reflects teamwork? Elaborate on it.
4   During the design phase what techniques concerning safety of operations are usually applied? List them.
5   How helpful is involvement of senior management in bringing about changes in an organization? (Hint. Use concept given by Severn, T.)
6   How should senor management focus on HSE management?
7   How would you embed HSE into the 'Business Plan' of your company?
8   Illustrate by a line diagram an Emergency Measures/Preparedness plan.
9   In what way should top/senior management be involved for efficient HSE management?
10   List the elements of HSE leadership and commitment.
11   List the causes of pollution. And prepare a line diagram to illustrate pollution caused by an industrial setup.
12   List the reasons for suffering from some occupational disease while working in civil, construction, mining and tunneling industries.
13   List the steps in a the Hazards and Effects Management Process.
14   Our ultimate aim is sustainable development, may it be in whatever industry – how it can be achieved?
15   Prepare a 'Checklist' to show your current status with regard to HSE Management – Leadership & Commitment.
16   Prepare a 'What if check list – for risk analyses'.
17   Prepare an 'Environment Management Plan' including a brief summary of steps that need to be taken during all the three phases; for Chemical, Petroleum, Mining, or any other industry that you belong to.
18   Tabulate the guidelines for control of hazards and effects.
19   Tabulate the measures used to reduce or limit the consequences arising from hazards.
20   What 'Safety leadership model – personal commitment' has been suggested by Latus?
21   What business activities do you consider as 'Critical' in achieving the set goals and objectives of any company?
22   What important aspects should be included while preparing an 'HSE Policy'?
23   What is an 'HSE Audit'? Outline the details for undertaking this for an in industrial setup.
24   What is the concept of 'Hazards and Effects Management Process (HEMP)'which is used in many industries? Describe it.
25   What is the most valuable asset (resource) to run any industry efficiently?
26   What strategy would you propose to achieve: zero-illness, zero-injury, zero-incident, zero-waste, zero-emissions and zero-rework?
27   What have you understood by: Recovery Measures, Planning and Procedures, Company Culture, Implementation and Monitoring, Review, Management Commitment? Prepare short notes.

REFERENCES

1 Arnold, R.M. (2001): Measurement techniques in safety management. In: *Mine health and safety management*, Michael Karmis (edt.), Littleton, Colorado, SME, AIME. pp. 51–57.

2 BHP Billiton Mitsubishi Alliance (2006): Creating a Mature Reporting Culture. *International conference focusing on safety and health on 14–16 November, Johannesburg, South Africa.* London ICMM

3 Canadian Standards Association (1995): *Emergency Planning for Industry*.

4 *Construction (Health Safety and Welfare) Regulations*; SI 1996/1592. HMSO London 1996.

5 *Good practice guidance on health impact assessment* (2010), London, International Council on Mining and Metals (ICMM).

6 *Guidelines for Tunnel Safety*, International Tunneling Association, Bron, France, 1991.

7 Gupta, J.P. (2000): Safety course at Sultan Qaboos University, Oman.

8 *Hazard Management in Contracts Guidelines*; Thomas Anwyl, TSE/53;

9 HSE Books, Sudbury, 1992; 1994; 1996; 2000; 2001.

10 International Organization for Standardization – *Organizations responsibilities and requirements to comply with ISO14001.*

11 Latus, M.: Leadership in Safety – One Company's Approach. *International conference focusing on safety and health on 14–16 November, Johannesburg, South Africa.* London ICMM. (Permission: Newmont Mining Corporation)

12 Martin, D.K. & Boling, H.L. (2001): Inspections and auditing. In: *Mine health and safety management*, Michael Karmis (edt.), Littleton, Colorado, SME, AIME pp. 175–18.

13 Omani legislation – general HSE obligations of organizations and individuals. Royal Decrees 34/73 and 10/82.

14 Pegg, M.J. (2000–03): Safety class-notes and course at Sultan Qaboos University, Oman.

15 Petroleum Development of Oman – HSE management (1996–2004).

16 Philip J. Stapleton, Principal, Margaret A. Glover, Principal: *Environmental Management Systems: An Implementation Guide for, Small and Medium-Sized Organizations.* S. Petie Davis, Project Manager. NSF ISR; 789 N. Dixboro Road; Ann Arbor, MI 48158; 1–888-NSF-9000.

17 Royal Dutch /Shell Group – requirements of HSE Management system (1996–2004).

18 Severn, T. (2006): Safe production. *International conference focusing on safety and health on 14–16 November, Johannesburg, South Africa.* London ICMM.

19 Shell International Exploration and Production Company – HSE Management system. (2000–04).

20 Standish, P.N. & Reardon, P.A. (2002): *Semi-Quantitative Risk Analysis for underground development projects*, International Tunnelling Association.

21 *Sustainable Development Framework – Indian Mining Initiative* (Draft Report), Federation of Indian Mining Institute (FIMI), 2009.

22 Tatiya, R.R. (2005): *Civil Excavations and Tunneling – A Practical Guide.* Institution of Civil Engineers/Thomas Telford Ltd. London, U.K.

23 Tatiya, R.R. (2000–04): *Course material for Petroleum and Natural Resources Engg; Chemical Engg. And Petroleum and Mineral Resources Engg.* Sultan Qaboos University, Oman.

24 Tatiya, R.R. (2006): Health, Safety and Environment (HSE) Management – Where do you stand? *Souvenir: 17th Environment Week Celebration, Indian Bureau of mines, Rajsthan region,* pp. 20-26.

Subject index

Abiotic xix, 19, 26, 29, 31, 34, 79
Abnormalities xxii, 177, 271, 316, 320,
   325, 332, 334
   Diagnosis xxii, 321, 322, 324
   Diagnosis technique: 5W – 2H
      analysis 320
   Remedial Measures to overcome them
      xxii, 321, 322, 324
Accidents xix, xxi, 1, 9, 12, 149, 206, 249,
   279, 281, 282, 287, 288, 290, 291,
   293–295, 297, 305, 325, 345, 353, 354
   A three-step event 290
   Causes xxi, 149, 290, 291, 294, 305, 345
   Consequences xxi, 279
   Costs xxi, 12, 279, 295, 354
   Frequency rate 293, 294
   Human failure 291
   Incidents responsible xxi, 197
   Losses 9, 282, 287, 297, 298, 325
   Prevention ratio 306
   Related calculations 293
   Remedial measures xxi, 295
   Severity rate 293, 294
   Strategies to minimize 282
Accumulation 90, 117, 149, 154, 160,
   201, 243
Acid Mine Drainage (AMD) xxi, 109,
   124, 133
   Chemistry 134
Acid Rain xix, xx, 8, 12, 14, 22, 59, 83–86,
   105, 376
   Formation 85
   Calculations related to acid rain 85
   Adverse impacts 85, 86
   Impacts of acid rain in Europe 86, 87
   pH 84–86
Acidity 109, 127
Air xix–xxi, 1, 3, 7, 12, 14, 19, 20, 22–24,
   26, 29, 32, 35, 40, 48, 53, 54, 59–61,
   69–71, 73, 74, 81–83, 86, 92, 98–103,
   105, 109–111, 127, 135, 149, 152, 157,
   159, 175, 177, 179, 180, 182, 189–195,
   197, 199, 201, 204, 207, 209, 219,
   227, 231, 233, 237, 238, 241–243, 253,
   276, 284, 300, 301, 310, 333, 353, 360,
   375–377
   Blast 192, 193
   Composition 59, 238
   Determination of (PSI) Value 73
   Disasters xx, 3
   Pollutants xx, 14, 59–61, 69–71,
      157, 377
   Pollution xx, 12, 48, 59, 71, 74, 83, 105,
      127, 152, 159, 175, 177, 180, 207,
      253, 377
   Pollution Standard Index (PSI) 73
   Quality Index xx, 59
   Quality Standards (AQI) 59, 69–71
   Toxics 59
Air conditioning and Refrigeration 238
Air pollutants xx, 59–61, 70, 157, 377
   Five major pollutants – comparison
   Receptors and adverse impacts 74, 75
Air samples 103
   Analyzing methods 103
   Collection methods 103
ALARP 273
Ammonium Nitrate Fuel Oil mixture
   (ANFO) 199
   Static hazards 199, 200
Analysis xxi, xxii, 106, 137, 143, 179,
   198, 206, 213, 271, 275, 291, 295,
   296, 305, 314, 334, 340, 353, 357,
   363, 366
Aqueous effluents 223, 226
   Discharge 245, 246
Asphyxiate gases 201, 219
Audit 169, 170, 274, 304, 333, 336–340,
   353, 354, 362, 372, 373, 382
   HSE Audit 372, 373
   Tools 372
Auto Ignition 179, 182

Behavior xxi, 20, 98, 261, 263, 267, 275, 279, 281–283, 287, 288, 304, 305, 345, 348, 373, 374
Benchmarking & standardization xxii, 309, 334
Best practices xx, xxii, 51, 53, 105, 205, 265, 270, 285, 286, 288, 298, 311, 312, 334, 341, 342, 380, 381
Biological Oxygen Demand (BOD) 109, 127
Biosphere 19, 22, 26, 48
Biotic Component of planet earth – Biosphere 22
Bottled Water xxi, 109, 145, 146
Breathing Apparatus 179, 217, 254, 300
Buyer's Market 1, 6

Carbon capture & storage xx, 46
Carbon Cycle 23, 24
Carnivores or secondary 27
Carrying capacity 29, 201
Case study xxi, xxii, 122, 135, 145, 265, 346, 349
  British sugar 304
  Computation of financial losses xxii, 332
  Three Pillars of equal strength for loss prevention xxii, 346, 349
  Waste management in petroleum industry xxi, 168
  Water pollution – Mining, Petroleum Products' handling & Industrial activities xxi, 135
Checklist 213, 215, 337, 362, 371
  Know your current status 359
  Legal, statutory & other compliances 337
Chemically Reactive Materials 207, 210
Chemicals in Motion: Cycles in the Ecosphere 33
Classification xxi, 1, 20, 99, 207, 301, 310
  Ecology 20
  Energy sources 36
  Hazards analysis techniques/methods 211
  Hazardous materials 207
  Toxic related hazards 204
  Waste disposal techniques 155
  Wastes xxi, 150
Clean Coal Technology (CCT) xx, 19, 45, 55
Climate Change 45, 79, 105
CO_2 emissions 44
  Global carbon dioxide emissions 44
  Risk due to CO_2 emissions 44

Coal xx, xxii, xxv, 7, 8, 19, 22, 23, 35–37, 40, 45, 47, 53, 55, 60, 74, 82, 83, 85, 86, 98, 99, 101, 133, 136, 180, 181, 193–195, 200, 242, 275, 376
  Clean coal technology xx, 19, 45, 55
  Converting coal-to-oil 47
  For energy security 45
Communication xxii, 5, 8, 35, 189, 253, 254, 283, 284, 286, 295, 298, 309–311, 334, 337, 341, 342, 346, 347, 355, 361, 382
Competency xxii, 309, 311, 315
Composting 149, 160, 175
Compressed Gases 207–209
Conceptual Planning 279, 295, 296
Consignment Note 149
Controlled Materials 210
Corrosive 121, 133, 179, 203, 207, 210, 223, 226, 245, 251
  Liquids 223, 226
  Material 210
Current Scenario 1, 2
Cycle xxi, 22, 23, 26, 27, 32, 34, 52, 109, 111, 128

Defects 294, 310, 316, 325, 333, 334, 336
Degree (Type) of injuries 294
Diesel Emissions 223, 226, 231
Disaster xix, xx, 12, 179–181, 202, 223, 287, 290
Diseases 8, 78, 98, 109, 125, 129, 207, 253, 267, 274, 275, 280, 353, 374, 375
  Notified 274
  Occupational diseases 8, 12, 207, 274, 275, 280
  Preventive measures 195, 274, 295
Disposal 40, 75, 124, 130, 137, 140, 149, 152, 156–159, 161, 170, 172, 174, 175, 177, 180, 205, 246, 250, 284, 285, 289, 331, 381
Dissolved Oxygen 109, 126, 127
Dow Index 179, 195, 196
Dumping site 149, 166, 168
Dust xxi, 48, 59, 71, 98–101, 132, 175, 180, 183, 194, 195, 203, 207, 223, 226, 233, 242, 251, 253, 254, 272, 300, 314, 316, 374, 375
  Conditions to become nuisance 98
  Control 100
  Exposure 98, 223
  Factors Affecting the Degree of Health Risk 98
  Physiological Effects 98, 99
  Properties 98, 195
  Sources 100

Earth's Great spheres 21
Ecological 13, 14, 19, 29, 31, 32, 40, 45,
   47, 69, 121, 126, 376
   Components 19
   Crisis 29, 31
   Pyramid 19, 28
Ecology xx, 19–21, 29, 55, 175
   Great Spheres 19
   Ecosphere 19
   Biosphere 19
   Biotic & Abiotic Components 19
Ecosystem 19–22, 25–27, 29, 31, 40, 48,
   53, 76, 79, 126, 127, 135
   Mountain 30
   Various combinations 29, 30
Education xxi, 4, 14, 175, 189, 252, 279,
   281, 282, 285, 287, 288, 295, 297, 298,
   305, 344, 361, 382
   Encouraging stewardship of Natural
      Resources 15
Effluent Discharges/disposal 248
   Marine 131, 248
Elimination 170, 273, 332, 334
Emergency xxii, 72, 185, 186, 189, 204,
   211, 217, 250, 251, 253, 254, 283, 290,
   291, 295, 300, 301, 311, 353, 354, 357,
   358, 362, 365, 371, 373, 382
   Measures and preparedness 254, 354, 371
   Plan development 369
   What if check list 370
Emission – Estimation of inventory 72, 74
Energy xx, 4, 7, 8, 15, 19, 24–27, 29, 32,
   35–37, 40, 44–48, 53–55, 59, 60, 76,
   81, 82, 84, 86, 92, 93, 105, 111, 127,
   134, 149, 154, 156, 158, 174, 177, 179,
   180–182, 190, 191, 196, 198, 200, 215,
   227, 238, 241, 243, 260, 264, 267, 279,
   283, 310, 312, 314, 325, 333, 342, 357,
   360, 376, 380, 381
   Conservation tips 54
   Consumption pattern 40
   Crisis xx, 19, 47
   Efficient lighting tips 53
   Green power and its purchasing
      options 37
   Promising alternative energies xx, 40, 45
   Sources xx, 19, 37, 40, 47, 53, 227
   Sources and their merits and
      limitations 38
   Top 7 promising alternative
      energies 40, 45
   Way out/solution to energy crisis xx
Enforcement xxi, 281, 282, 285, 305
Engagement xxi, 282, 305
Engineering xx, xxi, 7, 169, 224, 273,
   280–282, 285, 291, 295, 296, 305, 363

Environment xix–xxii, xxvi, 1, 4–8, 12–14,
   19, 21, 27, 29, 30, 47, 48, 54, 55, 69,
   96, 98, 105, 109, 114, 115, 117, 126,
   130, 131, 133, 145, 149–154, 156–159,
   167–169, 175, 182, 204, 207, 223, 245,
   246, 248, 250, 252, 255, 260, 271, 279,
   280, 287, 289–291, 298, 301, 311, 314,
   320, 331, 339, 341, 343, 347, 353, 354,
   356, 360, 362–365, 375, 376, 382
   Base-line information 378
   Degradation in an industrial set-up 376
   Impact assessment xx, xxii, 291, 382
   Management 14, 172, 175, 290,
      291, 301, 311, 335–339, 353, 354,
      373, 377
   Management plan xx, xxii, 170, 175,
      291, 377
   Policy 1, 14, 337, 338
   Related issues 4, 19, 153
   Why pollution also see pollution 8, 48,
      59, 152, 157, 280, 338, 353, 356, 360
Environmental Hazards 151, 180, 205
   Due to solid, aqueous and gaseous
      wastes 151
Equipment xxi, 1, 7, 9, 12, 15, 37, 54, 92,
   98, 132, 133, 149, 155, 170, 172–174,
   177, 179, 182, 184, 185, 187–190,
   198–200, 205, 206, 213–216, 223, 227,
   233, 250–252, 254, 255, 273, 274, 279,
   283–285, 289, 294–296, 300, 309,
   314–316, 320, 325, 334, 340, 345, 357,
   358, 360, 363, 365, 371, 372
   Availability and utilization 317, 320
   Effective maintenance 314, 316
   Efficient utilization 314
   Failure 98, 198, 216, 320
   Modifications 149, 173
   Overall equipment effectiveness 316
   Percentage utilization 315
   Percentage availability 315
   Preventive maintenance 320
Ergonomic fitness 274
Ergonomics xxi, 223, 224, 254–257, 259,
   260, 283, 315, 318
   Aims at 255
   Better working posture 257
   Good ergonomics xxi, 256
   Impacts of poor ergonomics 256
   Improved labor relations 256
   Introduction 254, 274
   Making things user-friendly xxi, 255
   Work in neutral postures 257
Eutrophication 109, 122, 126, 129
Explodable dusts 195
Explosions xix, xxi, 170, 179, 180, 190,
   191, 192, 194, 195, 198, 295

Boiling Liquid, Expanding Vapor
    Explosion (BLEVE) 193
  Classification 190
  Confined & Unconfined 179, 192
  Deflagration 190, 191
  Detonation 190, 191, 208
  Dust & Methane 190
  Ingredients 191, 192
  Mechanical 190
  Physical eruption 190
  Pressure Vessel 190
  Shock wave 190, 191, 208
  Sulfide dust 195
  Underground – Rock Bursts &
    Bumps 190
  Vapor Cloud 190, 193, 196
Exposure to 69, 72, 92, 99, 170, 198, 223,
  226, 243, 254, 271–273, 347, 375
  Aqueous Effluents 223, 225
  Corrosive Liquids 223, 225
  Diesel Emissions 223
  Dusts 223
  Extreme Temperatures 223, 225
  Fibers 223, 225, 226
  Heat & Humidity 223, 225
  Noise 94, 95, 223, 225, 375
  Radiations 225, 244
  Salts 226
  Toxic Gases 223, 225
  Vibrations 223, 225
  Welding 223, 225

Factors 8, 15, 32, 180, 203, 204, 214, 223,
  224, 242, 266, 267, 289, 290, 291, 314,
  315
Fatality 279, 287
Fibers 99, 181, 200, 207, 223, 226, 227
Fire Point 179, 182
Fire fighting 179, 185, 187, 250, 291
  Equipment 179, 185, 187, 250
  Department 179, 187, 188
Fires xix, xxi, 101, 170, 179–181, 183, 184,
  187, 189, 190, 198, 217, 295
  Classification 184, 185
  Concepts – mechanism 182
  Conduction, convection and
    radiation 183
  Fire and emergency 186
  Major ignition sources 183
  Protection 184, 185, 189, 289
  Self-oxidation 183
  Triangle concept 181
Flammable 179–183, 185, 186, 191,
  193–195, 198, 199, 207–210, 215, 249,
  291, 362

Gases 179, 181–183, 191, 194,
    195, 198, 208
  Liquids 179, 181–183, 185, 186, 198,
    199, 207, 209
  Liquids and solids 207, 209
Flash Point 179, 182, 219
Food 3, 4, 7, 19, 21, 22, 26, 27, 29, 32, 35,
  59, 69, 105, 111, 121, 127, 160, 176,
  197, 203, 250, 275, 325
  Chains 19, 27, 121, 127
  Webs 26
Foul gases 8, 227, 231, 331, 376
Fuels xix, 8, 23, 24, 35, 37, 40, 43–45, 48,
  53, 55, 56, 60, 81, 83, 84, 105, 127, 158,
  179, 181, 182, 309, 333, 376, 380

Gas Detection Techniques 59, 103
Gaseous Emissions xix, 149, 168
GDP 14, 40
  Energy Consumption Pattern and CO_2
    Emissions 40
Generation xix, 6, 14, 19, 37, 42, 44, 45,
  48, 120, 149, 152, 154, 155, 161, 162,
  175, 199, 200, 214, 226, 325, 344,
  376, 380
Global carbon dioxide emissions 44
Global Warming xix, xx, 12, 14, 24, 30,
  45, 59, 105
  Calculation of CO_2 emission from
    hydrocarbons 80
  Changing climate 79
  Greenhouse impacts 77
  The greenhouse effect 76
Global issues xix, 12, 14
Globalization 1, 5, 6
  Buyer's market 1, 6
  Competition xxii, 8
  Prevalent scenario 6
Green house gases 49, 78, 380
Green power 19, 37
  Purchasing options 37
Greenhouse Effect 8, 59, 76, 376
Groundwater 109, 110, 111, 117–119, 122,
  135, 137, 139, 141–143, 145, 160
Groundwater Contamination 109, 119, 143
  Sources and routes 118, 119
Growth xix, xxi, 1, 14, 16, 19, 24, 25, 29,
  32, 40, 43, 47, 99, 111, 121, 126, 128,
  152, 203, 204, 243, 307, 312, 336

Hardness 127, 333
Hazards xix–xxi, 1, 11, 135, 179, 180,
  195, 196, 200, 202, 205, 206, 210, 211,
  213, 223, 225–227, 250, 271, 272, 275,
  279, 280, 283, 286–290, 294, 296, 300,

301, 305, 341, 345, 353, 354, 357, 358, 362–365, 371–373, 380
  Analysis methods/techniques xxi, 179, 211, 213, 296, 301, 363, 365, 380
  Classification 1, 10, 207, 211
  Incidents responsible xxi, 197
  Summary and overview 212
  Surface or subsurface (underground) mines xxi, 166, 206
  Toxicity related 204
  Welding process 227, 228, 230
  While using machinery 205
Hazards and Effects Management Process (HEMP) 362
  Control of 364
  Lifecycle's phases 363
  Management 354, 362, 365
  Steps 362
  Tools (Risk analysis) 365
Hazards identification 280
Hazardous Materials xxi, 179, 196, 211, 213, 214, 216, 288, 294
  Biologically Active 207
  Chemically Reactive 207, 210
  Compressed Gases 207, 208
  Explosive Materials 207, 208
  Flammable liquids and solids 207, 209
  Radioactive Materials (See radioactive also) 99, 122, 151, 160, 207, 242
  Toxic materials 152, 207, 381
Hazardous salts 231
Hazardous waste xxi, 48, 119, 122, 124, 149–151, 161, 168, 249
Health & physique 353, 354, 374
Health Promotion Management (HPM) 263, 266
Health Risk 98, 179, 181, 227, 267
Health Risks and Behavior 390
Health, safety and loss prevention (HSLP) 345
Health surveillance xxi, 224, 261, 271
Healthcare Impact Assessment 381, 382
Heat and humidity 206, 223, 226, 233
Homeostasis 27
House Keeping 223, 249, 251
HSE xix–xxii, 8, 12, 13, 15, 223, 286, 287, 291, 298, 301, 309, 312, 325, 353–358, 360–362, 371–373, 382
  Accountabilities 356
  Interrelationship critical business activities 13, 354
  Leadership and commitment 353, 355
  Management xxii, 13, 291, 298, 353–357, 360, 361, 371–373, 382
  Policy 353, 360, 361, 373

HSE accountabilities as 286, 356
  Advisor 358
  Employee 354–356, 360, 362, 372
  Manager 361, 373
  Supervisor 358
HSE Leadership and commitment 355
  Accountabilities 356
  Checklist 359
  Culture 356, 380
  Informed Involvement 355, 356
  Target Settings 355, 356, 359
  Visibility 355, 359
HSE Management xxii, 353, 355–357, 360, 361, 372, 373, 382
  Audit 353, 354, 362, 372, 373, 382
  Checklist 359
  Documents 353
  Implementation and monitoring 353, 354, 371
  Organization 353, 372
  Recovery measures 365
  Resources required 360, 361
  Responsibilities 353, 356, 360–362, 371
  Role and responsibilities 360, 361
  Standards 353, 359, 360
  Training needs 298, 360, 361
Human xx, 8, 12, 14, 15, 21, 23, 27, 29, 32, 34, 36, 40, 45, 48, 59, 69, 70, 76, 93, 96, 99, 102, 105, 109, 110, 111, 117, 124–126, 128, 150–153, 159, 168, 170, 176, 181, 204, 210, 213, 224, 231, 238, 245, 255, 260, 280, 282, 283, 287, 288, 290, 291, 296, 309–311, 325, 333, 353, 360, 361, 376, 380, 382
Human Efficiency 325

Ignition Source 179, 181–183, 192
Incidents xxi, 15, 180, 286, 287, 295, 301, 304, 305, 345, 353, 366, 371, 372
  Causes xxi, 290, 306
  Human failure 293
Incineration 149, 158, 169, 174, 175, 219
Industrial (Occupational) Health, Safety & Environment (HSE) 8
Industrial evils xx, 1, 15
Industrial Hygiene see hygiene identification also xxi, 224, 225
Industrial Hygiene Identified/Recognized 226
  Dust/Fibers 223
  Radiations xix
  Noise xix, 48, 92, 223
  Fumes (Welding & Diesel) 233
  Heat & Humidity 223
  Toxic gases (NO_X, SO_X, CO) 204, 223, 233
  Extreme Temperature 223

Corrosive liquids (Acids) 223
Toxic Salts 225
Vibrations 223
Aqueous Effluents 223
Industrial xix–xxi, 1, 5, 7, 8, 12–14, 44, 45,
    48, 59, 86, 92, 94, 97, 99, 100, 109, 111,
    117, 120, 126, 128, 130, 132, 135, 137,
    141, 143, 145, 149, 150, 154, 156, 160,
    161, 177, 180, 204, 211, 215, 217, 223,
    224, 226, 245, 246, 249–251, 261, 271,
    275, 279–281, 283, 291, 296, 297, 300,
    309, 310, 312, 331–334, 341, 342, 344,
    349, 353, 354, 360, 362, 363, 365, 371,
    374–376, 380, 382
  Gains and impacts 2
  Hygiene see industrial hygiene also xxi,
    223–225
  Noise see noise also 48, 59, 92,
    97, 331
  Projects lifecycle 381
  Safety xxi, 1, 8, 12, 223, 279–281, 295,
    304, 364, 371
  Technology 14
Industrial workers xxi, 8, 223, 261, 271,
    280, 283
  Working life xx, xxi, 15, 270, 271
Industrialization xix, 1, 8, 11, 12,
    14, 32, 153
  And safety 223, 280, 291, 364
  Brief History 1
  Consequences xx, xxi, 15, 296, 363
  Damage Due to Industrialization 12
  Impacts 299
  Technological developments and
    renovations 6
Industries and related Issues 1
Industry & Environment 1, 8
Industry & Safety 1, 8
Industry/industries xix–xxii, xxv, 1, 5,
    7–9, 12, 14, 15, 42, 45, 46, 60, 79, 84,
    92, 100, 101, 109, 111, 119–121, 128,
    149–151, 153, 154, 156, 157, 175, 180,
    193, 195, 198, 200, 204, 205, 207, 211,
    218, 223, 226, 227, 233, 245, 259, 270,
    275, 279–281, 283, 285, 287, 288, 291,
    294–296, 301, 304, 305, 311, 331, 333,
    338, 339, 341, 343, 344, 353, 354,
    360–364, 366, 375, 380, 382
  Adverse impacts 2, 99
  Air polluting 1, 3
  Resources to run industries 2
Information Technology (IT) 7, 35, 310
  Impacts 7
Inherent Safer Design Strategies 213, 288
  Minimize 213
  Moderate 215

Simplify 216
  Substitute/Elimination 215
Inherent Safety features 289, 315
Inherently Safe 179, 279, 289–291
  Active 287
  Passive 287
  Procedural 287
Input resources 280, 309
Injury xxii, 189, 204, 205, 250, 255, 257,
    266, 279, 280, 287, 294, 295, 375
  Case study also see case study
  Change in injury rate w.r.t time 348
  Fatal injury frequency 347, 348
  Rate 287, 348, 383
Inventory Management 172
Irritant gases 201, 219
ISO 14000 336, 338
ISO 14001 336, 338, 339, 373
ISO 9000 xxii, 335, 336, 338

Land Degradation xxi, 48, 149, 160, 161,
    175, 207, 376
Landfill 37, 149, 156–161, 169, 174, 175
Layers of Protection xxi, 279, 290, 301, 365
Leadership xxii, 265, 381, 386, 398, 312,
    314, 334, 353, 355, 358, 373, 374, 380
  And commitment 353, 355
  Basics 312
Lead's adverse 237
Legal compliances xxi, 349
  Checklist
Leakage 197, 208, 334
Life style xix, 1, 32, 253, 267
Lifecycle xxi, 279, 288, 338
Lifecycle approach xxi, 288
Losses/loss xix–xxii, 9, 15, 27, 53, 54, 92,
    99, 132, 149, 152, 155, 160, 172, 181,
    194, 198, 201, 203, 216, 219, 227, 256,
    260, 266, 267, 275, 279, 280, 282, 286,
    287, 289, 290, 294, 295, 297, 298, 305,
    309–311, 315, 325, 331–333, 337,
    344, 349
  Case study xxii, 149, 332
  Classification 310, 325
  Cost loss matrix 331
  Direct xxii, 325
  In a manufacturing plant 325, 326
  In Chemical Industry 198
  Indirect xxii, 325
  Measures to minimize them xxii,
    325, 326
  Prevention xix–xxii, 279, 280, 309, 310,
    337, 344, 349
  Reasons 325
  Tree 332
  Types 309, 325, 326, 332, 333

Machinery hazards 179
Making it work 345, 346
   Implementation Strategy 345
   The Complete System 346
   Zero Tolerance to Risk Behavior 345
Management xx–xxii, 13, 14, 122, 124,
   149, 154, 161, 169, 170, 172, 175,
   249–251, 256, 261, 264–266, 270,
   272–274, 276, 281, 289–291, 297, 298,
   301, 304, 309–312, 314, 320, 325, 333,
   335–342, 344–347, 353–355, 371–373,
   376, 377, 380–382
   Commitment xxii, 55, 223, 254, 261,
      262, 270, 274, 291, 336, 338, 341,
      342, 353–356, 358, 373, 380, 382
   Impact Assessment xx, xxii, 291, 382
   Integrated Environment, Social and
      Healthcare xx, xxii, 291, 382
   Occupational Hazards (Health &
      Physique) 354
Management Plan xx, xxii, 169, 170, 291
Management System xxii, 311, 335–339,
   353, 355–357, 371–373, 380, 382
Marine Pollution 109, 130
Mass Balance 1, 11, 353, 376
   System/Equation 11
Material xix, xx, xxii, xxiii, xxv, 1, 7–9,
   11, 22, 29, 47, 54, 59, 76, 83, 100,
   101, 105, 122, 123, 130, 149–151, 154,
   156–159, 170, 172–175, 177, 179–183,
   186, 189, 193, 194, 196–200, 207, 208,
   215, 216, 226, 227, 241–243, 250–252,
   275, 279, 208, 285, 287–289, 291, 294,
   295, 297, 316, 325, 331, 333, 342, 357,
   360, 375, 376, 380, 381
Material Safety Data Sheet (MSDS)
   170, 210
Material substitution 149
Medical surveillance 223, 254
Metals 8, 83, 101, 121, 127, 130, 137, 143,
   145, 153, 159, 227, 233, 246, 249
   And their concentration 233, 237
   Intakes 236
Minerals xix, xx, 4–8, 15, 20, 21, 29, 35,
   36, 53, 99, 109, 114, 120, 133, 134, 149,
   153, 177, 226, 227, 238, 242, 246, 309,
   333, 342, 360, 376
   Contribution xx, 34
   Production and Consumption Trends 1, 7
   Rapid Resources Depletion 7
   The Nonrenewable Resources 19
   Use in Energy, Goods and Services
      Production 34
Minimization 170, 172
Models of standards 336
Moderate 4, 32, 71, 208, 215, 288

Natural Cycles 19, 22
   Carbon cycle 23
   Nitrogen cycle 22
   Sulfur cycle 22
Natural gases 233
Natural Pollution 126
Neutral Postures 257
Nitrogen Cycle 22
Noise xix, xx, 48, 59, 92, 94–97, 205, 219,
   223, 226–229, 272, 274, 301, 314, 317,
   321, 331, 376
   Control techniques 7, 122
   Induced hearing loss 227, 275, 303
   Industrial noise 59, 92
   Measurement 7, 96, 137, 224, 244, 245,
      270, 272, 286, 299, 320, 334
   Pollution ix, xx, 57, 59, 92, 97
   Related calculations 94
   Sources 92, 378
   Threshold limits 224, 303
Non-hazardous Waste 150, 158, 177
Non-renewable resources viii, 34, 53
Nonrenewable and renewable 19
Notified Diseases 274

Occupational hazards
   (health & physique) 354
Occupational Health (OH) xxi, 223
   And safety (OHS) xx, 223, 380
Occupational health, safety & environment
   (Interrelationship) xix, xx, 1, 13, 14, 298,
   343, 354
Occupational Hygienic Risk – exposure
   assessment and control 261
Occupational lung disease 66
Open pit elements (stripping ratio) 162
Organic pollution 126
Organizational Culture and
   Commitment 223
Organizational Culture and workplace
   stresses xxi, 261
Oxidizer 179, 181, 199
Ozone xix, xx, 8, 12, 14, 19, 60, 71, 72, 76,
   88–90, 105, 151, 205, 228–231, 237, 376
   Depletion process 90
   Hole 89, 90, 132, 133, 200
   Layer and formation of smog 89
   Worsening ozone hole 90
Ozone Depletion xix, xx, 12, 14, 59,
   90, 105
   Mechanism/process involved 90

Particulate matter (Grouping) xix, xxi, 45,
   59, 60, 63, 66, 71, 83, 101, 149,
   151, 231
Performance Monitoring 74

Periodic health surveillance 276
Periodical Medical Examination 275
Personal Protective Equipment (PPE) 279,
    285, 289
pH 30, 84–86, 88, 127, 134, 135,
    137, 143, 145
Photochemical Smog xix, xx, 89, 90, 105
Plan xx, xxii, 19–27, 29, 34, 36, 40, 46,
    48, 54, 83, 84, 92, 97, 101, 110, 111,
    120–122, 124, 126–129, 134, 135, 139,
    141–143, 151, 153, 158, 160, 161, 169,
    170, 174, 175, 178, 185, 204, 206, 274,
    291, 296, 340, 341, 363, 365, 371, 373,
    377, 382
Planning and Procedures 353, 354
Poisonous gases 202, 219
Policy and Strategic Objectives 353, 373
Pollutants xx, xxi, 14, 45, 47, 59, 60, 69,
    71, 74, 84, 89, 109, 121–124, 126, 128,
    129, 157
Pollution xix, xxi, 6, 8, 9, 12, 13, 16,
    36, 48, 53, 55, 59, 71, 72, 85, 86, 101,
    102, 118, 119, 121–123, 127, 130, 135,
    145, 146, 152, 156, 157, 160, 175, 180,
    233, 254, 279, 280, 283, 309, 325, 368,
    372, 376
Polluted Municipality water 130
Pollution see air pollution also xix–xxi, 6, 8,
    9, 12, 13
    Main sources 100
    Why pollution? 375
Pollution Standard Index (PSI) see air
    pollution xx, 12, 73
    Determination PSI Value 73
Population xix, 1–5, 11
    Crude Birth rate 31
    Doubling period 31, 111
    Growth curve 32
    Growth rate 2, 4, 31, 32
    Human population important aspects 32
    Water resources and world
        population 33
Population Growth 29, 31, 32
Portal of Entry 202
Primary 27, 29, 36, 84, 124, 130, 156, 170,
    174, 175, 320
Proactive Outlook 149, 177
Problems xix, xxi, 4, 20, 40, 101, 133,
    152, 154, 157, 160, 200, 242, 260,
    296, 298, 336
Producers 7, 26, 27, 29, 291

Quality xx–xxii, 1, 6, 14, 53, 59, 69, 70,
    71, 73, 101, 109, 114, 115, 117, 121,
    124, 125, 129, 131, 141, 170, 175, 180,
    206, 223, 224, 238, 246, 248, 260, 280,

301, 309, 311, 316, 325, 333–338, 342,
    345
    Control tools xxii
    Reasons for Defects 333

Radiation hazards in Mining 242
    Airborne radiation 242
    Radon and Radon-daughters 242
Radiations xix, 227, 241–243
    Electromagnetic Radiation 241, 243
Recycle 54, 154, 156–160, 169, 170, 176
Red Tide 109, 128
Remedial measures xxi, xxii, 14, 105, 225,
    279, 291, 371
Resources xix, xxi, 4, 5, 8, 13–15, 19, 29,
    32, 36, 37, 44–47, 53, 105, 109, 111,
    149, 153, 156, 177, 179, 181, 233, 251,
    262, 274, 279–281, 283, 290, 296, 300,
    301, 309–311, 333, 334, 337, 339, 342,
    344, 353, 356, 357, 360, 361, 363, 365,
    374, 376, 380, 381
Reuse 49, 156, 159, 161, 169, 170, 173,
    174, 246
Review 13, 172, 213, 231, 252, 271, 274,
    333, 337–340, 346, 354–356, 361–363,
    365, 371–373
Risk xxi, xxii, 43, 44, 48, 70–72, 98, 133,
    146, 179, 181, 185, 187, 198, 206, 207,
    214, 224, 225, 227, 251, 260, 261, 266,
    267, 269, 271–274, 279, 290, 295, 296,
    300, 301, 304, 345, 353, 356, 360, 362,
    364, 365, 371–373, 375, 380
    Assessment criteria 273
    Matrix 272, 274, 279, 301, 365, 380
    Of CO_2 emissions 44
    Ranking (matrix) 366
Rock burst and Bumps 193

Safe Handling Of Chemicals (SHOC) 208
Safety xix, xxi, xxv, 1, 8, 12–15, 61, 92, 97,
    99, 103, 113, 149, 170, 175, 184, 187,
    200, 204, 205, 207, 210, 213–215, 217,
    223, 227, 251–253, 255, 260–265, 270,
    279–291, 294, 295–301, 304, 305, 309,
    311, 333, 341, 343–373, 380, 382
    Awareness xxii, 52, 105, 149, 157, 279,
        280, 291, 297, 309, 311, 341, 361
    Leadership model 358
Salts 84, 86, 127, 128, 160, 223,
    226, 231
Self-Oxidation 179, 183
Severity Rating 279, 301
Sewage 109, 115, 126, 128–131, 245, 246,
    248, 249
Simplify 215, 279
Six Sigma xxii, 334

Soil Degradation 149, 161, 178
Solar Energy's Contribution 25
Spillage 197, 251
Standardization 309, 334, 338, 349
   Benchmarking & Standardization xxii,
     309
   ISO 14000 and ISO 14001 336,
     338, 373
   ISO 9000 xxii, 335, 336, 338
   Other Models of Standards 336
   Procedure to develop 338
Static hazards *see (ANFO)* 199
Substandard xxi, 279, 284, 291, 305, 325
Substitute 172, 213, 215, 288, 295
Subsurface (underground)
   mines' hazards 206
Subsurface Water 109
Sulfide dust explosions 195
Sulfur Cycle 22
Surface Excavations/Mining 161
   Classification xxi, 1, 99, 207, 299,
     301, 310
   Quarrying 163, 166, 167
Surface Mining 161–163
Suspended or sedimentary solids 109, 120,
   129, 130, 195
Sustainable Development xviii, xix, xxii,
   1, 6, 14, 15, 48, 288, 293, 344, 353, 354,
   360, 376, 380—382
Systems xxi, xxii, 5–7, 69, 109, 120, 122,
   133, 154, 161, 169, 172, 174, 179, 189,
   233, 248, 255, 260, 280, 283, 286, 287,
   289–291, 295, 309, 311, 331, 334–338,
   340, 348, 349, 363, 373, 380
   Developed to run the show 280

Teamwork xxii, 346, 356, 382
Technological Developments and
   Renovations 6
Temperature rise 79
Tertiary 27, 29, 130
The 5S Concept 251
Thermal Pollution 121, 126
Thermal stresses 333
Threshold Limit Values (TLVs) 245
Time management schedule 264
Toxic Gases 201, 202, 204, 223, 226,
   233, 374
Toxicity 70, 155, 168, 170, 174, 176, 179,
   202, 203, 301
Toxicology 202
   Factors influencing 203
Training and Education xxi, 279, 297,
   298, 361
Training needs 281, 297, 298, 305, 312,
   339, 340, 360, 361

Training, competency and awareness
   309, 311
Treatment xxi, 109, 120, 122, 128–130,
   132, 149, 154, 174, 175, 238, 248, 249,
   289, 295

Ultimate Goal 1, 310
Underground mines' hazards xxi, 121, 163,
   192, 193, 200, 207
Unintended venting 197
Units of measuring concentration 69, 70
   Ideal gas law 70
   Molecular weight (mol.wt) 70
   Parts per million (ppm) 61, 70
Unsafe Acts 12, 279, 281, 282, 287, 356,
   358
Unsafe Conditions 12, 279, 281, 287

Vapors 182, 183, 191, 207–209
Vibrations 59, 97, 105, 223, 226,
   274, 314
Vocabulary 219

Wastage 4, 15, 149, 177, 309, 325,
   331, 342
   Impacts of wastage 331
Waste xix, xxi, xxii, 4, 8, 15, 19, 23, 40,
   47, 48, 53, 75, 105, 119–122, 124,
   127–130, 147, 149–164, 169, 170,
   172–175, 205, 214, 238, 249–252,
   289, 297, 325, 331, 336,
   342, 375
   Classification xxi, 1, 20, 99, 207, 299, 301
   Recycling/Recovery 149, 154
   Responsible Disposal 149, 154
   Reuse 49, 156, 159, 161, 169, 170, 173,
     174, 246
   Source Reduction 149, 154, 168, 169,
     175, 176
   Treatment xxi, 109, 122, 149, 154, 174,
     238, 249, 295
Waste disposal in Mining 149, 156, 157,
   160, 177, 180, 205, 289
   Dumping site 149, 168
   Procedure of dumping 167
   Types of dumping sites 168
   Various schemes of waste rock dumps 163
Waste management in petroleum industry
   xxi, 168
Water xix, xxi, 3, 4, 7, 8, 14, 15, 19–21,
   24–27, 29, 30, 32, 35, 40, 47, 48, 53,
   56, 57, 59, 76, 77, 83–86, 98, 100, 101,
   103, 105, 109–111, 113–115, 117–135,
   137, 139, 141, 143, 145, 149, 151, 152,
   159–161, 166, 172–175, 179, 180,
   187–191, 194, 195, 199, 207, 208, 236,

243, 245, 246, 248, 249, 251, 253, 254,
    275, 279, 301, 333, 360, 375, 376
Water-bodies 110, 111, 117, 126, 127, 146,
    246, 248
Water pollutants xxi, 14, 146
    Biological 109
    Biological Agents 125
    Chemical 123
    Natural 119
    Organic Substance 126
    Physical 123
    Radio active 123
    Thermal 123
    Toxic Substance 66, 122, 126, 160, 203,
        249, 362
Water pollution
    Non-point source 123
    Point source 123
Water Quality Discharge limits to the
    Marine Environment 131, 248
Water Quality Standards (WQS) 114, 246
Water resources and world population 33
Welding 223, 225, 230, 231
Wellness in workplace 271
What if check list 370
Work culture 261, 282
    Negative 261
    Positive 263

Working xx–xxiii, 13, 15, 45, 91, 121,
    131–133, 166, 193, 201, 204–208,
    215, 217, 223, 233, 243, 244,
    250–253, 255, 256, 258–261, 270,
    274, 279, 280, 283–285, 287–289,
    293, 294, 297, 306, 307, 317, 322,
    327, 335, 342, 348, 349, 355, 358–360,
    371, 374–377
    Environment xix, xxi, xxii, 8, 12, 45,
        255, 279, 280, 287, 322
    Pressure 91, 215, 229, 266, 289
Working Conditions xxi, 133, 206,
    223, 283, 287, 288, 293, 297,
    349, 376
Workplace Hazardous Materials
    Information System (WHMIS) 210
Workplace Stress xxi, 223, 254, 260–262,
    265, 276, 315, 320
Workplace Risk assessment 226
Workplace Stress xxi, 223, 254, 260–262,
    265, 276, 315, 320
World population 33, 111, 153
    Human population 32
    Impacts of population growth 31

Yellow Boy 109, 134, 135, 146